Integrated Researches in Immunology, Physiology and Genetics

Integrated Researches in Immunology, Physiology and Genetics

Edited by **Jim Wang**

SYRAWOOD
PUBLISHING HOUSE
New York

Published by Syrawood Publishing House,
750 Third Avenue, 9th Floor,
New York, NY 10017, USA
www.syrawoodpublishinghouse.com

Integrated Researches in Immunology, Physiology and Genetics
Edited by Jim Wang

International Standard Book Number: 978-1-68286-052-6 (Hardback)

Contents

Preface

Over the recent decade, advancements and applications have progressed exponentially. This has led to the increased interest in this field and projects are being conducted to enhance knowledge. The main objective of this book is to present some of the critical challenges and provide insights into possible solutions. This book will answer the varied questions that arise in the field and also provide an increased scope for furthering studies.

The exploration of human body and its processes has been the focus of many experiments and studies worldwide. Different approaches, evaluations, methodologies and advanced studies on a diverse array of topics such as genetics, immunology, evolution, physiology, epidemiology, etc. have been included in this book. It is a collective contribution of a renowned group of international experts and scientists. Students actively engaged in this field will find this book full of crucial and unexplored concepts.

I hope that this book, with its visionary approach, will be a valuable addition and will promote interest among readers. Each of the authors has provided their extraordinary competence in their specific fields by providing different perspectives as they come from diverse nations and regions. I thank them for their contributions.

Editor

Sperm DNA Fragmentation as a Factor of Male Low Reproductive Function in IVF Practice

Ievgeniia Zhylkova[1], Olexandr Feskov[1], Iryna Feskova[1], Olena Fedota[2] & Vladyslav Feskov[1]

[1] Centre of Human Reproduction "Clinic of Professor Feskov A. M.", Kharkov, Ukraine

[2] Kharkov National University named after V. N. Karazin, Kharkov, Ukraine

Correspondence: Ievgeniia Zhylkova, Centre of Human Reproduction "Clinic of Professor Feskov A. M.", 61098 Kharkov, Yelizarova Str. 15, Ukraine. E-mail: zhilkova@feskov.com.ua

Abstract

The correlation between semen DNA fragmentation level and the male age was investigated. The dependence for the blastocyst formation rates on the sperm DNA fragmentation has been examined in patients with low reproductive function. The increase of DNA fragmentation level correlates with the decrease of the blastocyst formation rates ($p < 0.05$). The significant negative relationship between sperm DNA fragmentation and the male age is proved ($p < 0.05$). The age of 35 years old could be discussed as clinically critical male age for the process of chromatin compaction during the process of spermatogenesis.

Keywords: sperm DNA fragmentation, blastocyst formation rates, IVF

1. Introduction

1.1 Sperm DNA Fragmentation as a Male Infertility Factor

Recently high attention in the reproductive medicine is paid to the paternal genome failures. Assisted reproductive techniques (ART) such as conventional *in vitro* fertilization (IVF), and especially intracytoplasmic sperm injection (ICSI), allow couples whose sperm characteristics are impaired to obtain a pregnancy, whereas a few years ago, these couples would have had to use sperm donation in order to obtain their child. One can nevertheless wonder about the capacity of poor quality sperm samples to generate embryos having normal capacities of development. Among the factors involved in the failure of obtaining embryos and/or pregnancies, the impaired sperm genome is frequently incriminated (Ahmadi, 1999). Sperm DNA fragmentation is rather new discussible reason of male infertility. There are contradictory data about the possible influence of sperm DNA fragmentation on the sperm fertilization ability and the process of embryo development, particularly on the process of the blastocyst formation (Agarwal, 2003; Findikli, 2004; Borini, 2006). There are data about the correlation of the sperm DNA fragmentation level and high percentage of the spontaneous miscarriages for patients included in the IVF treatment (Benchaib, 2003; Seli, 2005; Oleszczuk, 2011). Sperm DNA fragmentation consists of single and doublestranded DNA breaks, frequently occurring in semen of subfertile patients (Lopes et al., 1998; Irvine et al., 2000; Muratori et al., 2000). Despite the origin and the mechanisms responsible for such genomic anomaly are not yet clarified, it has been proposed that sperm DNA fragmentation could be a good parameter to predict the male fertility status as an alternative or in addition to poorly predictive standard parameters presently determined in routine semen analysis (Lewis, 2007; Erenpreiss et al., 2006). Spermatogenesis is a complicated process that includes spermatozoa development and maturation. It depends on such factors as genetics, hormonal statement, environmental conditions etc. Spermatogenesis failures could lead to the formation of aneuploid sperm or sperm with the DNA damage (Cayli, 2004; Tesarik, 2002).

1.2 The Etiology of Chromatin Compactization Failures

The level of sperm DNA fragmentation reflects the integrity of genetic material of the gamete. This parameter is important since DNA lesions of many types induce mutations commonly observed in mutated oncogenes and tumour suppressor genes (Marnett, 2000). The possible reason of sperm DNA fragmentation are changing of chromatin structure during the process of spermatogenesis and apoptosis (Brugnon, 2006; Schlegel, 2005; Calle, 2008). Defects of the process of spermatozoa maturation should also be discussed. In average about 20% of spermatozoa in ejaculate could be found as apoptotic in men with different deviations in the semen analysis

(Mehdi, 2003; Muratori, 2008). It is also found out that the sperm DNA fragmentation level is statistically higher for patients with spermatogenesis failures comparing with men with normal semen parameters (Speyer, 2012). On the next side there are data that DNA fragmentation level has no correlation with such semen parameters as motility, concentration or sperm morphology (Luke, 2010). There is still discussible question about the dependence of sperm DNA fragmentation level on the male age (Bronet, 2012). Results of studies aimed to establish whether the amount of sperm DNA fragmentation could predict the outcome of Assisted Reproduction Techniques (ARTs) are conflicting.

1.3 The Research Design

The aim of the presence work was to investigate the correlation for sperm DNA fragmentation level on the male age and to examine if there is any dependence in the process of the early embryo development on the sperm DNA fragmentation in patients with low reproductive function during the infertility treatment using the methods of ART. Taking into account that the most important criteria of the normal embryo development is the blastocyst formation, the affect on sperm DNA fragmentation level on the blastocyst formation rates were examined for infertile patients during the infertility treatment using the IVF procedure including the manipulation of ICSI. It was also very important to compare sperm DNA fragmentation level in the group of infertile patients and in men with normal reproductive function.

2. Method

2.1 Sperm DNA Fragmentation Level Examination

Sperm DNA fragmentation level was examined using the method of the sperm chromatin dispersion (SCD) (HaloSperm, Halotech, Spain). The method is based on the chromatin dispersion around the spermatozoa nuclei that allows finding the spermatozoa with fragmented DNA. The SCD test is based on the principle that sperm with fragmented DNA fail to produce the characteristic halo of dispersed DNA loops that is observed in sperm with non-fragmented DNA, following acid denaturation and removal of nuclear proteins. The SCD test is a simple, accurate, highly reproducible. and inexpensive method for the analysis of sperm DNA fragmentation in semen and processed sperm. It could potentially be used as a routine test for the screening of sperm DNA fragmentation in the andrology laboratory (Fernández, 2003). The examination was carried out using the fluorescent microscope Nikon Eclipse 80i. The obtained photos were documented with the cytogenetic program Lucia FISH (LIM, Czech Republic). Normally the sperm DNA fragmentation level should not be higher then 20.0%.

2.2 Participants Characteristics

During the investigation the group consisted of 33 infertile patients with the male low reproductive function were formatted. The middle male age in the group was 38.0 ± 5.7 years old. The IVF procedure and the examination of the sperm DNA fragmentation level were carried out for all the participants of the mentioned group. Simultaniously the control experiment with a normal group consisted of 20 fertile men was carried out to compare the value of DNA fragmentation level. The middle male age in the control group was 34.0 ± 6.9 years old. Basic sperm parameters (concentration, motility and morphology) showed a high variability among individual patients, ranging between normal values and severe oligoasthenoteratozoospermia. Blastocyst formation rates were variable for the patients (Table 1).

Table 1. Basic sperm parameters of examined group of infertile patients

Patient	Basic sperm parameters			Sperm DNA fragmentation, %	BFR, %
	Concentration ($\times 10^6$/ml)	Motility, %	Normal forms, %		
1	15	52	22	38.0	18.1
2	34	26	31	3.0	12.5
3	23	17	12	14.5	56.2
4	54	70	37	27.5	0
5	72	31	41	14.5	69.5
6	12	34	39	3.8	50.0
7	43	63	19	5.2	100.0
8	7	13	15	20.0	0
9	32	49	34	10.0	50.0
10	8	35	21	33.4	62.5
11	19	41	36	11.7	25.0
12	21	15	45	13.2	78.2
13	59	48	42	22.3	42.1
14	37	52	39	42.2	44.4
15	29	24	17	12.3	33.3
16	3	23	5	10.5	47.1
17	51	63	39	21.3	33.3
18	26	17	14	61.5	14.3
19	38	46	34	18.3	54.5
20	6	10	11	37.5	14.2
21	42	23	13	12.4	77.8
22	40	39	25	9.8	37.5
23	74	57	57	12.0	61.5
24	67	68	54	2.0	80.0
25	31	20	35	3.5	58.3
26	18	23	13	60.0	20.0
27	4	16	10	5.5	60.0
28	22	46	24	1.5	61.5
29	65	58	51	7.5	28.5
30	11	29	9	14.5	50.0
31	9	14	35	5.2	50.0
32	20	36	21	46.7	38.4
33	38	48	18	38.0	55.5

2.3 IVF Procedure

During the IVF programs the protocols with a-GnRH (antagonist-Gonadotropin Releasing Hormone) were used for controlled ovary stimulation for oocytes retrieval. Ovary stimulation took not less then 10 days in every case. In the day of the transvaginal puncture the size of follicles achieved 18 mm. To maintain the luteal phase the progesterone consisted medicines were used. Oocyte retrieval was carried out under general anaesthesia by a vaginal ultrasonography-guided aspiration. At 16-18 h after insemination or microinjection, as previously described (Borini et al., 2004a, 2004b), oocytes were assessed for two PN presence.The fertilization of oocytes was carried out using the procedure of ICSI taking into account the male factor of infertility. ICSI manipulation was done on the inverted microscope Nikon Eclise TI-u with the Eppendorf micromanipulators. Spermatozoa were immobilezed using the laser system OCTAX (MTG, Germany). The obtained oocyted and embryos were cultured in the Universal IVF Medium and ISM-1 Medium (Medicult) during the first three days and in the Blast Assist (Medicult) till the fifth day of culture. The gametes and embryos were cultured in temperature 36.8 °C - 37.1 °C and in 5.5% - 5.8% CO_2 level.

2.4 Statistics

The obtained data for abnormal distribution were statistically checked using nonparametric methods of statistic analysis. The correlation was examined by the method of analysis with the Spearman coefficient. Calculation was done using the program module Statistica-6.

3. Results

The positive correlation between the sperm DNA fragmentation level and male age was proved for the examined group. Total Spearman coefficient is 0.35 ($p < 0.05$). The critical male age for the sperm DNA fragmentation level growth is 35 years old. Spearman coefficient for patients elder then 35 years old is 0.45 ($p < 0.05$). There is no any correlation for DNA fragmentation and male age in the group of patients younger then 35 years old. Spearman coefficient for the group of patients younger then 35 years old is -0.04 ($p > 0.05$). The obtained results are mentioned in Table 2.

Table 2. Correlation between the sperm DNA fragmentation level and male age

Age group	N	Age, years	DNA fragmentation, %	r_s	$r_{critical}$	p
≤ 35 years	11	30.7±2.7	13.5±6.6	-0.04	0.58	$p > 0.05$
> 35 years	22	41.6±4.4	21.2±15.0	0.45	0.43	$p < 0.05$
Total group	33	38.0±5.7	18.6±12.5	0.35	0.34	$p < 0.05$

The significant negative correlation between the semen DNA fragmentation level and blastocyst formation rates is proved. However, the DNA fragmentation less then 5.0% did not impair the blastocysts formation rates. The critical DNA fragmentation level for blastocysts formation rates is 5.0%. The Spearmen coefficient for the group of patients with the DNA fragmentation level higher then 5.0% is -0.44 ($p < 0.05$). The Spearmen coefficient for the total examined group is -0.41 ($p < 0.05$). The obtained results are mentioned in Table 3.

Table 3. Correlation between the semen DNA fragmentation level and blastocyst formation rates

Critical DNA fragmentation level	N	DNA fragmentation, %	BFR, %	r_s	$r_{critical}$	p
> 5.0%	28	21.4±12.6	43.2±18.6	-0.44	0.38	$p < 0.05$
< 5.0%	5	2.8±0.8	52.5±17.0	-0.60	0.94	$p > 0.05$
Total group	33	18.6±12.5	44.9±18.9	-0.41	0.34	$p < 0.05$

Photo of spermatozoa with the fragmented DNA and with normal compactization of chromatin is present on Figure 1.

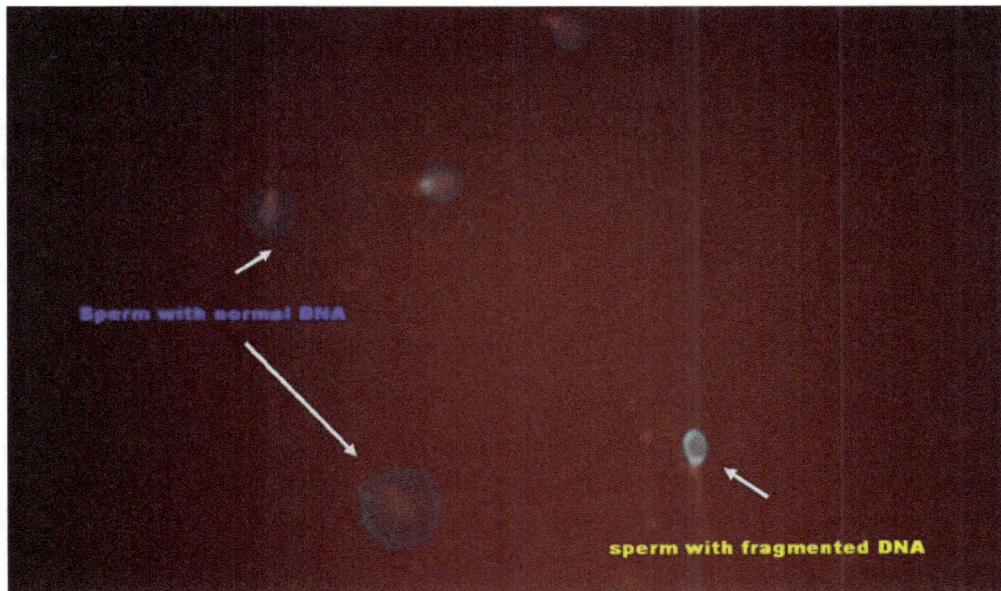

Figure 1. Sperm DNA fragmentation by the method of SCD

The average semen DNA fragmentation level in the control group of fertile men was 5.48 ± 2.03. The level of DNA fragmentation level was significantaly higher in infertile patients comparing with the control ($p < 0.05$). Such fact could prove the influence of the sperm chromatin compactization failures on the semen parameters and male reproductive function. Further investigations should be performed.

4. Discussion

In studies investigating the impact of sperm DNA fragmentation on reproduction, the prevailing idea is that sperm with damaged DNA, even if retaining the ability to fertilize the oocyte (Ahmadi, 1999), affect the subsequent steps resulting in increased failure of embryo development and miscarriage (Agarwal, 2004; Lewis, 2005; Li, 2006). However, data on the relationship between DNA damage and ART outcome are very conflicting (O'Brien, 2005; Li, 2006). The obtained data confirmed that DNA fragmentation could be one of the male infertility factor that effects the embryo development. The increase of DNA fragmentation showed negative correlation with the blastocyst formation rates (BFR) ($p < 0.05$). The level of sperm DNA fragmentation 5.0% could be reported as a critical one for the process of blastocyst formation in ART (assisted reproductive techniques). Such conclusion is practically important because of an absence of the strict recommendations about normal or abnormal level of semen DNA damage levels. The discussible range of acceptable level of semen DNA fragmentation varies in diapasone 10%-40% in different studies (Benchaib, 2003). The studies by previous workers reported above, supported by the present study, indicate clearly that strand breaks in the sperm DNAhave little or no effect on fertilization and early embryo growth but begin to have an effect at the stage of blastocyst development, and then have a very marked effect on implantation of the embryo, but the exact clinically important DNA fragmentation level that could impair the blastocyst development is still discussible (Ahmadi, 1999; Fatehi, 2006). There was no significant effect on fertilization rate or early cleavage, but the effects on blastocyst development and implantation rate were very marked. The negative correlation between the sperm DNA fragmentation on the male age is proved ($p < 0.05$). The patients' age 35 years old could be discussed as clinically critical male age for the process of chromatin compaction during the spermatogenesis. Such conclusion is clinically important, taking into account that nowadays more attention to the female age is paid and there is a little information about the age of male fertility. As both males and females decide to conceive later, the question of whether this may impact their fertility individually and as a couple becomes even more crucial. A paternal age of over 40 years at the time of conception is a frequently quoted male age threshold, however, currently there is no clearly accepted definition of advanced paternal age or even a consensus on the implications of advancing male age (Humm, 2013). Further study is also needed to determine if there is any dependence of sperm DNA damage on the basic sperm parameters. Several studies have stressed the importance of traditional sperm parameters as predictors of fertility potential (Nallella et al., 2006). Because of the evidence of correlations between sperm DNA fragmentation and clinical pregnancy and pregnancy loss rates, it would be practically important to find a relationship between traditional sperm evaluation parameters (concentration, motility and morphology) and pregnancy and pregnancy loss rates in IVF groups.

Acknowledgements

The clinical research was performed at Feskov Center of Human Reproduction in Kharkov, Ukraine. Authors are grateful to Mr. Gakh Andrey (Kharkov National University named after V. N. Karazin) for the support in the statictics analisis.

References

Agarwal, A., & Said, T. M. (2003). Role of sperm chromatin abnormalities and DNA damage in male infertility. *Hum. Reprod., 19*(4), 331-345. http://dx.doi.org/10.1093/humupd/dmg027

Ahmadi, A., & Soon-Chye, Ng. (1999). Fertilizing ability of DNA-damaged spermatozoa. *J. Exp. Zool., 284,* 696-704. http://dx.doi.org/10.1002/(SICI)1097-010X(19991101)284:6%3C696::AID-JEZ11%3E3.0.CO;2-E

Benchaib, M., Braun, V., & Lornage, J. (2003). Sperm DNA fragmentation decreases the pregnancy rate in an assisted reproductive technique. *Hum. Reprod., 18*(5), 1023-1028. http://dx.doi.org/10.1093/humrep/deg228

Borini A, Bonu, M. A., Coticchio, G, Bianchi, V., Cattoli, M., & Flamigni, C. (2004b). Pregnancies and births after oocyte cryopreservation. *Fertil Steril, 82*(3), 601-605. http://dx.doi.org/10.1016/j.fertnstert.2004.04.025

Borini, A., Tallarini, A., Maccolini, A., Dal Prato, L., & Flamigni, C. (2004a). Perifollicular vascularity monitoring and scoring; a clinical tool for selecting the best oocyte. *Eur. J. Obstet Gynecol Reprod Biol., 115,* 102-105. http://dx.doi.org/10.1016/j.ejogrb.2004.01.021

Borini, A., Tarozzi, N., Bizzaro, D., Bonu, M., Fava, L., Flamigni, C., & Coticchio, G.. (2006). Sperm DNA fragmentation: paternal effect on early post-implantation embryo development in ART. *Hum. Reprod., 21*(11), 2876-2881. http://dx.doi.org/10.1093/humrep/del251

Bronet, F., Martırnez, E., Gaytarn, M., Lin~arn, A., Cernuda, D., Ariza, M., ... Nogales, M. (2012). Sperm DNA fragmentation index does not correlate with the sperm or embryo aneuploidy rate in recurrent miscarriage or implantation failure patients. *Hum. Reprod., 27*(7), 1922-1929. http://dx.doi.org/10.1093/humrep/des148

Brugnon, F., Van Assche, E., & Verheyen, G. (2006). Study of two markers of apoptosis and meiotic segregation in ejaculated sperm of chromosomal translocation carrier patients. *Hum. Reprod., 21*(3), 683-685.

Calle, J. F., Muller, A., & Walschaerts, M. (2008). Sperm deoxyribonucleic acid fragmentation as assessed by the sperm chromatin dispersion test in assisted reproductive technology programs: results of a large prospective multicenter study. *Fertility and Sterility, 19*(6), 671-682.

Cayli, S., Sakkas, D., Vigue, L., Demir, R., & Huszar, G. (2004). Cellular maturity and apoptosis in human sperm: creatin kinase, caspase-3 and Bcl_{XL} levels in mature and diminished maturiry sperm. *Mol. Hum. Reprod., 10*(5), 365-372. http://dx.doi.org/10.1093/molehr/gah050

Fatehi, A. N., Bevers, M. M., Schoevers, E., Roelen, B. A. J., Colenbrander, B., & Gadella, B. M. (2006). DNA damage in bovine sperm does not block fertilization and early embryonic development but induces apoptosis after the first cleavages. *J. Androl., 27,* 176-188. http://dx.doi.org/10.2164/jandrol.04152

Fernández, J. L., Muriel, L, Rivero, M. T., Goyanes, V., Vazquez, R., & Alvarez, J. G. (2003). The sperm chromatin dispersion test: a simple method for the determination of sperm DNA fragmentation. *J. Androl., 24,* 59-66.

Findikli, N., Kahraman, S., & Kumtepe, Y. (2004). Assessment of DNA fragmentation and aneuploidy on poor quality human embryos. *Reprod Biomed Online, 8*(2), 196-206. http://dx.doi.org/10.1016/S1472-6483(10)60516-0

Henkel, R., Kierspel, E., & Hajimohammad, M. (2003). DNA fragmentation of spermatozoa and assisted reproduction technology. *Reprod Biomed Online, 7,* 477-484. http://dx.doi.org/10.1016/S1472-6483(10)61893-7

Hong Y., Guo-ning, H., Yang, G., & De, Y. L. (2006). Relationship between human sperm-hyaluronan binding assay and fertilization rate in conventional in vitro fertilization. *Hum. Reprod., 21*(6), 1545-1550. http://dx.doi.org/10.1093/humrep/del008

Humm, K. C., & Sakkas, D. (2013). Role of increased male age in IVF and egg donation: is sperm DNA fragmentation responsible? *Fertility Sterility, 99*(1), 30-6. http://dx.doi.org/10.1016/j.fertnstert.2012.11.024

Lewis, S. E., & Aitken, R. J. (2005). DNA damage to spermatozoa has impacts on fertilization and pregnancy. *Cell Tissue Res., 322*, 33-41. http://dx.doi.org/10.1007/s00441-005-1097-5

Li, Z., Wang, L., Cai, J., & Huang, H. (2006). Correlation of sperm DNA damage with IVF and ICSI outcomes: a systematic review and meta-analysis. *J. Assist Reprod Genet., 23*, 367-376. http://dx.doi.org/10.1007/s10815-006-9066-9

Luke, S., Gunnar, B., Michael, S., & Deborah, L. (2010). Clinical significance of sperm DNA damage in assisted reproduction outcome. *Hum. Reprod., 25*(7), 1594-1608. http://dx.doi.org/10.1093/humrep/deq103

Marnett, L. J. (2000). Oxyradicals and DNA damage. *Carcinogenesis, 21*, 361-370. http://dx.doi.org/10.1093/carcin/21.3.361

Mehdi, B., ValeÂrie, B., Jacqueline, L., & Samia, H. (2003). Sperm DNA fragmentation decreases the pregnancy rate in an assisted reproductive technique. *Hum. Reprod., 18*(5), 1023-1028. http://dx.doi.org/10.1093/humrep/deg228

Muratori, M., Marchiani, S., Tamburrino, L., Tocci, V., Failli, P., Forti, G. & Baldi, E. (2008). Nuclear staining identifies two populations of human sperm with different DNA fragmentation extent and relationship with semen parameters. *Hum. Reprod., 23*(5), 1035-1043. http://dx.doi.org/10.1093/humrep/den058

Nallella, K. P., Sharm, R. K., Aziz, N., & Agarwal, A. (2006). Significance of sperm characteristics in the evaluation of male infertility. *Fertil Steril., 85*(3), 629-634. http://dx.doi.org/10.1016/j.fertnstert.2005.08.024

O'Brien, J., & Zini, A. (2005). Sperm DNA integrity and male infertility. *Urology, 65*, 16-22. http://dx.doi.org/10.1016/j.urology.2004.07.015

Oehninger, S., Chaturvedi, S., & Toner, J. (1998). Semen quality is there a paternal effect on pregnancy outcome in in-vitro fertilization/intracytoplasmic sperm injection? *Hum. Reprod., 13*, 2161-2164. http://dx.doi.org/10.1093/humrep/13.8.2161

Oleszczuk, K., Giwercman, A., & Bungum, M. (2011). Intra-individual variation of the sperm chromatin structure assay DNA fragmentation index in men from infertile couples. *Hum. Reprod., 26*(12), 3244-3248. http://dx.doi.org/10.1093/humrep/der328

Schlegel, P. N., & Paduch, D. A. (2005). Yet another test of sperm chromatin structure. *Fertil Steril., 84*(4), 854-859. http://dx.doi.org/10.1016/j.fertnstert.2005.04.050

Seli, E., & Sakkas, D. (2005). Spermatozoal nuclear determinants of reproductive outcome: implications for ART. *Hum. Reprod. Update, 11*(4), 337-349. http://dx.doi.org/10.1093/humupd/dmi011

Speyer, B. E., Pizzey, A. R., Ranieri, M., Joshi, R., Delhanty, J. D. A., & Serhal, P. (2012). Fall in implantation rates following ICSI with sperm with high DNA fragmentation. *Hum. Reprod., 25*(7), 1609-1618. http://dx.doi.org/10.1093/humrep/deq116

Tesarik, J., Mendoza, C., & Greco, E. (2002). Paternal effects acting during the first cell cycle of human preimplantation development after ICSI. *Hum. Reprod., 17*, 184-189. http://dx.doi.org/10.1093/humrep/17.1.184

2

Physiological and Ethological Effects of Caffeine, Theophylline, Cocaine and Atropine; Study Using the Ant *Myrmica sabuleti* (Hymenoptera, Formicidae) as A Biological Model

Marie-Claire Cammaerts[1], Zoheir Rachidi[1] & Geoffrey Gosset[1]

[1] Faculté des Sciences, Université Libre de Bruxelles, Bruxelles, Belgium

Correspondence: Marie-Claire Cammaerts, DBO, CP 160/12, Université Libre de Bruxelles, 50, Av. F.D. Roosevelt, 1050 Bruxelles, Belgium. E-mail: mtricot@ulb.ac.be

Abstract

Aiming to check if ants can be used as biological models, we studied the effects of four alkaloids on the ant *Myrmica sabuleti*. Caffeine and theophylline increased the ants' linear speed, decreased their precision of reaction and their food consumption, did not affect their response to pheromones, nor their audacity, largely increased their conditioning ability and their memory; ants did not become habituated to nor dependent on these alkaloids. The effects of caffeine exponentially decreased in the course of time; those of theophylline slowly decreased sigmoidally after a latency period. Cocaine decreased the ants' linear speed, their precision of reaction, their response to pheromones and their consumption of food; it increased their audacity; under its consumption, ants became unable to acquire conditioning. Ants became habituated to and dependent on such an alkaloid. The effects of cocaine rapidly vanished in the course of time. Atropine did not affect the ants' locomotion, decreased their precision of reaction and their response to pheromones because it reduced their olfactory perception; it did not affect their audacity, decreased their food consumption, increased the speed in acquiring conditioning but not the quality of the resulting conditioning; it slightly increased their memory. Ants did not become habituated to, nor dependent on this alkaloid. The effects of atropine vanished in about 20 hrs - 30 hrs. These effects are in agreement with those known by physiologists, psychologists and doctors in medicine; observations made on ants even lead to more precise deductions. Consequently, ants can efficiently be used as biological models.

Keywords: conditioning, dependence, habituation, locomotion, memory

1. Introduction

Animals can be used as biological models, for studying several physiological functions, because most of the biological processes are similar for all animals, also for human beings (i.e. genetics, metabolism, nervous cells functioning). Invertebrates are more and more used as biological models because they offer scientists many advantages, among others a short life cycle and a simple anatomy, being available in large numbers, presenting few ethical problems (Sovik & Barron, 2013; Wolf & Heberlein, 2003). Some invertebrates are become famous as biological models, for instance, *Dendrocelium lacteum, Caenorhabdotes elegans, Aplysia californica, Tribolim castaneum, Drosophila melanogaster. Apis mellifera* (Kolb & Wishaw, 2002; Wehner & Gehring, 1999; Deutsch, 1994; Devineni & Heberlein, 2013). Among the invertebrates, insects, and especially social hymenoptera, are advantageously used as biological models (Andre, Wirtz, & Das, 1989; Doring & Chittka, 2011). Bees, for instance, are often used as biological models (for instance, Abramson, Wells, & Janko, 2007). Thousands of ants can easily be maintained in laboratories, at low cost and very conveniently, throughout the entire year. Despite being invertebrates, ants are among the most evolved organisms as for their morphology, their physiology, their social organization and their behavior. They are the most morphologically elaborate hymenoptera, having indeed a unique resting position of their labium, mandibles and maxilla (Keller, 2011a, 2011b), as well as a lot of glands emitting numerous, efficient compounds (Billen & Morgan, 1998). Their societies are highly organized with a strong division of labor, an age-based polyethism and social regulation (Hölldobler & Wilson, 1990). Their behavior is incomparable: they care for their brood, build sophisticated nests, chemically mark the inside of their nest, and, differently, their nest entrances, their nest surroundings and their foraging area (Passera & Aron, 2005). They generally use an alarm signal, a trail pheromone, and a recruitment signal (Passera & Aron, 2005); they are able to navigate using memorized visual and olfactory cues (cf. Cammaerts, 2012a, 2012b, and references therein);

they efficiently recruit nestmates where, when and as long as it is necessary (Passera, 2006), and, finally, they provide their area with cemeteries (Keller & Gordon, 2006). Mankind has probably not yet discovered all the ants' capabilities. The biology of some ant species is now very well known: for instance *Solenopsis invicta, Cataglyphis cursor, Atta sexdens,* and *Gigantiops destructor* (Passera & Aron, 2005).

We have largely studied species of the genus *Myrmica,* and above all *Myrmica sabuleti* Meinert 1861 (Cammaerts, 2004, 2008; Cammaerts & Rachidi, 2009). We became aware that *M. sabuleti* may be a good biological model after studies concerning the impact of age, activity and diet on the ants' conditioning capability (Cammaerts & Gosset, 2014). At that time, we presumed that several other factors may also impact the ants' (and any animals') learning as well as other ethological and physiological functions. Among these factors, some drugs may influence memorization (this has been observed in bees: Chittka & Peng, 2013) and may also impact animals' general activity. Works of Bateson (for instance, 1979) are also in favor of such a presumption. Therefore, in the present study, we aimed to define the effects of four alkaloids (caffeine, theophylline, cocaine, atropine) on several ant biological functions in order 1) to have an idea of the impact, on human beings, of these four commonly consumed alkaloids, 2) to check if ants could be efficiently used as models for collecting information about effects of substances or treatments on living organisms, 3) to estimate if ants may be used for detecting the presence of small amount of substances in collected samples.

In this study, we focused on ten ethological and physiological parameters:

➤ 1) The general activity of the individuals: by assessing their linear and angular speed

➤ 2) The precision of their response: by examining their orientation towards a source of alarm pheromone.

➤ 3) Their response to their specific pheromones: by quantifying their response to their trail pheromone.

➤ 4) Their food consumption: by assessing the ants present on a fresh meat food provided after two days of starvation.

➤ 5) Their tendency in accomplishing risky tasks: by assessing the ants coming onto a tower set on their foraging area.

➤ 6) Their ability in being conditioned: by assessing their acquisition of a visual operant conditioning, and comparing the results with those previously obtained on ants having never consumed any alkaloid.

➤ 7) Their memory: by assessing, as in a previous experiment on ants having never received any alkaloid, the time period during which they retain their acquired visual conditioning after training ended.

➤ 8) The 'habituation' to the consumed alkaloid: by checking if, after having consumed the alkaloid during 6 days, the revealed effects on the ants' locomotion or precision of response decreased or remained unchanged.

➤ 9) The 'dependence': by looking if individuals having consumed an alkaloid, preferred sugar water containing the alkaloid to sugar water free of it.

➤ 10) The time period during which the effects of a given alkaloid persist after its consumption ended: by assessing the ants' locomotion or precision of response, several times, in the course of time, after having removed the alkaloid from the ants' food.

The obtained results were compared to what is known about the effects of the alkaloids used, and conclusions were drawn as for 1) the impact of these alkaloids on living organisms; 2) the potential use of ants as models for examining physiological and ethological effects of substances or treatments; 3) the possible use of ants for detecting substances in collected material.

2. Material and Methods

2.1 Collection and Maintenance of Ants

The study was made on eight colonies of *M. sabuleti* collected, in summer 2013, in two old quarries, one located at Treigne, and one in the Aise valley (Ardenne, Belgium). The ants were nesting under stones, on a field covered with small, often odorous plants. *Myrmica sabuleti* is a well known species: its ecology, eye morphology, visual perception, visual and olfactory conditioning, navigation system, recruitment strategy, responses to pheromones, acquisition by callow ants of adults' cognitive abilities have already been studied (M.-C. Cammaerts & D. Cammaerts, 2012; Cammaerts, Rachidi, Bellens, & De Doncker, 2013; Cammaerts, 2013a, 2013b, 2014, in press). The eight collected colonies were demographically identical, containing about 600 workers, one or two queens and brood (at larval stage). They were maintained in the laboratory in artificial nests made of one to three glass tubes half-filled with water, with a cotton-plug separating the ants from the water. The glass tubes were deposited in trays (34 cm × 23 cm × 4 cm), the sides of which were covered with talc to prevent the ants from escaping. The trays served as foraging areas, food being delivered into them. The ants were fed with sugar water provided *ad libitum* in a small glass tube plugged with cotton, and with cut *Tenebrio molitor* larvae (Linnaeus 1758) provided twice a week on a glass slide. Temperature was maintained between 18 °C and 22 °C, humidity at about 80%, these conditions remaining constant over the course of the study. Lighting had a constant intensity of 330 lux while caring for the ants, training and testing them. During other time periods, the lighting was dimmed to 110 lux. The electromagnetic field had an intensity of 2-3 $\mu W/m^2$.

2.2 Acquisition of the Alkaloids; Realization of Aqueous Solutions for Ants

One gram of caffeine and of theophylline, as well as 100 mg of cocaine and of atropine sulfate, was provided, via the pharmacist Mr. Cardon J. (Brussels), by the manufacturer CERTA. The alkaloids were provided as a white bright powder, at the highest level of purity possible. Using a precise balance, 15 mg of each alkaloid was prevailed and dissolved in 15 ml of a saturated solution of brown sugar, the ants' common liquid food. The concentration in alkaloid of the solution was thus 1/1,000. These solutions were given to the ants, like their usual liquid food, in a small glass tube plugged with cotton, this cotton being refreshed each two days.

2.3 Assessment of the Ants' Linear Speed, Angular Speed and Orientation

Ants' linear and angular speed was assessed for detecting excitation in the animals. This assessment was made on ants freely moving on their foraging area. Ants' orientation towards an isolated congener's head (the source of the ants' alarm pheromone) was assessed for examining the ants' precision of reaction. An isolated worker's head, with widely open mandibles, is a source of alarm pheromone identical to that of an alarmed worker, in terms of the dimensions of the emitting source (the mandibular glands opening) and the quantity of pheromone emitted (Cammaerts-Tricot, 1973).

Each time, such assessment was made firstly before giving alkaloid to the ants, secondly one day after the ants had drunk sugar water containing the alkaloid. For each assessment, the movement of 10 ants of each two colonies (N = 20 ants) was analyzed.

Trajectories were recorded manually, using a water-proof marker pen, on a glass slide placed on the top of the experimental tray, set horizontally 3 cm above the area where the tested individuals were moving. A metronome set at 1 second was used as a timer for assessing the total time of each trajectory. Each trajectory was recorded during 5 to 10 seconds or until the ant reached the stimulus. All the trajectories were then traced (copied) with a water-proof marker pen onto transparent polyvinyl sheets (Figures 1E and 1F). These sheets could then be affixed to a PC monitor screen. The trajectories were then analyzed using specifically designed software (Cammaerts, Morel, Martino, & Warzée, 2012). Briefly, each trajectory was defined in the software by clicking as many points as needed with the mouse. Then, the total time of the trajectory (assessed using the metronome) was entered, and feature of the trajectory could be measured (linear speed, angular speed, orientation).

The three variables used to characterize the trajectories were defined as follows:

The linear speed (V) of an animal is the length of its trajectory divided by the time spent moving along this trajectory. It was measured in mm/s.

The angular speed (S) (i.e. the sinuosity) of an animal's trajectory is the sum of the angles, measured at each successive point of the trajectory, made by the segment 'point i → point i – 1' and the segment 'point i → point i + 1', divided by the length of the trajectory. This variable was measured in angular degrees/cm.

The orientation (O) of an animal towards a given point (here an ant's head) is the sum of the angles, measured at each successive point of the registered trajectory, made by the segment 'point i of the trajectory → given point' and the segment 'point i → point i + 1' divided by the number of measured angles. This variable was measured in angular degrees. When such a variable (O) equals 0°, the observed animal perfectly orients itself towards the point; when O equals 180°, the animal fully avoids the point; when O is lower than 90°, the animal has a tendency to orient itself towards the point; when O is larger than 90°, the animal has a tendency to avoid the point.

Each distribution of 20 variables was characterized by its median and its quartiles since it was not Gaussian (Tables 1, 3, 5, 7, lines 1, 2), and the distribution of values obtained for ants having consumed alkaloid was statistically compared to that previously obtained for the same ants before they consumed alkaloid using the non-parametric χ^2 test (Siegel & Castellan, 1989, pp. 111-116). The significance threshold was set to $\alpha = 0.05$.

2.4 Assessment of the Ants' Trail Following Behavior

This behavior was assessed for examining the ants' response to their pheromones. The trail pheromone of *Myrmica* ants is produced by the workers' poison gland. So, ten of these glands were isolated in 0.5 ml (500μl) hexane and stored for 15 min at -25 °C. To perform one experiment, 0.05 ml (50 μl) of the solution was deposited, using a normograph pen (a pen used for drawing, hexane extract being poured inside the pen instead of ink), on a circle (R = 5 cm) pencil drawn on a piece of white paper and divided into 10 ang. deg. arcs. One minute after being prepared, the piece of paper with the artificial trail was placed in the ants' foraging area. When an ant came into contact with the trail, its movement was observed. Its response was assessed by the number of 10 ang. deg. arcs it walked without departing from the trail, even if it turned back on the trail. If an ant turned back when being in front of the trail, its response was assessed as "zero arcs walked"; when an ant crossed the trail without following it, its response equaled "one walked arc". Before testing the ants on a trail, they were observed on a circumference imbibed with 50 μl of pure hexane and the control numbers of walked arcs were so obtained. For each control and test experiment, 20 individuals of each two used colonies were observed; 40 numbers of walked arcs were so each time recorded. Each distribution of values was characterized by its median and its quartiles, since it was not Gaussian (Tables 1, 3, 5, 7, line 3) and those obtained for ants having consumed alkaloid were statistically compared to the corresponding ones obtained for the same ants before they consumed alkaloid, this by using the non parametric χ^2 test (Siegel & Castellan, 1989). On such experimental trails, *Myrmica* workers do not deposit their trail pheromone because they do so only after having found food or a new nest site.

2.5 Assessment of the Ants' Audacity

Before the ants consumed alkaloid and three days after they had consumed such alkaloid, a tower built in strong white paper (Steinbach ®) (h = 4 cm; diam = 1.5 cm) was set on the ants' foraging area (Figure 1A), and the ants present on it were counted 10 times. The mean and the extreme values of the obtained values were established each time and the two series of values were compared using the non parametric Mann-Whitney U test (Siegel & Castellan, 1989, pp. 128-137; Tables 1, 3, 5, 7, line 4).

2.6 Assessment of the Ants' Consumption of Food

Before the ants consumed a given alkaloid, and after they had consumed that alkaloid for three days, the workers present on the *T. molitor* larva were counted 10 times. The numbers obtained under the two kinds of food consumption were statistically compared using the Mann-Whitney U test (same reference as above) and the mean as well as the extreme values of the recorded numbers were established (Tables 1, 3, 5, 7, line 5).

2.7 Assessment of the Ants' Ability in Acquiring Operant Visual Conditioning and of the Ants' Memory

Briefly, at a given time, a green hollow cube was set above the *T. molitor* larva, this time tied to the supporting piece of glass. The ants so underwent visual operant conditioning. Tests were performed in the course of the ants conditioning acquisition, then, after having removed the green cube, of their conditioning loss.

In detail, ants were collectively visually trained to a hollow green cube constructed of strong paper (Canson ®) according to the instructions given in Cammaerts and Nemeghaire (2012) and set over the meat food which served as a reward (Figure 1B). The color has been analyzed to determine its wavelengths reflection (Cammaerts, 2007). The ceiling of each cube was filled unlike the four vertical faces, this allowing the ants entering the cubes. The green cube was considered to be the 'correct' choice when the ants were tested as explained below.

Ants were individually tested in a Y-shaped apparatus constructed of strong white paper according to the instructions given in M.-C. Cammaerts, Rachidi, and D. Cammaerts (2011), and set in a small tray (30 cm × 15 cm × 4 cm), apart from the experimental colony's tray (Figure 1C). Each colony had its own testing design. The apparatus had its own bottom and the sides were covered with talc to prevent the ants from escaping. In the Y-apparatus, the ants deposited no trail since they were not rewarded. However, they may utilize other chemical secretions as traces. As a precaution, the floor of each Y-apparatus was changed between tests. The Y-apparatus was provided with a green cube in one or the other branch (Figure 1C). Half of the tests were conducted with the cube in the left branch and the other half with the cube in the right branch of the Y maze, and this was randomly chosen. Control experiments had previously been made on never conditioned ants, and conditioning experiments had also been previously performed on ants having never received alkaloid (Tables 1, 3, 5, 7, lines 6, 7; Cammaerts et al., 2011). Indeed, once any animals are conditioned to a given stimulus, they become no longer naïve for such an experiment. It was thus impossible, in the present study, to perform, on the same colonies, the ants' conditioning without then with the alkaloid.

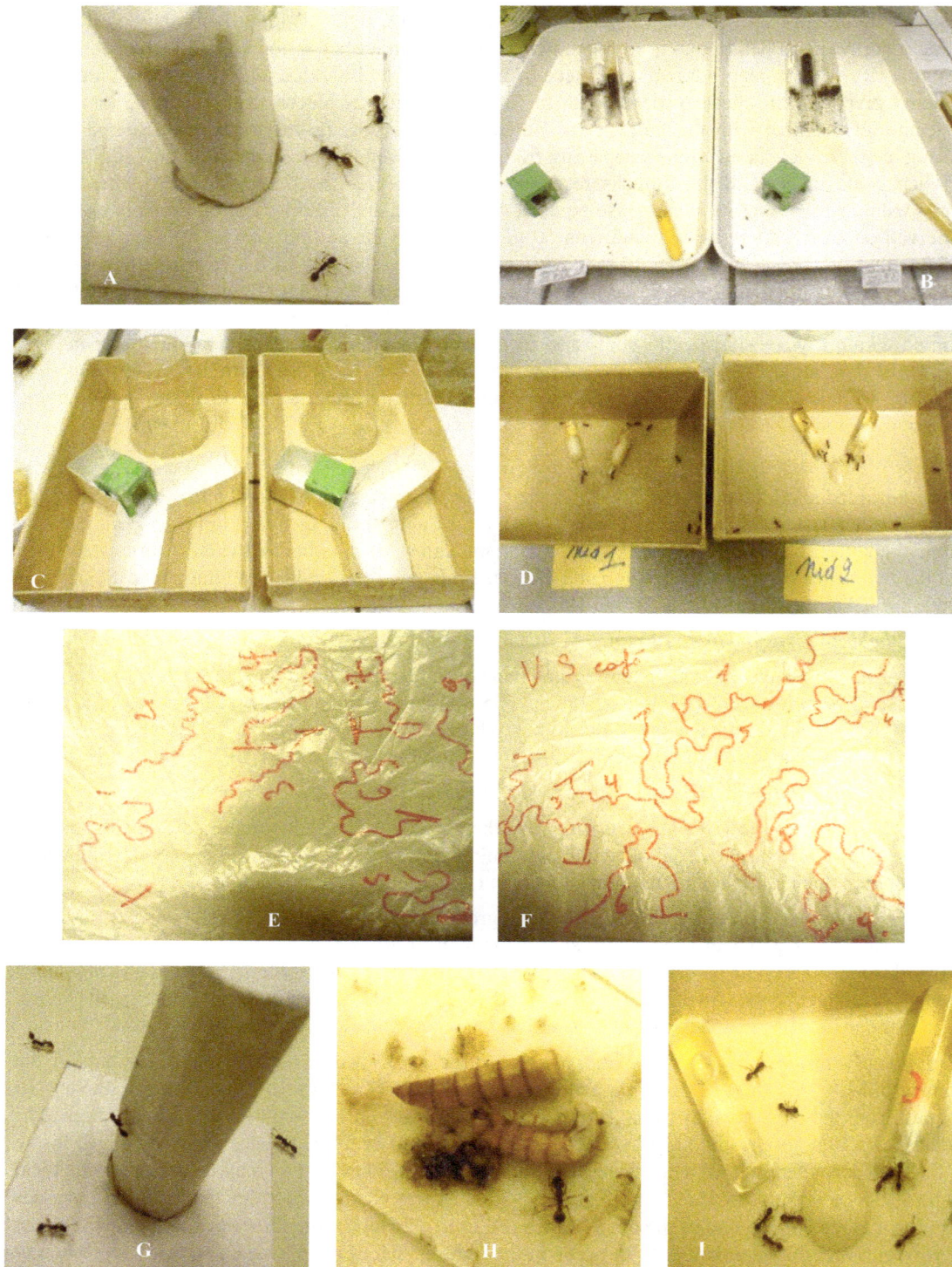

Figure 1. Selected views of the experiments

A: the tower used for assessing the ants' daring. B: two ant colonies under training to a visual cue, a green hollow cube. C: experimental apparatus used for testing ants being under operant visual conditioning. D: experimental design used for assessing the ants' dependence on an alkaloid, this one being present in the right tube for nest 1 and in the left tube for nest 2; in the present case, the alkaloid was theophylline and ants did not exhibit any dependence. E, F: recorded ants' trajectories E: under normal diet and F: after caffeine consumption. G: ants having consumed cocaine and coming onto the tower. H: ants having consumed cocaine and staying near the meat food without eating. I: ants having consumed cocaine and choosing sugar water containing this alkaloid during the test allowing revealing such dependence.

To conduct a test on a colony, 10 workers of that colony - randomly chosen from the workers of that colony - were transferred one by one to the area at the entrance of the Y-apparatus. Each transferred ant was observed until it turned either to the left or to the right in the Y-tube, and its choice was recorded. Only the first choice of the ant was recorded and this only when the ant was entirely under the cube, i.e. beyond a pencil drawn thin line indicating the entrance of a branch (Figure 1C). Afterwards, the ant was removed and transferred to a polyacetate cup, in which the rim was covered with talc, until 10 ants were so tested, this avoiding testing the same ant twice. All the tested ants were then placed back on their foraging area. For each experiment, the numbers of ants, among 10 + 10 = 20, which turned towards the "correct" green cube, or went to the "wrong" empty branch of the Y were recorded. The percentage of correct responses for the tested ant population was so established (Tabs 1, 3, 5, 7, lines 6, 7). The results obtained for ants that have consumed alkaloid were compared to those obtained for ants that have never consumed such substances (Tables 1, 3, 5, 7, lines 6, 7; Figure 2) using the non parametric test of Wilcoxon (Siegel & Castellan, 1989, pp. 87-95). For such a test, the value of N, T, and P, according to the nomenclature of the here above cited authors, are given in the results section.

2.8 Ants' Habituation to the Consumed Alkaloid

Four to seven days after ants had continuously consumed a given alkaloid, their linear and angular speed or their orientation towards an isolated worker's head, were assessed. The results were compared to the control ones and to those obtained after one day of alkaloid consumption using the non parametric χ^2 test (Siegel & Castellan, 1989; Tables 2, 4, 6, 8, line 1).

2.9 Ants' Dependence on the Consumed Alkaloid

After the ants had continuously consumed a given alkaloid during five to nine days, an experiment was performed to examine if they have acquired some dependence on the consumed alkaloid. Fifteen ants of each two used colonies were transferred into a small tray (15 cm × 7 cm × 5 cm), the borders of which had been covered with talc and in which laid two tubes (h = 2.5 cm, diam. = 0.5 cm), one containing sugar water, the other sugar water + a given alkaloid (at the concentration 1/1,000), each tube being plugged with cotton (Figure 1D). In one of the trays, the tube containing the alkaloid was located on the right; in the other tray, it was located on the left. The ants drinking each liquid food were counted 12 times, the mean values being then established for each kind of food (Tables 2, 4, 6, 8, line 2). They were statistically compared to the values expected if ants randomly went drinking each kind of food, using the non parametric goodness of fit χ^2 test (Siegel & Castellan, 1989, pp. 45-51).

2.10 Duration of the Effect of the Alkaloid

Eight to twenty days after that the ants had continuously consumed a given alkaloid, the liquid food containing this alkaloid was removed from the ants' tray and replaced by sugar water free of alkaloid. This change was made at a given recorded time. After that, the ants' linear and angular speed or their orientation towards an isolated worker's head, were assessed after successive given time periods. The results revealed the decrease of the effects, on ants, of the consumed alkaloid (Tables 2, 4, 6, 8, line 3; Figure 3). Their statistical significance could be estimated via the non parametric χ^2 test (Siegel & Castellan, 1989, pp. 111-124).

3. Results

3.1 Concerning Caffeine

3.1.1 Effect on Ants' Locomotion

After having consumed caffeine for one day, the ants moved more quickly and less sinuously. These changes in locomotion were statistically significant (Table 1, line 1; linear speed: $\chi^2 = 21.51$, df = 1, P < 0.001; angular speed: $\chi^2 = 17.22$, df = 2, P < 0.001), and obvious while observing the ants, and their recorded trajectories (Figures 1E and 1F).

3.1.2 Effect on Ants' Precision When Reacting

The ants' reaction to an isolated worker's head (a source of their alarm pheromone) consists in a positive orthokinesis due to the attractiveness of the alarm pheromone. Under normal diet, ants oriented themselves very well towards an isolated worker's head (median value = 38.5 ang. deg., far lower than 90 ang. deg.). After having consumed caffeine for one day, the ants went on being attracted by their alarm pheromone (median value = 57.2 ang. deg, still lower than 90 and deg.) but their positive orthokinesis was of lower quality than before consuming caffeine, this difference being statistically significant (Table 1, line 2; $\chi^2 = 9.50$, df = 2, P < 0.01).

3.1.3 Effect on Ants' Response to Their Pheromones

After a two day time period of caffeine consumption, the ants' trail following behavior (= a response to one of their pheromones) was similar to that exhibited before such a consumption (Table 1, line 3; $\chi^2 = 1.68$, df = 2, NS). Caffeine did not affect the ants' response to their pheromones.

3.1.4 Effect on Ants' Audacity

No difference could be revealed between the numbers of ants coming onto a tower (a 'risky' apparatus) before consumption of caffeine and after three days of such consumption (Table 1, line 4; Figure 1A; U = 30.5, Z = 0.105, P = 0.878).

3.1.5 Effect on Ants' Consumption of Food

While consuming caffeine, statistically fewer ants eat the provided *T. molitor* larva (Table 1, line 5; U = 11, Z = 2.91, P = 0.002). The ants did not stop eating; they only spent less time in performing this task.

3.1.6 Effect on Ants' Conditioning Capability

While under normal diet, *M. sabuleti* workers acquired 80% of operant visual conditioning in 167 hrs (= 7 days), after having consumed caffeine for 5 days, they acquired a conditioning score of 80% in 23hrs and an exceptional score of 90% after a total of only 47 hrs (2 days). Such a difference was significant (Table 1, line 6; N = 4, T = 10, P = 0.06; Figure 2, left part, empty and black circles).

3.1.7 Effect on Ants' Memory

Under normal diet, visually conditioned ants lost nearly all their conditioning in about 90 hrs (2 1/3 days) and kept 10% of it during several days, even weeks (Cammaerts et al., 2011). After having consumed caffeine for seven days, the conditioned ants never lost their conditioning, still presenting a score of 80% after 46hrs and retaining 30% of their conditioning for 10 or more days or weeks. Such a difference was significant (Table 1, line 7; N = 8, T = 36, P = 0.004; Figure 2, right part, empty and black circles).

3.1.8 Ants' Habituation to Caffeine

The ants' locomotion was examined one day and seven days after they had consumed caffeine. No difference in linear speed could be statistically detected (the ants' linear speed was even slightly larger after 7 days); the ants' sinuosity was statistically larger after 7 days than after one day (Table 2, line 1; linear speed: $\chi^2 = 1.75$, df = 2, NS; angular speed: $\chi^2 = 7.03$, df = 1, P < 0.01). So, no decrease of effect occurred in the course of caffeine consumption.

Table 1. Effect of caffeine on seven physiological and ethological functions

Functions	Variable Assessed	Normal Diet	Diet + Caffeine	Statistics
Activity n = 20 (locomotion)	Linear speed mm/sec	10.5(9.3-12.1)	17.8(15.8-20.9)	P<0.001
	Sinuosity ang.deg./cm	142(130-165)	113(93-123)	P<0.001
Precision of a reaction n=20	Orientation to an alarm signal ang.deg.	38.5(32.6-58.9)	57.2(51.2-68.4)	P < 0.01
Response to pheromones n = 40	Trail following behavior (n° of arcs walked)	C: 1 (1-2) T:10.5(7.8-18.5)	C: 1 (1 – 2) T:9.0(6.0-15.0)	NS NS
Audacity n = 10	n° of ants on a tower	7.7 (6 – 9)	7.5 (2 – 13)	P = 0.878
Food consumtion n = 10	n° of ants eating meat (mean, extrema)	3.5 (1 – 6)	1.8 (0 – 3)	P = 0.002

Learning ability n=120 + 20=140	% of correct responses	* C: 61/59 50% 7 hrs 61/59 51% 23 hrs 68/52 57% 30 hrs 68/52 57% 47 hrs 76/44 63% 55 hrs 75/45 63% 71 hrs 80/40 67% 79hrs 79/41 66% 95 hrs 84/36 70% 103 hrs 82/38 68% 118 hrs 87/33 73% 127 hrs 83/37 69% 145 hrs 94/26 78% 167 hrs 97/23 81% 215 hrs 96/24 80%	7 hrs 15/5 75% 23 hrs 16/4 80% 30 hrs 17/3 85% 47 hrs 18/2 90%	N=4, T= 10 P = 0.06
Memory n=120 + 20=140	% of correct responses	*6 hrs 87/33 73% 20 hrs 92/28 77% 46 hrs 81/39 68% 71 hrs 76/44 63% 95 hrs 71/49 59% 116 hrs 72/48 60% 143 hrs 69/51 58% 166 hrs 72/48 60%	6 hrs 17/3 85% 20 hrs 17/3 85% 46 hrs 18/2 90% 71 hrs 16/4 80% 95 hrs 15/5 75% 116 hrs 15/5 75% 143 hrs 17/3 85% 166 hrs 16/4 80% 189 hrs 16/4 80% 214 hrs 16/4 80% 237 hrs 16/4 80%	N=8, T=36 P = 0.004

Details are given in the text. Briefly, caffeine increased ants' linear speed, learning ability and memory, unchanged their audacity and their response to pheromone, decreased the precision of their response and their food consumption. Results about conditioning and memory are graphically presented in Figure 2 (black circles). * = results obtained by Cammaerts et al. (2011).

3.1.9 Ants' Dependence on Caffeine

The appropriate test was performed after that the ants had consumed caffeine for nine days. Meanly 48.1% of ants chose the sugar food containing caffeine and 51.9% the food free of the alkaloid. The numbers obtained (79 vs 81) did not statistically differ from those expected if ants chose the provided foods at random (80 vs 80) (Table 2, line 2; $\chi^2 = 0.006$, df = 1, 0.90 < P < 0.95, NS). Thus, ants did not become dependent on caffeine consumption.

3.1.10 Time Period During Which Consumed Caffeine Affected the Ants

Twenty days after ants had consumed caffeine, their sugar food containing the alkaloid was removed and replaced by sugar food free of this substance. The ants' trajectories were recorded 1, 2½, 4½, and 8 hrs later. The ants' linear speed decreased in the course of time, revealing so a decrease of the effect of caffeine (Table 2, line 3). This decrease may be a negative exponential function of the time (Figure 3, black circles and dotted line). After one day, the ants' linear speed equaled 11.4 mm/sec and their angular speed 120 ang.deg./cm, values which were not identical to though statistically not different from the control ones (linear speed: $\chi^2 = 0.42$, df = 2, P ≈ 0.80, NS; angular speed: $\chi^2 = 1.75$, df = 2, 0.30 < P < 0.50, NS).

3.1.11 Other Observed Effects

One hour after ants had consumed caffeine, a few ones appeared to suffer just like if they were very thirsty. We gave tap water to the ants, each day, on their foraging area. Ants came drinking and no longer suffered.

Table 2. Habituation to caffeine, dependence on its consumption, and duration of its effect on ants

Effect Studied	Variable Assessed	Numerical Results	
Habituation	Linear speed mm/sec	18.7 (17.8 – 20.2) *vs* 1 day: NS	
(after 7 days)	Sinuosity ang.deg./cm	132 (105 – 153) *vs* 1 day: 0.001<P<001	
Dependence	choices of sugar water + caffeine	nest 1: 51 ants *vs* 44 ants = 51/96 = 53.1%	
(12 counts)	*vs* sugar water	nest 2: 28 ants *vs* 37 ants = 28/65 = 43.1%	
		T 1hrs 20.3(17.8-21.8)	125(102-133)
Duration of effect	Linear (mm/sec) and angular	*T 2½hrs* 18.7(17.2-20.7)	125(103-138)
	(ang.deg./cm) speed	*T 4½hrs* 16.9(14.8-19.2)	126(108-144)
		T 8hrs 13.8(12.6-16.3)	133(117-154)

Details are given in the text. Briefly, ants did not become habituated to nor dependent on caffeine and the effects of the alkaloid exponentially decreased in the course of time, ending in about 11 – 13 hours. The latter observation is also graphically presented in Figure 3 (black circles).

3.2 Concerning Theophylline

3.2.1 Effect on Ants' Locomotion

After having drunk, for one day, sugar water containing theophylline, *M. sabuleti* workers obviously and statistically walked more rapidly than usually (21.4 mm/sec *vs* 13.0 mm/sec; p < 0.001) but not statistically less sinuously (Table 3, line 1; linear speed: $\chi^2 = 18.09$, df = 1, P < 0.0001; angular speed: $\chi^2 = 1.79$, df = 2, 0.30 < P < 0.50, NS). In fact, the ants moved very quickly, then stopped, then moved again rapidly.

3.2.2 Effect on Ants' Precision When Reacting

Under theophylline consumption, the ants' orientation by orthokinesis towards a source of alarm pheromone was only slightly and not statistically lower than without consuming the alkaloid (Table 3, line 2; $\chi^2 = 2.09$, df = 2, 0.30 < P 0.50, NS).

3.2.3 Effect on Ants' Response to Their Pheromones

Under theophylline consumption, the ants less efficiently followed a trail (9.5 *vs* 16.5 arcs walked along the circular trail) but the variability of the ants' responses was so large that this decrease in efficiency was not statistically significant (Table 3, line 3; $\chi^2 = 3.23$, df = 3, 0.30 < P 0.50, NS).

3.2.4 Effect on Ants' Audacity

After having consumed theophylline for five days, as many ants moved onto a tower (a risky apparatus) as before consuming such an alkaloid (Table 3, line 4; U = 602, Z = -0.66, P = 0.514). The ants' behavior in front of the apparatus was exactly the same before and after consumption of theophylline: hesitation, coming then moving away, stopping, and cleaning antennae.

3.2.5 Effect on Ants' Consumption of Food

Under consumption of theophylline, the ants eating the *T. molitor* larva were less, but not statistically lees, numerous than those eating that meat food before receiving the alkaloid (meanly 1.8 *vs* 2.3) (Table 3, line 5; U = 38.52, Z = 0.832, P = 0.393). In fact, the ants spent less time on the food site, this being probably due to their increase of locomotion.

3.2.6 Effect on Ants' Conditioning Capability

Under normal diet, after 30hrs of training, the ants presented a conditioning score of 57%; under theophylline consumption, they reached, in the same time period, the exceptional conditioning score of 90% (Table 3, line 6; N = 4, T = 10, P = 0.06; Figure 2, left part, black triangles *vs* empty circles). The tested ants' behavior clearly reflected such a result: they scarcely hesitated and went rapidly under the hollow green cube.

Table 3. Effect of theophylline on seven physiological and ethological functions

Functions	Variable Assessed	Normal Diet	Diet+Theophylline	Statistics
Activity n = 20	Linear speed mm/sec	13.0(11.9-14.3)	21.4(19.2-23.6)	P < 0.001
(locomotion)	Sinuosity ang.deg./cm	115(99-151)	111(77-137)	NS
Precision of a reaction n=20	Orientation to an alarm signal ang.deg.	47.3(35.7-52.7)	55.2(43.2-62.7)	NS
Response to pheromones n = 40	Trail following behavior	C: 1 (1-2)	C: 1 (1 − 2)	NS
	(n° of arcs walked)	T:16.5(8.0-23.5)	T:9.5(3.8-22.5)	NS
Audacity n = 10	n° of ants on a tower	1.4 (0 − 3)	1.3 (0 − 2)	P = 0.514
Food consumption n = 10	n° of ants eating meat (mean, extrema)	2.3 (0 − 3)	1.8 (0 − 3)	P = 0.393
Learning ability n=120 + 20=140	% of correct responses	* C: 61/59 50% 7 hrs 61/59 51% 23 hrs 68/52 57% 30 hrs 68/52 57% 47 hrs 76/44 63% 55 hrs 75/45 63% 71 hrs 80/40 67% 79hrs 79/41 66% 95 hrs 84/36 70% 103hrs 82/38 68% 118hrs 87/33 73% 127hrs 83/37 69% 145hrs 94/26 78% 167hrs 97/23 81% 215hrs 96/24 80%	7 hrs 15/5 75% 23 hrs 17/3 85% 30 hrs 18/2 90% 47 hrs 17/3 85%	N=4, T=10 P = 0.06
Memory n=120 + 20=140	% of correct responses	*6 hrs 87/33 73% 20 hrs 92/28 77% 46 hrs 81/39 68% 71 hrs 76/44 63% 95 hrs 71/49 59% 116hrs 72/48 60% 143hrs 69/51 58% 166hrs 72/48 60%	6 hrs 17/3 85% 20 hrs 17/3 85% 46 hrs 18/2 90% 71 hrs 18/2 90% 95 hrs 16/4 80% 116 hrs 17/3 85% 143 hrs 17/3 85% 166 hrs 17/3 85% 189 hrs 17/3 85% 214 hrs 17/3 85% 237 hrs 18/2 90%	N=8, T=36 P = 0.004

Details are given in the text. Briefly, theophylline increased the ants' linear speed, learning ability and memory, unchanged their response to pheromones and their audacity, slightly decreased the precision of their reaction and their food consumption. Results relative to the conditioning and the memory are graphically presented in Figure 2 (black triangles). * = results obtained by Cammaerts et al. (2011).

3.2.7 Effect on Ants' Memory

While under normal diet, the ants presented a conditioning score of 59% 95hrs after training ended, under theophylline consumption, the ants still presented a score of 80% after the same time period without training (Table 3, line 7; N = 8, T = 36, P = 0.004). Under normal diet, the ants retained 10% of their conditioning after a few days; under theophylline consumption, they retained 35% – 40 % of their conditioning for at least 10 days (Figure 2, right part, black triangles *vs* empty circles).

3.2.8 Ants' Habituation to Theophylline

After seven days of theophylline consumption, the ants' linear speed was a little higher than and the ants' angular speed was similar to those presented after one day of this alkaloid consumption (Table 4, line 1; linear speed: χ^2 = 8.01, df = 2, P < 0.02; angular speed: χ^2 = 0.92, df = 2, 0.50 < P < 0.70, NS). No ants' habituation to theophylline could so be detected.

3.2.9 Ants' Dependence on Theophylline

During the appropriate test performed after the ants had consumed theophylline for nine days, similar numbers of ants went drinking the sugar water containing the alkaloid and the sugar water free of it (Table 4, line 2). The numbers obtained (37 *vs* 40) did not statistically differ from those expected if ants chose the two kinds of sugar food at random (38.5 *vs* 38.5) (χ^2 = 0.03, df = 1, 0.80 < P < 0.90, NS). So, no dependence on theophylline consumption could be detected.

3.2.10 Time Period During Which Consumed Theophylline Affected the Ants

Theophylline was removed from the ants' food after fifteen days of consumption. No decrease of effect was observed during the first four hours. Then, a slow decrease was revealed, the effect of the alkaloid being yet detectable after 15 hours (Table 4, line 3; linear speed after 15 hrs: χ^2 = 8.22, df = 2, P < 0.02). An entire loss of this effect could take about 24 hrs to 30 hrs (Figure 3, black triangles). The curve of the loss of effect might be sigmoid.

3.2.11 Other Observed Effects

To avoid ants' suffering, we gave them tap water each day, on their foraging area.

Table 4. Habituation to theophylline, dependence on its consumption, and duration of its effect on ants

Effect Studied	Variable Assessed	Numerical Results	
Habituation	Linear speed mm/sec	23.2 (20.7 – 26.5) *vs* 1 day: NS	
(after 7 days)	Sinuosity ang.deg./cm	106 (92 – 122) *vs* 1 day: NS	
Dependence	ants on sugar water +	nest 1: 19 ants *vs* 22 ants = 19/41 = 46.3%	
(12 counts)	theophylline *vs* sugar water	nest 2: 18 ants *vs* 18 ants = 18/36 = 50.0%	
		T 1hrs 22.5(21.7-23.2)	98(85-117)
		T 2½hrs 23.3(20.7-24.9)	95(85-113)
		T 4½hrs 21.3(19.0-22.8)	101(85-110)
Duration of effect	Linear (mm/sec) and angular (ang.deg./cm) speed	*T 7½hrs* 17.2(15.5-18.5)	132(117-145)
		T 10hrs 17.8(14.5-20.0)	106(90-151)
		T12½hrs 16.7(14.3-17.8)	114(104-128)
		T 15hrs 15.6(14.0-18.5)	113(105-123)
		T 28hrs 14.4(13.3-17.3)	99(89-115)

Details are given in the text. Briefly, ants did not become habituated to nor dependent on theophylline and the effects of this alkaloid decreased sigmoidally, after a latency period of four hours, in about twenty hours. The latter observation is also graphically presented in Figure 3 (black triangles).

3.3 Concerning Cocaine

3.3.1 Effect on Ants' Locomotion

After having consumed cocaine for one day, the ants moved more slowly, what was obvious and statistically significant (Table 5, line 1; $\chi^2 = 15$, df = 2, P < 0.001).

3.3.2 Effect on Ants' Precision When Reacting

Under normal diet, ants oriented themselves very well towards an isolated worker's head (median value = 36.1 ang. deg., far lower than 90 ang. deg., Table 5, line 2). After having consumed cocaine for one day, the ants went on being attracted by their alarm pheromone since the median value of their orientation was still lower than 90 ang. deg., but their positive orthokinesis was of lower quality since the median value of their orientation equaled 61.1 ang. deg. This difference in the ants' precision of reaction was statistically significant (Table 5, line 2; $\chi^2 = 7.04$, df = 2, P ≈ 0.02).

3.3.3 Effect on Ants' Response to Their Pheromones

After having consumed cocaine for two days, the ants walked less efficiently along a trail (the median value equaled 5.5 arcs while, under normal diet, it equaled 11.5 arcs). Such a difference was statistically significant (Table 5, line 3; $\chi^2 = 13.34$, df = 3, P = 0.001).

3.3.4 Effect on Ants' Audacity

The numbers of ants coming onto a tower (a 'risky' apparatus) before consumption of cocaine was low (1.3), but after three days of such consumption, it was statistically larger (4.5; Table 5, line 4; U = 266.5, Z = -2.71, P = 0.006; Figure 1G).

3.3.5 Effect on Ants' Consumption of Food

The number of ants eating the provided T. molitor larva was statistically lower while ants consumed cocaine (Table 5, line 5; U = 150, Z = 4.43, P = 000003). Generally, the ants came onto the meat food site but stayed there, motionless, without eating (Figure 1H).

3.3.6 Effect on Ants' Conditioning Capability

While under normal diet, M. sabuleti workers acquired about 63% of operant visual conditioning in 50 hrs and reach a score of 80% in about 167 hrs (= 7 days), after having consumed cocaine for six days, they could never acquire any visual conditioning. The ants presented a typical behavior in the testing apparatus: when reaching the choice space, they generally stopped, turned their head to the right then to the left, walked a little forward, then turned back; in other words, they hesitated during several minutes, before going either to their right or to their left. Their mean conditioning score equaled 40% - 55% even after 127 hrs of training. Such a result was highly significant (Table 5, line 6; N = 11, T = 66, P = 0.0005; Figure 2, left part, empty circles and squares).

3.3.7 Effect on Ants' Memory

Since after having consumed cocaine for six days, the ants could never retain (memorized) the presented visual cue, the study of their memory was impossible. The conclusion could be that eventually only very short term memory occurred, while long-lasting memory could no longer be acquired under cocaine consumption (Table 5, line 7). Such a result could be estimated as being significant.

3.3.8 Ants' Habituation to Cocaine

The ants' locomotion was examined one day (see above) and four days after they had consumed the alkaloid. After cocaine consumption, the ants' linear speed was no longer lower than under normal diet (Table 6, line 1; linear speed compared to the control one: $\chi^2 = 1.82$, df = 2, 0.30 < P < 0.50, NS). Consequently, the effect of cocaine on ants decreased in the course of that alkaloid consumption. This was obvious while looking to the ants, and this means that the ants became habituated to such alkaloid consumption.

3.3.9 Ants' Dependence on Cocaine

The appropriate test was performed after the ants had consumed cocaine for five days. Meanly 80.5% of ants chose the sugar food containing cocaine, while only 19.5% chose the food free of the alkaloid. The numbers obtained (52 vs 11) statistically differed from those expected if ants chose the provided foods at random (31.5 vs 31.5) (Table 6, line 2; $\chi^2 = 26.68$, df = 1, P < 0.001). Such a result was obvious while testing the ants (Figure 1I). The ants having consumed cocaine actively searched for food containing this alkaloid; they thus became dependent on cocaine consumption.

Table 5. Effect of cocaine on seven physiological and ethological functions

Functions	Variable Assessed	Normal Diet	Diet + Cocaine	Statistics
Activity n = 20 (locomotion)	Linear speed mm/sec	13.0(11.6-14.1)	10.9(9.9-11.6)	P<0.001
	Sinuosity ang.deg./cm	138(108-164)	126(113-143)	NS
Precision of a reaction n=20	Orientation to an alarm signal ang.deg.	36.1(27.7-45.3)	61.1(37.5-80.9)	P≈0.02
Response to pheromones n = 40	Trail following behavior (n° of arcs walked)	C: 1 (1-2) T:11.5(8.0-18.0)	C: 1 (1 – 1) T:5.5(3.0-12.3)	NS P = 0.001
Audacity n = 15	n° of ants on a tower	1.3 (0 – 3)	4.5 (1 – 5)	P = 0.006
Food consumption n = 15	n° of ants eating meat (mean, extrema)	2.2 (1 – 3)	1.6 (0 – 2)	P = 0.000003
Learning ability n=120 + 20=140	% of correct responses	* C: 61/59 50% *7 hrs 61/59 51% *23 hrs 68/52 57% *30 hrs 68/52 57% *47 hrs 76/44 63% *55 hrs 75/45 63% *71 hrs 80/40 67% *79hrs 79/41 66% *95 hrs 84/36 70% *103 hrs 82/38 68% *118 hrs 87/33 73% *127 hrs 83/37 69% *145 hrs 94/26 78% *167 hrs 97/23 81% *215 hrs 96/24 80%	7 hrs 10/10 50% 23 hrs 11/9 55% 30 hrs 9/11 45% 47 hrs 10/10 50% 55 hrs 11/9 55% 71 hrs 11/9 55% 79hrs 8/12 40% 95 hrs 10/10 50% 103 hrs 9/11 45% 118 hrs 10/10 50% 127 hrs 10/10 50%	N=11,T=66 P = 0.0005
Memory n=120 + 20=150	% of correct responses	*6 hrs 87/33 73% 20 hrs 92/28 77% 46 hrs 81/39 68% 71 hrs 76/44 63% 95 hrs 71/49 59% 116 hrs 72/48 60% 143 hrs 69/51 58% 166 hrs 72/48 60%	*This could not be studied since ants never retained the presented cue*	

Details are given in the text. Briefly, cocaine decreased ants' linear speed, the precision of their response, their response to pheromone, and their food consumption; it increased their audacity. Under this alkaloid consumption, ants became unable to memorize a visual cue. The latter result is also graphically presented in Figure 2 (empty squares). * = results obtained by Cammaerts et al. (2011).

3.3.10 Time Period during Which Consumed Cocaine Affected the Ants

After having received cocaine for six days, the ants received no cocaine during one day, then again sugar water containing this alkaloid. They were numerous in coming drinking this liquid food. One hour later, the ants' locomotion was assessed, and just after, at t = 0 hrs, the food containing cocaine was replaced by sugar water free of this alkaloid. The ants' locomotion was then assessed after 1hrs, 2½hrs, 4½hrs, 7½hrs and 10hrs. During the first hour, effects of cocaine slightly decreased. Then, they rapidly decreased during the 2 – 3 following hours, and went on decreasing during 1 - 2 more hours. After 2½hrs, the ants' linear speed was already statistically similar to the control one (χ^2 = 0.15, df = 2, 0.90 < P < 0.95, NS). So, the effects of cocaine vanished (probably sigmoidally) in 4 to 5 hours (Table 6, line 3; Figure 3, empty squares).

Table 6. Habituation to cocaine, dependence on its consumption, and duration of its effect on ants

Effect Studied	Variable Assessed	Numerical Results	
Habituation (after 7 days)	Linear speed mm/sec	14.6 (12.7 – 17.3) *vs* 1 day: p<0.001	
		vs control: NS	
	Sinuosity ang.deg./cm	115 (109 – 128) *vs* 1 day: NS	
		vs control: NS	
Dependence (20 counts)	choices of sugar water +cocaine *vs* sugar water	nest 1: 18 ants *vs* 8 ants = 18/26 = 69.2%	
		nest 2: 34 ants *vs* 3 ants = 34/37 = 91.9%	
Duration of effect	Linear (mm/sec) and angular (ang.deg./cm) speed	*T 0hrs* 8.8(8.1-9.7) 130(117-149)	
		T 1hrs 9.4(8.3-9.9) 137(129-155)	
		T 2½hrs 12.0(11.5-14.0) 124(107-135)	
		T 4½hrs 13.0(11.8-14.6) 123(97-154)	
		T 7½hrs 14.0(12.0-15.2) 119(107-128)	
		T 10hrs 12.9(11.6-14.8) 132(112-146)	

Details are given in the text. Briefly, ants became habituated to cocaine, and dependent on its consumption. The effects of the alkaloid rapidly and sigmoidally decreased in the course of time, ending in about 4 - 5 hours; this observation is also presented in Figure 3 (empty squares).

3.4 Concerning Atropine

3.4.1 Effect on Ants' Locomotion

Atropine did not affect the ants' locomotion: the insects went on walking at the same linear speed (14.5 mm/sec *vs* 14.0 mm/sec; χ^2 = 2.78, df = 2, P > 0.20) with a nearly identical sinuosity (101 ang deg./cm *vs* 114 ang. deg./cm; χ^2 = 1.62, df = 2, P > 0.30; Table 7, line 1).

3.4.2 Effect on Ants' Precision When Reacting

Under atropine consumption, ants seemed having difficulties in olfactorily perceiving the source of alarm pheromone. When they sufficiently perceived this source, they moved towards it, but often, they badly perceived it and did not well orient themselves towards it. Finally, their orientation had a median value of 55 ang. deg. while during the control experiment, this value equaled 33.8 ang. deg. (χ^2 = 9.56, df = 2, P < 0.01; Table 7, line 2).

3.4.3 Effect on Ants' Response to Their Pheromones

Once more, under atropine consumption, the ants' olfactory perception seemed to be affected. The ants moved their antennae, cleaned them, moved slowly, stopped, and followed again the trail with difficulties. Their trail following behavior had a media value of 6 arcs, while the control value was 12.5 arcs (χ^2 =8.38, df = 3, 0.02 < P < 0.05; Table 7, line 3).

3.4.4 Effect on Ants' Audacity

After having consumed atropine for three days, no more or fewer ants came onto the presented tower (Table 7, line 4; U = 137, Z = 1.69, P = 0.09). Thus, the ants' audacity was not affected by atropine consumption.

3.4.5 Effect on Ants' Consumption of Food

After having consumed atropine for four days, the ants statistically and obviously consumed less meat food. They were four times less numerous on the presented *T. molitor* larvae (Table 7, line 5; U = 0, Z = 5.19, P = 0.000).

Table 7. Effect of atropine on seven physiological and ethological functions

Functions	Variable Assessed	Normal Diet	Diet + Atropine	Statistics
Activity n = 20	Linear speed mm/sec	14.0(12.7-15.0)	14.5(14.0-16.1)	NS
(locomotion)	Sinuosity ang.deg./cm	114(105-126)	101(86-121)	NS
Precision of a reaction n=20	Orientation to an alarm signal ang.deg.	33.8(27.9-45.2)	55.0(41.3-77.8)	P<0.01
Response to pheromones n = 40	Trail following behavior	C: 1 (1-1)	C: 1 (1 – 1.3)	NS
	(n° of arcs walked)	T:12.5(8.0-18.5)	T:6.0(3.0-9.3)	P < 0.05
Audacity n = 15	n° of ants on a tower	1.0 (0 – 2)	0.8 (0 – 2)	P = 0.09
Food consumption n = 15	n° of ants eating meat (mean, extrema)	3.2 (2 – 4)	0.8 (0 – 1)	P = 0.000
Learning ability n=120 + 20=140	% of correct responses	* C: 61/59 50% 7 hrs 61/59 51% 23 hrs 68/52 57% 30 hrs 68/52 57% 47 hrs 76/44 63% 55 hrs 75/45 63% 71 hrs 80/40 67% 79hrs 79/41 66% 95 hrs 84/36 70% 103 hrs 82/38 68% 118 hrs 87/33 73% 127 hrs 83/37 69% 145 hrs 94/26 78% 167 hrs 97/23 81% 215 hrs 96/24 80%	7 hrs 13/7 65% 23 hrs 14/6 70% 30 hrs 15/5 75% 47 hrs 16/4 80% 55 hrs 16/4 80% 71 hrs 15/5 75% 79hrs 15/5 75% 95 hrs 15/5 (75%) NS	N=8, T=36 P = 0.004
Memory n=120 + 20=150	% of correct responses	*6 hrs 87/33 73% 20 hrs 92/28 77% 46 hrs 81/39 68% 71 hrs 76/44 63% 95 hrs 71/49 59% 116 hrs 72/48 60% 143 hrs 69/51 58% 166 hrs 72/48 60%	6 hrs 15/5 75% 20 hrs 15/5 75% 46 hrs 14/6 70% 71 hrs 14/6 70% 95 hrs 14/6 70% 116 hrs 14/6 70% 143 hrs 14/6 70% 166 hrs 14/6 70%	N=8,T=34.5 0.008<P<0.012

Details are given in the text. Briefly, atropine did not affect the ants' linear and angular speed nor their audacity; it affected the ants' olfactory perception and consequently decreased the precision of their response to the alarm pheromone as well as their response to the trail pheromone; it decreased their food consumption. Under this alkaloid consumption, ants soon acquired visual conditioning but did not reach a higher score, and better memorized the learned visual cue. The latter result is graphically presented in Figure 2 (empty triangles). * = results obtained by Cammaerts et al. (2011).

3.4.6 Effect on Ants' Conditioning Capability

Ants having consumed atropine for seven days soon acquired operant visual conditioning: after 7, 23, 30, and 47 hrs of training, they reached the conditioning scores of 65%, 70%, 75%, and 80% respectively, while under normal diet, their score then equaled 51%, 57%, 57%, and 63% and reached 80% after 167 hrs of training. This result was statistically significant (N = 8, T = 36, P = 0.004). After having reached 80% of conditioning in 47 hrs, the ants having consumed atropine no longer increased their conditioning score and maintained themselves at the 75% - 80% level (Table 7, line 6; Figure 2, left part, empty circles and triangles). They so more quickly acquired conditioning but did not reach a better score.

3.4.7 Effect on Ants' Memory

After that the green follow cube was removed from the ants' food site, the ants' conditioning score (under atropine consumption) first decreased from 75% to 70%, like it decreased when ants received no atropine, then stayed at that rather high value for at least 200 hrs (Table 7, line 7; N = 8, T = 34.5, 0.008 < P < 0.012; Figure 2, right part, empty circles and triangles), the ants retaining so 20% of their learning instead of 10% when consuming no atropine. Thus, atropine somewhat increased the ants' long-lasting visual memory.

3.4.8 Ants' Habituation to Atropine

Under atropine consumption, the ants' locomotion was not affected (see above). Another parameter was thus used for examining the ants' habituation to atropine: the ants' orientation towards an isolated worker's head, a behavior affected by the alkaloid. After seven days of atropine consumption, the ants' orientation towards an alarm signal was still affected, even somewhat more than after one day of this alkaloid consumption (Table 8, line 1; χ^2 = 12.38, df = 1, P < 0.001). Ants' olfactory perception went on being affected by atropine, and so ants did not become habituated to that alkaloid consumption.

3.4.9 Ants' Dependence on Atropine

After seven days of atropine consumption, ants were presented with sugar food free of atropine and sugar food containing this alkaloid. They came onto the food free of the alkaloid and drunk, or went to the food containing atropine, tasted it, then went away and moved towards the food free of the alkaloid. Meanly 8.3% ants drunk the food containing atropine, while 91.7 drunk the food free of that alkaloid (Table 8, line 2; χ^2 = 44.8, df = 1, P < 0.001). Thus, ants did not become dependent on atropine consumption, and even avoided it.

Table 8. Habituation to atropine, dependence on its consumption, and duration of its effect on ants

Effect Studied	Variable Assessed	Numerical Results	
Habituation (after 7 days)	Orientation towards an alarm signal ang.deg.	76.6 (50.3 – 92.4)	*vs* control: p<0.001
			vs one day: NS
Dependence (20 counts)	choices of sugar water +atropine *vs* sugar water	nest 1: 2 ants *vs*	18 ants = 2/20 = 10%
		nest 2: 3 ants *vs*	45 ants = 3/48 = 6.6%
Duration of effect	Orientation towards an alarm signal ang.deg.	*T 0hrs*	67.7 (53.6 - 82.5)
		T 1hrs	60.4 (43.5 - 75.6)
		T 2½hrs	59.4 (43.1 - 76.9)
		T 4½hrs	55.7 (50 3 - 66.4)
		T 7½hrs	45.4 (34.6 - 65.1)
		T 10hrs	43.1 (37.4 - 54.0)
		T 24hrs	41.1 (33.0 - 48.0)

Details are given in the text. Briefly, ants did not become habituated to atropine, and did not become dependent on its consumption (they obviously preferred food free of atropine). The effects of this alkaloid first rapidly then slowly (so exponentially?) decreased in the course of time, ending in 20 - 30 hours; this observation is also presented in Figure 3 (empty triangles).

3.4.10 Time Period during Which Consumed Atropine Affected the Ants

Fourteen days after that the ants had consumed atropine, they no longer received this alkaloid for one day. Then, they once more received sugar water containing atropine and came drinking this liquid food. One hour later, at a given recorded time, this food was removed from the ants' tray and replaced by sugar water free of atropine. From that time, the ants' orientation towards an isolated worker's head was examined at successive given and recorded times (Table 8, line 3; Figure 3, empty triangles). At t = 4½hrs, the ants' orientation was still statistically affected (χ^2 = 8.11, df = 1, P < 0.01); at t = 10hrs, the effect of atropine was only slightly perceptible and at the limit of statistical significance (χ^2 = 4.59, df = 2, 0.05 < P < 0.10). So, the effects of this alkaloid decreased rapidly during one or two hours, then went on vanishing slowly for 24 -30 hrs.

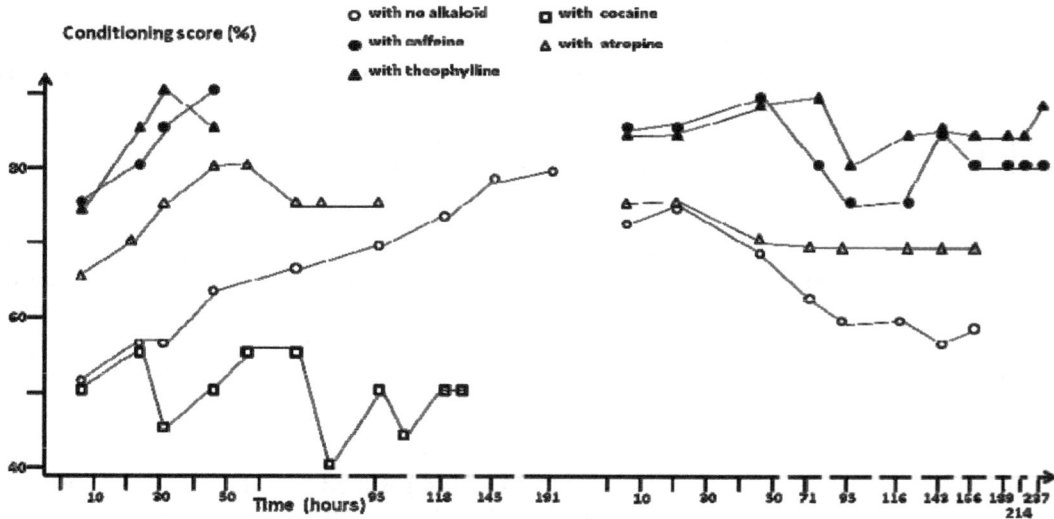

Figure 2. Kinetics of the acquisition (left graphs) and the loss (right graphs) of operant visual conditioning by ants consuming normal diet (empty circles), caffeine (black circles), theophylline (black triangles), cocaine (empty squares) or atropine (empty triangles)

Obviously, the two first alkaloids increased the ants' ability in acquiring conditioning and improved their memory, theophylline being the most efficient as for these effects; cocaine had a negative effect on these abilities and atropine, a moderated positive effect.

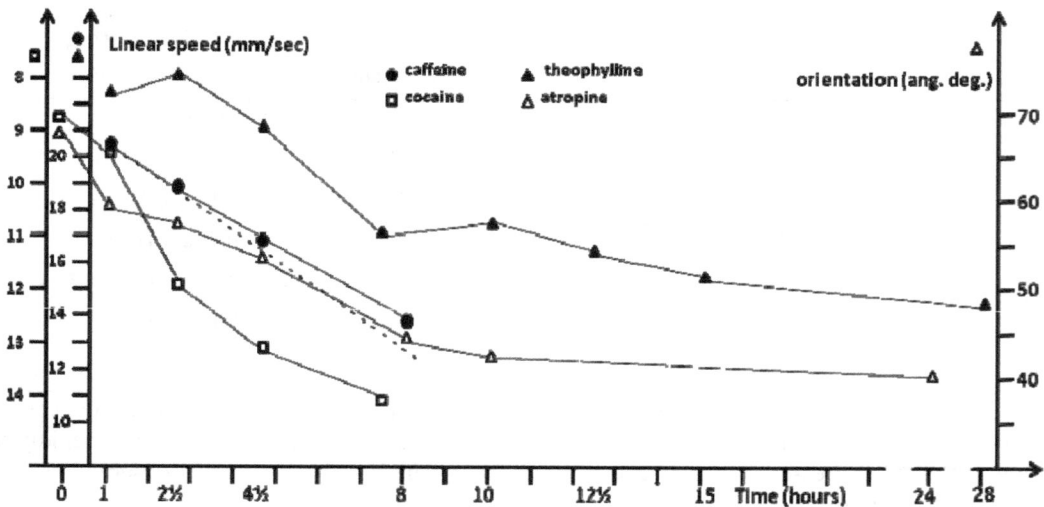

Figure 3. Loss of the effects induced by caffeine (black circles), theophylline (black triangles), and cocaine (empty squares) which Y axis is on the left (linear speed) as well as by atropine (empty triangles) which Y axis is on the right (orientation) after consumption of these substances ended

For caffeine, the loss was immediate and exponential; for theophylline, it was delayed and sigmoid; for cocaine, it was immediate, rapid and sigmoid; for atropine, it was first rapid then rather slow, so probably exponential

4. Discussion

We examined the effects, on ants, of four commonly used alkaloids, caffeine, theophylline, cocaine and atropine (Figure 4). We found that caffeine and theophylline increase the ants' speed of locomotion, slightly decrease the precision of their response, do not affect their response to pheromones, nor their audacity, and slightly reduce their consumption of food. Each of these two alkaloids, especially theophylline, spectacularly increases the ants' ability in acquiring conditioning and memory. The ants do not develop habituation to or dependence on each of these two alkaloids. At the end of consuming each of these two alkaloids, their effects decrease in the course of time: the effects of caffeine exponentially vanish in 11 – 16 hours; the effects of theophylline remain intact for 4 hours, and then sigmoidally vanish in about 28 hours. Cocaine decreases the ants' speed of locomotion, their precision of reaction, and their response to pheromone; it largely increases their audacity and reduces their consumption of food. Under cocaine consumption, ants retain nothing of a visual conditioning; they might still have a very short-lasting memory but no longer a long-lasting one. In a few days, the ants become habituated to the consumption of cocaine, and develop a strong dependence. The effects of cocaine vanish in 4 to 5 hours. Atropine does not affect the ants' locomotion but seems to reduce their olfactory perception and consequently decreases their orientation towards an alarm signal as well as their trail following behavior. This alkaloid does not affect the ants' audacity but decreases their food consumption. It increases the ants' rapidness in acquiring visual conditioning but not their conditioning score; it slightly increases the ants' visual long-lasting memory. Ants do not become habituated to nor dependent on, such alkaloid consumption.

All the effects revealed using ants as biological models are in agreement with those previously observed in mammals (Arnaud, 1987; Juliano & Griffiths, 2004; Brady, Lydiard, Malcolm, & Ballenger, 1991; Lupica & Berman, 1988; Vaubourdolle, 2007 and references therein) and in bees (Chittka & Peng, 2013). Indeed, these observations are, briefly, the following ones:

Caffeine is present in several plants such as coffee tree, tea plant, guarana, yerba mate, and cacaoyer; it is commonly consumed by human beings since it is present in coffee, tea, cacao, ice tea, and energizing drinks. It is rapidly and nearly completely assimilated by the organism, and quickly reaches the brain. Caffeine increases the individuals' awakening and decreases their physical and intellectual tiredness. This alkaloid increases the level of blood glucose, having a positive impact on the functioning of the brain as well as the muscles. Among others, nerve cells become more active and the cognitive performances of higher quality. As for the effect of this alkaloid on memory capacity, some kinds of memorization would be increased, but not all. Chittka and Peng (2013) showed that caffeine increases bees' memory. Caffeine is also known for having a tendency in reducing the individuals' appetite and for increasing their water elimination. Theophylline is present in several plants such as the tea plant, coffee tree, cacaoyer, guarana, yerba mate; it is commonly consumed by human beings via coffee, tea, cacao, and energizing drinks. This alkaloid increases the level of blood calcium, and consequently improves the functioning of the muscles. It increases the level of cyclic AMP, and consequently the quantity of available glucose, which induces a better general neuronal functioning. Theophylline increases the respiratory function, and is commonly used for helping people suffering from respiratory deficiencies. However, this last effect is still a matter of experimental investigation in order to confirm and elucidate the medical use of theophylline. Cocaine is a tropane alkaloid extracted from the plant 'coca'. It is known that persons consuming cocaine progressively become accustomed to and soon dependent on such consumption. These persons eat less, present unusual behavior, are no longer inhibited in performing risky tasks, and are less sensitive to pain. Atropine is a tropane alkaloid present in several Solanaceae plants such as belladonna, datura, jusquiame, and mandragore. Atropine sulfate is used in medicine for its negative effect on the cholinergic neurotransmissions, and consequently its positive effect on the adrenergic neurotransmissions. It can also be used as an antidote, but the quantity used must be limited for avoiding atropine poisoning. This alkaloid slows olfactory perception while diminished cholinergic neuronal activity does not. High amounts of atropine inhibit several physiological secretions, and increase the sense of pain.

In the present study, ants revealed several of the known effects of caffeine, theophylline, cocaine and atropine. These insects may thus be adequate biological models, at least initially, for studying effects of substances or treatments, before experimenting on mammals. The observations made on ants may be even more precise, more complete, and obtained in a shorter time period than those performed on mammals. For instance, the effect of theophylline on learning and memorization was not previously known as we could make using ants. Also, the decrease, in the course of time, of the alkaloid effects could be more precisely described when using ants. By doing so, it appeared that caffeine and theophylline are differently eliminated: the first alkaloid is immediately and exponentially eliminated while the second one is sigmoidally eliminated after some delay. Cocaine is rapidly and sigmoidally eliminated in a few hours, while atropine is first quickly, and then very slowly, eliminated. In the same way, ants, adequately tested, can objectively reveal the impact of substances (or treatments) on consumption of food, audacity, habituation and dependence.

Let us remark that cocaine decreased the ants' food consumption and this might have had a negative impact on the ants' conditioning ability, since food (meat food without cocaine) was used as a reward during training. But theophylline also slightly reduced the ants' food consumption though largely increasing the ants' ability in acquiring conditioning and improving their visual memory. Consequently, the adverse effect of cocaine on memorization must be seen as true, independent of the ants' lower consumption of food.

Concerning dependence, none was observed, in ants, for caffeine and theophylline. However, it is commonly accepted that people become dependent on these alkaloids. They are bitter and generally consumed at a concentration of about 1/1,000, which cannot be perceived. People might become dependent on other substances contained in or linked to food containing these alkaloids. On the contrary, a strong dependence on cocaine consumption was observed in ants, what also occurs in human beings.

Alkaloids	loc.	O	trail	audacity	food	conditioning	memory	hab.	Dependence	↘ effects
Caffeine	↗	↘	↘	—	↘	↗	↗	—	—	16hrs
Theoph.	↗	↘	↘	—	↘	↗↗	↗↗	—	—	28hrs
Cocaine	↘	↘	↘	↗	↘↘	↘↘	o	+	+	4-5hrs
Atropine	—	↘	—	—	↘	↗	↗	—	—	8-10hrs

Figure 4. Effects of four alkaloids, on basis of studies made on ants

Theoph.: theophylline; loc.: locomotion; O: orientation (→ precision of responses); trail: trail following behavior (→ responses to pheromones); food: food consumption; hab.: habituation to the alkaloid; ↘ effects: vanishing of the effects after the alkaloid consumption ended. The studies confirmed known effects (blue), lead to more precise information (green), bring new ones (orange), and reveal that dependence occurs when habituation exists and effects quickly vanished (red).

In ants, caffeine, theophylline and atropine did not lead to habituation, nor to dependence, while cocaine lead to such ethological effects. The effects of caffeine, theophylline and atropine slowly vanished in the course of time, whereas the effects of cocaine quickly vanished. It might be postulated that substances leading to dependence are those which primarily affect the nervous system, to which habituation soon occurs and the effects of which rapidly vanish in the course of time.

So, in summary, ants have been satisfactorily used as biological models. Such use of animals as biological models, for studying genetic, physiological, ethological, or embryological problems, is necessary and common in sciences and medicine. Invertebrates are actually more and more employed for unfolding such issues, and some of them are become famous (see references in the introduction section). We here demonstrate that ants could also be used in this manner; they are available during the entire year, they require small housing areas, their maintenance is easy and very cheap. As a conclusion of the present work, ants may be used, initially, to analyze the effects of substances and other influences on living organisms, before investigations are carried out on animals whose maintenance and manipulation is expensive and time consuming. One elegant feature of using ants as experimental animals is that these insects do not lend themselves to psychological explanatory models; they do react to the actual adverse or beneficial effects of a given treatment. They might also serve to reveal adverse effects of large amounts of substances consumed as well as to approximately determine their lethal dose. Finally, it shall be noted that ants are very sensitive to chemical compounds; thus, they may also be used for detecting the presence of presumed elements in collected samples, as already demonstrated by Cammaerts et al. (2012).

Acknowledgments

We are very grateful to Dr R. Cammaerts who helped us with the Mann-Whitney U test analyses, and was indulgent while we made the long-lasting experiments. We genuinely thank two anonymous referees whose comments allowed us improving our paper. We feel indebted to Associate Professor Olle Johansson, at the Karolinska Institute, Stockholm, Sweden, who patiently corrected the English of our paper. Our greatest thanks are for the Editor in chief and the Editorial Assistant for their attention and their excellent work.

References

Abramson, C. I., Wells, H., & Janko, B. (2007). A social insect model for the study of ethanol induced behavior: the honey bee. In R. Yoshida (Ed.), *Trends in Alcohol Abuse and Alcoholism Research* (pp. 197-218). Nova Sciences Publishers, Inc.

Andre, R. G., Wirtz, R. A., & Das, Y. T. (1989). Insect Models for Biomedical Research. In A. D. Woodhead (Ed.), *Nonmammalian Animal Models for Biomedical Research* (November 13, 2008). Boca Raton, FL: CRC Press.

Arnaud, A. M. (1987). The pharmacology of caffeine. *Progress in Drug Research, 31*, 273.

Bateson, G. (1979). *Mind and Nature: A Necessary Unity (Advances in Systems Theory, Complexity, and the Human Sciences)*. Hampton Press. ISBN 1-57273-434-5.

Billen, J., & Morgan, E. D. (1998) *Pheromone communication in social insects - sources and secretions*. In R. K. Vander Meer, M. D. Breed, K. E. Espelie, & M. L. Winston (Eds.), *Pheromone Communication in Social Insects : Ants, Wasps, Bees, and Termites* (pp. 3-33). Boulder, Oxford: Westview Press.

Brady, K. T., Lydiard, R. B., Malcolm, R., & Ballenger, J. C. (1991). Cocaine induced psychosis. *Journal of Clinical Psychiatry, 52*, 509-512.

Cammaerts, M.-C. (2012b). Olfactory and visual operant conditioning in the ant *Myrmica rubra* (Hymenoptera, Formicidae). *Bulletin de la Société Royale Belge d'Entomologie, 148*, 199-208.

Cammaerts, M.-C., & Cammaerts, D. (2012). Know-how of three *Myrmica* species foragers. *Symposium de la Société Royale Belge d'Entomologie* (December 7, 2012).

Cammaerts, M.-C., & Gosset, G. (2014). Impact of age, activity and diet on the conditioning performance in the ant *Myrmica ruginodis* used as a biological model. *International Journal of Biology, 6*(2), 10-20. http://dx.doi.org/10.5539/ijb.v6n2p10

Cammaerts, M.-C. (2004). Operant conditioning in the ant *Myrmica sabuleti*. *Behavioral Processes, 67*, 417-425. http://dx.doi.org/10.1016/j.beproc.2004.07.002

Cammaerts, M.-C. (2007). Colour vision in the ant *Myrmica sabuleti* MEINERT, 1861 (Hymenoptera: Formicidae). *Myrmecological News, 10*, 41-50.

Cammaerts, M.-C. (2008). Visual discrimination of cues differing as for their number of elements, their shape or their orientation, by the ant *Myrmica sabuleti*. *Biologia, 63*, 1169-1180. http://dx.doi.org/10.2478/s11756-008-0172-2

Cammaerts, M.-C. (2012a). Navigation system of the ant *Myrmica rubra* (Hymenoptera, Formicidae). *Myrmecological News, 16*, 111-121.

Cammaerts, M.-C. (2013a). Ants' learning of nest entrance characteristics (Hymenoptera, Formicidae). *Bulletin of Entomological Research, 6*.

Cammaerts, M.-C. (2013b). Learning of trail following behaviour by young *Myrmica rubra* workers (Hymenoptera, Formicidae). *ISRN Entomology*, Article ID: 792891, p. 6.

Cammaerts, M.-C. (2014). Performance of the species-typical alarm response in young workers of the ant *Myrmica sabuleti* is induced by interactions with mature workers. *Journal of Insect Sciences* (In press).

Cammaerts, M.-C. (2014). Learning of foraging area specific marking odor by ants (Hymenoptera, Formicidae). *Journal of Entomological Research* (In press).

Cammaerts, M.-C., & Nemeghaire, S. (2012). Why do workers of *Myrmica ruginodis* (Hymenoptera, Formicidae) navigate by relying mainly on their vision? *Bulletin de la Société Royale Belge d'Entomologie, 148*, 42-52.

Cammaerts, M.-C., & Rachidi, Z. (2009). Olfactive conditioning and use of visual and odorous elements for movement in the ant *Myrmica sabuleti* (Hymenoptera, Formicidae). *Myrmecological News, 12*, 117-127.

Cammaerts, M.-C., Morel, F., Martino, F., & Warzée, N. (2012). An easy and cheap software-based method to assess two-dimensional trajectories parameters. *Belgian Journal of Zoology, 142*, 145-151.

Cammaerts, M.-C., Rachidi, Z., & Cammaerts, D. (2011). Collective operant conditioning and circadian rhythms in the ant *Myrmica sabuleti* (Hymenoptera, Formicidae). *Bulletin de la Société Royale Belge d'Entomologie, 147*, 142-154.

Cammaerts, M.-C., Rachidi, Z., Bellens, F., & De Doncker, P. (2013). Food collection and responses to pheromones in an ant species exposed to electromagnetic radiation. *Electromagnetic Biology and Medicine, 32*(3), 315-332. http://dx.doi.org/10.3109/15368378.2012.712877

Cammaerts-Tricot, M.-C. (1973). Phéromone agrégeant les ouvrières de *Myrmica rubra*. *Journal of Insect Physiology, 19*, 1299-1315. http://dx.doi.org/10.1016/0022-1910(73)90213-8

Chittka, L., & Peng, F. (2013). Caffeine boosts bees' memories. *Science, 339*, 8. http://dx.doi.org/10.1126/science.1234411

Deutsch, J. (1994). *La drosophile: des chromosomes aux molécules* (p. 112). In J. Libbey (Ed.). Paris.

Devineni, A. V., & Heberlein, U. (2013). The evolution of *Drosophila melanogaster* as a model for alcohol addiction. *Annual Review of Neurosciences, 36*, 121-138. http://dx.doi.org/10.1146/annurev-neuro-062012-170256

Døring, T. D., & Chittka, L. (2011). How human are insects and does in matter? *Formosan Entomologist, 31*, 85-99.

Hölldobler, B., & Wilson, E. O. (1990). *The ants* (p. 732). Harvard University Press: Springer-Verlag Berlin. http://dx.doi.org/10.1007/978-3-662-10306-7

Juliano, L. M., & Griffiths, R. R. (2004). A critical review of caffeine withdrawal: empirical validation of symptoms and signs, incidence, severity, and associated features. *Psychopharmacology, 176*, 1-29. http://dx.doi.org/10.1007/s00213-004-2000-x

Keller, L., & Gordon, E. (2006). *La vie des fourmis* (p. 204). Odile Jacob, Paris.

Keller, R. A. (2011a). A phylogenetic analysis of ant morphology (Hymenoptera: Formicidae) with special reference to the Poneromorph subfamilies. *Bulletin of the American Museum of Natural History, 355*, 99. http://dx.doi.org/10.1206/355.1

Keller, R. A. (2011b). Ants protect their mouthparts by locking them in place. *Colloque organized at Banyuls/mer* (April, 2011).

Kolb, B., & Whishaw, I. Q. (2002). *Neuroscience & cognition: cerveau et comportement* (p. 635). New York, Basing Stoke: Worth Publishers.

Lupica, C. R., & Berman, R. F. (1988). Atropine slows olfactory bulb kindling while diminished cholinergic innervation does not. *Brain Research Bulletin, 20*, 203-209. http://dx.doi.org/10.1016/0361-9230(88)90180-3

Passera, L. (2006). *La véritable histoire des fourmis* (p. 340). Librairie Fayard.

Passera, L., & Aron, S. (2005). *Les fourmis: comportement, organisation sociale et evolution* (p. 480). Ottawa, Canada: Les Presses Scientifiques du CNRC.

Siegel, S., & Castellan, N. J. (1989). *Nonparametric statistics for the behavioural sciences* (p. 396). Singapore: McGraw-Hill Book Company.

Søvik, E., & Barron, A. B. (2013). Invertebrate models in addiction research. *Brain Behavior and Evolution, 82*, 153-165. http://dx.doi.org/10.1159/000355506

Vaubourdolle, M. (2007). *Médicaments* (Tome 2, Ed. 3, p. 867). France: Editeurs Wolters Kluwer.

Wehner, R., & Gehring, W. (1999). *Biologie et physiologie animals* (p. 844). De Boek Université, Thieme Berlag, Paris, Bruxelles.

Wolf, F. W., & Heberlein, U. (2003). Invertebrate models of drug abuse. *Journal of Neurobiology, 54*, 161-178. http://dx.doi.org/10.1002/neu.10166

3

Discovery, Modification and Production of T4 Lysozyme for Industrial and Medical Uses

Alaa Alhazmi[1]*, Johnathan Warren Stevenson[1]*, Samuel Amartey[2] & Wensheng Qin[1]

[1] Department of Biology, Lakehead University, 955 Oliver Road, Thunder Bay, ON, P7B 5E1, Canada

[2] Division of Biology, Imperial College of Science, Technology and Medicine, London, E8 1PQ, UK

Correspondence: Wensheng Qin, Department of Biology, Lakehead University, Thunder Bay, ON., P7B 5E1, Canada. Email: wqin@lakeheadu.ca

*The authors contribute equally to the work

Abstract

Lysozyme has attracted immense attention as an antimicrobial agent because of its ability to lyse the bacterial cell wall. It is found in a wide variety of body fluids and in cells of the innate immune system. Lysozyme can act as muramidase or as a Cationic Antimicrobial Peptide (CAMP). Lysozyme has many applications in the medical and industrial fields. Based on enzyme nomenclature, lysozyme is classified as a glycosylase under the group hydrolases. This manuscript covers a fundamental review of lysozyme in terms of discovery, history, functions and various sources and types of lysozyme. The biological and molecular structure is discussed as well as notable bioengineering and protein modifications. Furthermore, the mechanisms of resistance to lysozyme in microorganisms have also been discussed. Lastly, different methods that have been developed for detecting and measuring the activity of lysozyme are outlined. Although, a recombinant lysozyme has not yet been produced, several studies have attempted to generate a modified lysozyme either for large-scale production or that which is more suitable for industrialization purposes.

Keywords: antimicrobial, applications, biomolecular structure and functions, detection, thermostability

1. History

Lysozyme (1, 4-N-acetylmuramidase, E.C.3.2.1.17) is a small cationic protein first reported by Laschtschenko in 1909 (Burgess, 1973). However, the discovery of lysozyme is attributed to Alexander Fleming who reported that lysozyme was a "powerful bacteriolytic element found in human tissues and secretions" (Fleming, 1922). Fleming was at the time suffering from a cold and he allowed drops of his nasal secretions to fall onto an agar culture plate that was thickly colonized with bacteria. The plates were then incubated at a certain temperature for a number of hours, after which radial inhibition beyond the nasal mucus on the plate was observed. This experiment was later termed the lysoplate and it was concluded that the nasal secretions contained an enzyme capable of bacterial lysis. This enzyme was named lysozyme.

With the success of his preliminary experiment, Fleming (1922) continued to work with lysozyme, testing its antibacterial properties with several different bacteria. He reported the discovery of a small round bacterium *Micrococcus lysodeikticus* (now referred to as *Micrococcus luteus*) which was particularly vulnerable to lysozyme. He later diluted the nasal secretions in saline and added the solutions to a thick suspension of the *M. luteus*. Within minutes of incubation at 37°C the opaque bacterial solution had cleared. This experiment was later termed the Turbidimetric test. Fleming and Allison also showed increased levels of lysozyme in patients with pyogenic infections (pus producing) such as meningitis, the first indication that lysozyme could be a marker for sepsis (Fleming & Allison, 1927).

Since Fleming's initial discovery of lysozyme in tears and nasal secretions, many early studies have explored its various characteristics. For example, Wolff (1927) reported on its chemistry; Hoder (1931) its relation to immunology; Anderson (1932) its importance in vitamin A deficiency; Corper (1932) its relation to tuberculosis; Meyer et al. (1936a) its purification and properties; and Meyer et al. (1936b) combined with Daly (1938) the mechanism of its action. It is a paramount component of innate immunity due to its antibacterial, antiviral, antitumor and immune modulatory activities, such as anti-inflammation and immunomodulation (Helal et al., 2010). There has also been interest in lysozyme as a "natural" antibiotic. Commercially, lysozyme is produced from chicken egg whites. While T4 lysozyme (T4L) originates from phage T4, until now, there has been no commercial production of T4L ("BRI of CAAS," 2011, p. 4). However, in the future, the development of advanced technology in molecular biology and fermentation could facilitate large scale production of T4L for industrial and medical purposes.

2. Biomolecular Structure and Function

2.1 Biological and Molecular Functions of Lysozyme

Lysozyme is a bacteriolytic enzyme that has the ability to hydrolyze glycosidic bonds of 1,4-beta-linkages between *N-acetylglucosamine* (*NAG*) and *N-acetylmuramic acid* (*NAM*) in peptidoglycan (PG), which is present in the cell walls of prokaryotes (e.g., bacteria) (Mir, 1977; Akinalp et al., 2007) (Figure 1). In addition, its lytic activity has also been reported to inhibit viruses and eukaryotes, including fungi and parasites, in the absence of typical PG in their cell walls. However, the lysis of yeast and mould has been explained by the presence of an important component of their cell walls called chitin, which has the same β-(1-4) glycosidic bonds as the bacterial PG, except that chitin links two *NAG* residues rather than *NAG* and *NAM*; this means that lysozyme also possesses chitinase activity (Figure 2). Chitin is present in insects, crustaceans and fungi cell walls. Furthermore, the inhibition of *Entamoeba histolytica* by lysozyme can be explained by the presence in its membrane of lipopeptidophosphoglycan (LPPG), which can react with the enzyme in a similar manner as the PG (Benkerroum et al., 2008).

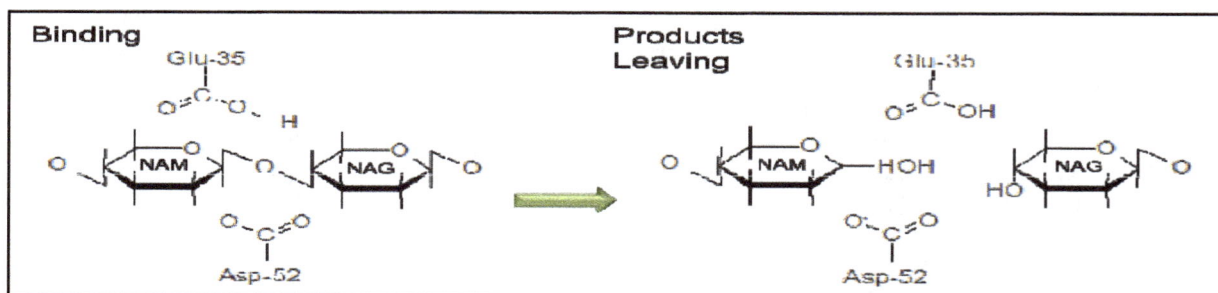

Figure 1. The catalytic mechanism of lysozyme. The substitution on the sugar rings have been omitted to improve the clarity of the diagram. The diagram was drawn by using ACD/ChemSketch Freeware software, a free comprehensive chemical drawing package (http://www.acdlabs.com). It was originally adapted from (Smith & Wood, 1991)

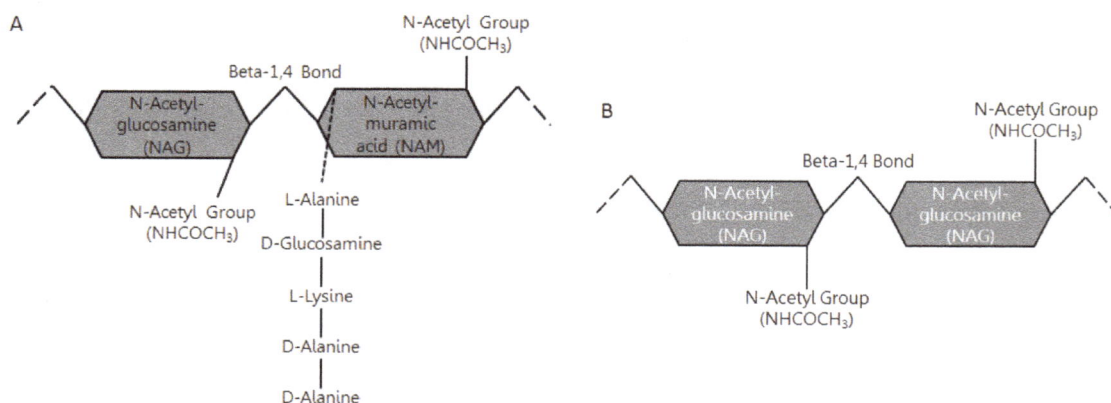

Figure 2. Simplistic representations of (A) peptidoglycan monomer and (B) chitin monomer. The beta-1,4 glycosidic bond between subunits is highlighted. Important substituents to the base glucosamine subunits are also highlighted, with other details left out to improve the clarity of this figure

Along with chitinase, lysozyme characterizes a significant class of polysaccharide-hydrolyzing enzymes. The chemical similarity between the two-polysaccharide substrates leads to some lysozymes being able to hydrolyze chitin, but they do so less effectively. Therefore, some lysozymes can be considered good chitinases, yet they have no obvious amino acid sequence similarities (Wohlkonig et al., 2010).

The antimicrobial activity of lysozyme has been extensively demonstrated *in vitro* or in physiological fluids and secretions, including milk, blood serum, saliva and urine. The antimicrobial activity of milk lysozyme, part of the non-specific innate defense mechanism, is also well established. Specifically, it acts independently by lysing sensitive bacteria or acts as a component of complex immunological reactions to enhance the phagocytosis of bacteria by macrophages thereby contributing to the innate protection from milk spoilage after drawing (i.e., bacteriostasis period) (Benkerroum et al., 2008). Lysozyme also has a Cationic Antimicrobial Peptide (CAMP) function that leads to bacterial death via the destabilization of the cytoplasmic membrane (Le Jeune et al., 2010).

Lysozyme is recognized to be non-dialyzable, soluble in water and a weak saline, insoluble in alcohol and ether, resistant to heat and desiccation and stable at room temperature (Burgess, 1973); it has also been demonstrated to be as stable at acidic pH and labile at alkaline pH (Jolles, 1969). Understanding the biochemistry and genetics of lysozyme enzymes, their phylogenetic relationships and methods of estimation will make them very useful in a variety of processes in the future (Patil et al., 2000).

2.2 The Structure of Lysozyme

Lysozyme is one of the first enzymes for which the X-ray structure was determined (Blake et al., 1965). Furthermore, several classes of lysozymes have been identified based on their sequence similarities. The best known include C-type (chicken-type), G-type (goose-type) and V-type (viral type) (Wohlkonig et al., 2010). Despite the variability in the genetic sequence of the classes of lysozyme, amino acids of the catalytic center of active sites are well conserved to maintain the hydrolytic function of the enzyme from different sources. In particular, glutamic acid (Glu35) and aspartic acid (Asp52) residues are directly involved in the breakdown of the glycosidic bond between *NAG* and *NAM*. As a result their presence in the catalytic center is crucial for the hydrolytic activity of the enzyme. Glutamic acid acts as a proton donor through the free carbonyl group of its side chain, while aspartic acid acts as a nucleophile that generates a glycosyl-enzyme intermediate, which then reacts with a water molecule to give the product of hydrolysis and release the enzyme unchanged (Figure 3). However, the amino acid sequence of known lysozyme has revealed that aspartic acid is not consistently present in the active sites of lysozyme molecules. In contrast, substitution of glutamic acid has resulted in a complete inactivation of the enzyme, which confirms the critical role of this amino acid in the enzymatic activity of lysozyme, regardless of its origin or the class to which it belongs (Benkerroum, 2008).

Furthermore, there are two structural domains and one polysaccharide ligand involved in the hydrolysis of glycosidic bonds. The structural domains are typically an α-helix and a β-sheet that form a cleft in which the ligand binds with polysaccharide at the active sites Glu35 and Asp52 (Figure 4) (Alberts et al., 2010). Once the polysaccharide binds to the active sites the enzyme begins to hydrolyze available glycosidic bonds within the polysaccharide.

The structure of T4 lysozyme is well characterized, the backbone of T4L has two separate domains, the N-terminal and the C-terminal, which are joined by a long α-helical chain (Baase et al., 2010). Structural studies have shown that the two domains form a pocket that opens and closes to facilitate ligand binding and release (Ramanathan et al., 2011). The native structure of this protein has been determined to 1.7Å resolution by X-ray crystallography (Poteete et al., 1997). Also, T4L is notably devoid of disulphide bridges since it only has two Cys residues, located in positions 48 and 97, locations that prevent interaction.

The potential of lysozyme has attracted considerable interest over many years (Gorin et al., 1971). For example, hen egg white lysozyme (HEWL) is very similar to human lysozyme; it has 129 amino acids, while human lysozyme has ~148 amino acids and is three to four times more reactive as tested by the turbidimetric test (Jolles, 1969; Lollike et al., 1995). However, due to its structural similarity, availability and inexpensiveness, HEWL has been used as a model for human lysozyme experiments, making it one of the most studied enzymes (Lollike et al., 1995).

Studies determining the structure of lysozymes showed that HEWL, T4 phage lysozyme (T4L) and goose egg white lysozyme (GEWL) have an overall similar three-dimensional structure, but drastically different amino acid sequences. Among the different lysozymes, the only highly conserved element within the active site is a glutamic acid residue (Glu-35 in HEWL, Glu-11 in T4L and Glu-73 in GEWL). As an example of the wide variability among the lysozymes, HEWL has a necessary Asp-52 residue, which does not correspond to Asp-20 of T4L, while GEWL has no equivalent for this acidic residue (Kuroki et al., 1999).

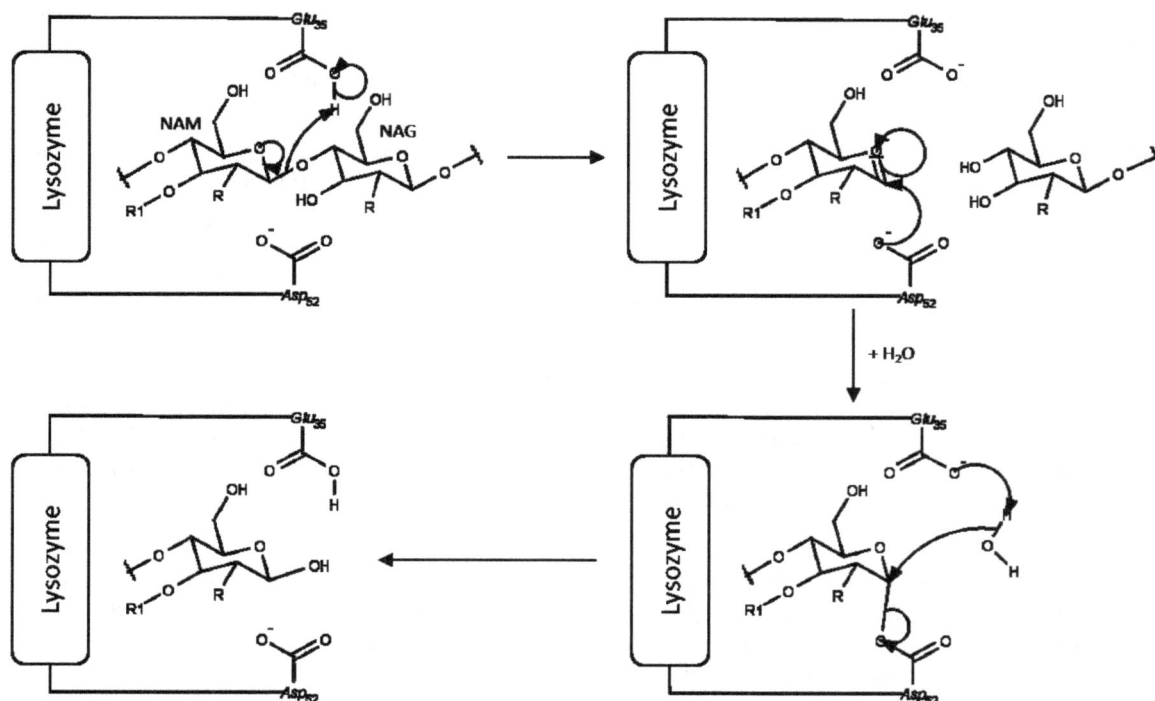

Figure 3. Chemical mechanism for enzymatic breakdown of a peptidoglycan monomer. R units denote the N-acetyl group and R1 denotes the tetrapeptide side chain to NAM outlined in Figure 2. This has been simplified to improve the clarity of the figure. This figure was sketched using MarvinSketch, a free program used for chemical sketching (http://www.chemaxon.com/products/marvin/marvinsketch/)

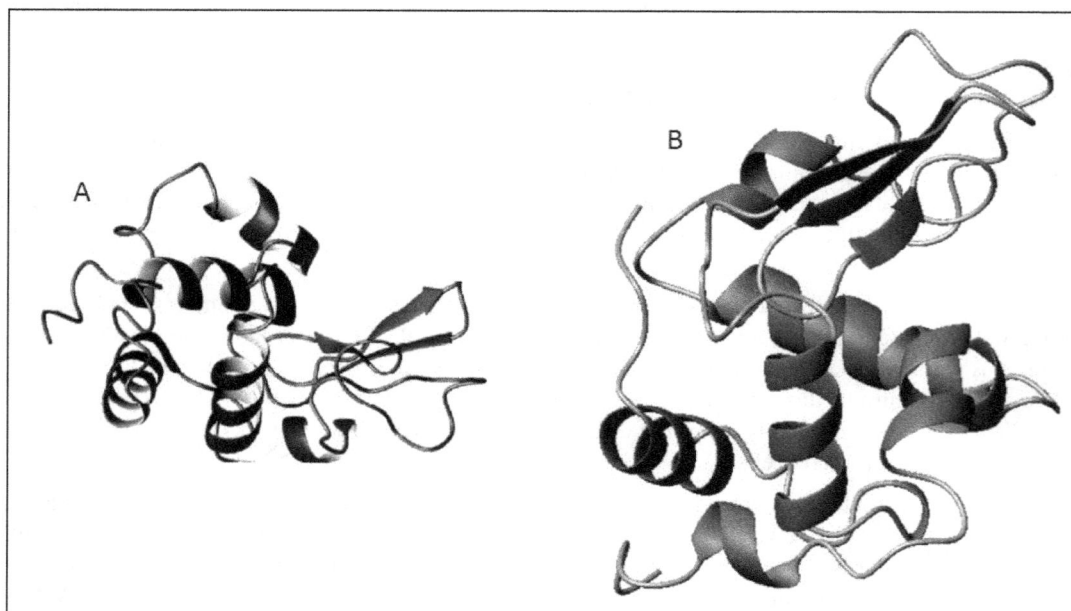

Figure 4. A ribbon model presenting (A) hen egg white lysozyme (HEWL) and (B) bovine milk lysozyme. The differences in structure among the two types of lysozymes are evident. Figure adapted from (Benkerroum, 2008)

2.3 Lysozyme from Different Species

While lysozyme is widely distributed in prokaryotes and eukaryotes, it maintains a consistent catalytic function even when grouped into various types based on its amino acid sequence, protein structure and catalytic characteristics (Akinalp et al., 2007; Dong et al., 2008). These types, or classes, are classified by lysozyme's wide variability in origin, structural, antigenic, chemical and enzymatic properties. The chicken-type (i.e. c-lysozyme), derived from the egg white of a domestic chicken (*Gallus gallus*), is the most studied and best known. Although c-lysozyme is typically found in the egg white of birds, it is also purified from various tissues and secretions of mammals. Other types of commonly studied lysozyme include the g-type, derived from the egg white of the domestic goose (*Anser anser*); h-type lysozyme from plants; i-type from invertebrates; b-type from bacteria (*Bacillus*); and v-type from viruses (Benkerroum, 2008).

Fleming (1922) showed that lysozyme from different sources affect different organisms: *Pneumococcus* was attacked only by the lysozyme of pus and *Streptococcus haemolyticus* only by the lysozyme of tears; *Staphylococcus* and *Streptococcus viridans* were more strongly affected by the lysozyme of nasal mucus than by that of tears or pus. Nevertheless, other researchers who worked with egg white lysozyme (which may act differently from lysozyme derived from body tissues) found that none of many *Staphylococcus albus* and *Staphylococcus aureus* strains tested, showed marked susceptibility to the lytic action of the lysozyme (Daly, 1938).

2.3.1 T4 Lysozyme

Phage T4 is a bacterial virus that infects only bacteria; specifically, it invades and controls bacterial metabolism, then terminates its incubation by lysing the infected-bacterial cell wall using T4L (Sulakvelidze et al., 2001; Trun & Trempy, 2004). T4L is similar to Hen Egg White Lysozyme (HEWL) in the structure of its active sites and sites of attack on the PG cell wall. Surprisingly, T4L lyses T4-infected cells by attacking the cell wall from the cytoplasmic side (Akinalp et al., 2007). In other words, it can have active sites on both sides of the bacterial cell wall. T4L is a small, well-characterized protein with a molecular weight of approximately 18 kDa and dimensions of 5.4 x 2.8 x 2.4 nm (Bower et al., 1998). It is a single polypeptide chain whose entire amino acid composition has been determined; it has a single NH_2-terminal methionine residue and a single COOH-terminal leucine residue (Tsugita et al., 1968).

Systematic probing of the T4L structure has also shown that 74 of the 164 positions in its sequence are sensitive to single amino acid substitutions. That is, at least one single amino acid substitution at one of these positions results in at least a 50-fold reduction in function. Not surprisingly, the most critical amino acid residues in the protein have been found to be buried or solvent-exposed in the active site cleft (Poteete et al., 1997).

Although lysozymes from T4 phage and hen egg whites have similar catalytic activity, they have non-homologous amino acid sequences. A comparison of the 3D structure of the phage enzyme with that of hen egg whites has been determined in other studies. The 3D structures of HEWL and T4L have also been seen to be quite different (Matthews & Remington, 1974); it is not clear whether their respective mechanisms of catalysis are related.

2.3.2 Hen Egg White Lysozyme (HEWL)

Lysozyme enzyme isolated from hen egg white is used extensively as a model in protein chemistry, enzymology, protein crystallography, molecular biology and genetics, immunology and evolution. The progress made by the study of these different aspects has been reviewed (Van Dael, 1998). However, HEWL remains, by far, the richest source of lysozyme, with a concentration that ranges between 3400 and 5840 mg/l (Benkerroum, 2008). HEWL is a 14,600 Da large monomeric protein that consists of 129 amino acid residues. It has been classified as a "hard" protein because the adsorption of HEWL to solid surfaces does not involve any major conformational changes (Malmsten, 2003).

2.3.3 Human Lysozyme

Lysozyme has also been identified in numerous human exocrine secretions such as gastroduodenal, genito-urinary, middle ear, salivary, lacrimal, upper and lower airway secretions and in colostrum (Hinnrasky et al., 1990). This enzyme is also found in a wide variety of body fluids, such as respiratory secretions and in cells of the innate immune system, including neutrophils, monocytes, macrophages and epithelial cells (Le Jeune et al., 2010).

Human lysozyme is a key component of the innate immune system and the production of recombinant forms of the enzyme could be promising in the search for therapeutic agents against drug-resistant infections. In addition to its catalytic hydrolysis of cell wall PG, human lysozyme also exhibits catalysis-independent antimicrobial properties. The dual functionality results in a protein that attacks Gram-positive and Gram-negative bacterial pathogens; it has been shown to be the most effective cationic anti-pseudomonal agent in human airway fluids (Gill et al., 2011). Lysozyme has also been found in high concentrations in human airway secretions collected from patients with chronic pulmonary diseases (Hinnrasky et al., 1990). In addition, human colon adenocarcinoma cells synthesize and secrete several specific proteins, including remarkably large amounts of lysozyme. The lack of diagnostic or prognostic markers, as well as the refractory nature of human colon carcinoma to standard chemoimmuno- and radio-therapy, makes this an extremely relevant system when investigating the chemistry and biology of such secreted macromolecules (Fett et al., 1985).

Human lysozyme is single polypeptide chain with a low molecular weight of around 14.6 kDa (Cabellero et al., 1999). Human lysozyme is present in the lysosomes of phagocytic cells, granulocytes and monocytes (Burgess et al., 1994). It is released as part of the non-specific immune response and exists among cells of the blood, especially leukocytes. Lysozyme is found in all stages of the maturation of the myelocytic series, but not in the myeloblast, eosinophil or basophil series (Davis, 1971). Monocytes contain large amounts of lysozyme, yet none are found in the lymphocytes. Moreover, in tissues, lysozyme is mainly found in bone marrow, lungs, intestines, spleen and kidneys. Lysozyme exists here due to the breakdown of neutrophilic granulocytes in these organs (Hansen et al., 1972). Tissue macrophages release lysozyme into serum, nasal and lacrimal secretions.

2.3.4 Lysozyme in Other Vertebrates

In higher vertebrates, lysozymes are involved in a broad battery of defense mechanisms and embrace several actions such as opsonization, immune response potentiation, restricted anti-viral and antineoplastic activities. Moreover, several studies have indicated that lysozyme may play a role as a defense mechanism against infectious disease in fish (Lie et al., 1989). In addition to their defensive role, lysozymes also play a role in digestive tasks; a dominant gene for high lysozyme activity has been detected in cattle (Lie et al., 1989). Also, high levels of intestinal lysozyme occur in some species of house mice in which the enzyme may play a digestive role. This trait appears in only one of two closely related lysozyme genes in the mouse genome; a Paneth cell (lysozyme P) and a macrophage-specific (lysozyme M) gene (Markart et al., 2004). This resulted in a specific over-expression of lysozyme P in the small intestine but a normal expression of lysozyme M in other tissues (Cross et al., 1988).

Furthermore, the milk of virtually all mammals contains lysozyme, either as a free soluble protein or within leucocytes and lysosomes. Although all lysozyme from milk has been reported as belonging to the c-type, it varies widely in terms of structure and physicochemical properties; this includes the folding and unfolding status, ability to bind calcium ions, stability to heat or pH and isoelectric point (Table 1). In regards to milk lysozyme content, human, equine and canine milks are the main representatives of high levels of lysozyme (averages of 200 to 1330 mg/L), while bovine, ovine and caprine milks (Table 2) represent low levels of lysozyme (3000 to 6000 times less than the milks of the first group) (Benkerroum, 2008). Additionally, the concentration of soluble lysozyme in milk varies considerably from one species to another, as well as within the same species, depending on various factors such as the breed, stage of lactation, parturition, nutrition, udder health and season of the year.

Table 1. Main chemical properties of milk lysozymes of different mammals: chicken egg-white lysozyme properties are also given as a reference for the c-type lysozyme (Benkerroum, 2008)

Origin	MW (kDa)	Catalytic center residues	Reference
Camel milk	14.4	NA	El Agamy et al. (1996)
Cow milk	14.4	Glu_{35} Asp_{53}	El Agamy et al. (2000)
Ewe milk	16.2	Glu_{53} Asp_{21}	Maroni et Cuccuri (2001)
Human	15	---	Parry et al. (1960)
Goat	14.4	Glu_{35} Asp_{53}	Jolles et al. (1990)
Buffalo	16.0	NA	El Agamy (2000)
Mare	14.7	Glu_{35} Asp_{53}	Jauregui-adell (1974); Sarwar et al. (2001)
Canine	14.5	Glu_{35} Asp_{53}	Grobler et al. (1994)
Egg-white	14.3	Glu_{35} Asp_{52}	Matagne and Dobson (1998)

Table 2. Reported concentrations (mg/L) of lysozyme in the milk of different mammals (Benkerroum, 2008)

Animal Species	Average concentration	References
Human	400	Mathur et al. (1990)
	320	Montagne et al. (1998)
	270-890	Montagne et al. (2001) and Chandan et al. (1968)
	224-426	Hennart et al. (2004)
Ass	1428	Salimei et al. (2004)
Mare	790	Jauregui-Adell (1975)
	1330	Sarwar et al. (2001)
Cow	0.13	Chandan et al. (1968)
	0.07	El Agamy et al. (1996)
	0.05 – 0.21	Piccinni et al. (2005)
Buffalo	0.0012	Priyadarshini and Kansal (2003)
Ewe	0.1	Chandan et al. (1968)
Goat	0.25	Chandan et al. (1968)
Sow	6.8	Schultz and Müller (1980)
Camel	0.15	El Agmay et al. (1996)

2.4 Bioengineering and Protein Modifications

The gene encoding lysozyme was cloned from mammalian tissues and secretions, insects, plants, protozoa, bacteria and viruses, then expressed in bacteria, fungi, yeast and plants (Akinalp et al., 2007). For example, T4L is one of the most thoroughly characterized enzymes through site-specific mutagenesis; more than 2,000 single-amino acid substitutions have been done (Malmsten, 2003). Studies using T4L variants, which differed from each other by at least one amino acid residue provided data that clearly showed a change in enzyme structure (Bower et al., 1998). Numerous variants of T4L have also been produced through site-directed mutagenesis, including a set of variants that differ by substitution of the isoleucine residue at position 3 (Ile 3). Producing variants help to test various physiological characteristics of T4L and the importance of various amino acids in regards to their input on overall protein structure and function. Since the Ile 3 residue contributes to a major hydrophobic core of T4L, any substitution can significantly alter the enzyme's stability (Bower et al., 1998). Notably, when Ile 3 is replaced by Cysteine (Cys) an increase in thermal stability is observed due to the formation of a disulphide bond with another Cys residue at position 97. Thermal unfolding of this mutant revealed that the disulphide bond stabilized the mutant by 1.2 kcal/mol compared to the wild-type. However, in another mutant, Ile 3 was replaced by the larger amino acid Tryptophan (Trp). This decreased the thermal stability of this mutant by 2.8 kcal/mol relative to the wild-type, one of the least stable T4L mutants characterized (Malmsten, 2003). Therefore, Ile 3 contributes to the major hydrophobic core of the C-terminal domain and helps link the N- and C-terminal domains.

In other studies, researchers examined the structure of a mutant T4L with a single amino acid substitution 2.5 nm from the active site. To put this in perspective, lysozyme dimensions are 3.0 nm x 3.0 nm x 4.5 nm (Kim et al., 2002). The change (replacement of glutamate-128 by a lysine residue) caused very little change in the structure of the lysozyme molecule. Nonetheless the enzyme had only 4% of the native enzyme's catalytic activity. These results suggest that glutamate-128 participates directly in substrate binding or catalysis. The glutamate residue is located on the C-terminal domain, an area that has no counterpart in the HEWL molecule. Therefore, it has been suggested that the role of this C-terminal domain is to bind the peptide crosslink that connects neighboring saccharide strands in the E. coli cell wall (Grutter & Matthews, 1982).

Generally, a major goal of protein engineering is to design proteins that demonstrate enhanced stability and activity. Among the physical forces that maintain the tertiary structure of proteins, disulphide bonds make a substantial contribution. However, the addition of new disulphides has not always increased stability. One aspect of this problem is the limited knowledge of the mechanism for which crosslinks such as disulphide bonds, stabilize or destabilize proteins (Mastsumura et al., 1989).

The main purposes for early structural studies of lysozyme were to screen for temperature-sensitive (TS) mutants; these studies did not have the benefit of site-directed mutagenesis as methods to generate specific substitutions had not yet been developed. As a result, classical protein chemistry had to be used to determine any sequence changes in TS lysozymes, rather than methods involving DNA sequencing. Nonetheless, five TS T4Ls were developed that reduced the protein's melting temperature by an average of 10°C. These mutants included the single amino acid changes: Arg 96 → His (R96H), Met 102 → Thr (M102T), Ala 146 → Thr (A146T), Gly 156 → Asp (G156D), and Thr 157 → Ile (T157I) (Baase et al., 2010). In a similar study, a multiple mutant was engineered from T4L: I3L/S38D/A41V/A82P/N116D/V131A/N144D (PDB Code 189L) and was shown to increase T_m by 8.3°C (Zhang et al., 1995).

Currently, entropic effects, hydrophobic stabilization, helical propensity, salt-bridge interactions, metal binding, disulphide bridges, methionine replacement, computational procedures and genetic selection methods are used to generate TS T4Ls. Disulphide bridges are particularly effective at increasing the stability of the protein; for example the formation of a Cys 21-Cys 142 bridge across the active site cleft increases melting temperature by 11°C at pH 2 because the protein bends to accommodate the disulphide bond (Baase et al., 2010). Furthermore, the genetics of T4L have been studied extensively and a variety of mutant enzymes have been isolated and characterized. A previous study suggested that the residues Asp 20, Glu 22, Glu 105, Trp 138, Asn 140 and Glu 141 are all essential for full catalytic activity on the grounds that changes in any of these amino acids drastically reduce the catalytic effectiveness of the enzyme (Matthews & Remington, 1974).

In contrast, the close proximity of the substituted residue Ile 3 to the active site Glu 11 may also contribute to the observed differences in activity among T4L variants because both residues are located on the same α-helix. The α-helix content of a protein in a solution is known to increase or decrease in response to changes in its environment (e.g., pH, ionic strength and hydrophobic nature of the solvent). A T4L variant containing a relatively large tryptophan residue in position 3 could make conformational changes in adjacent residues. If these effects are propagated along the α-helix, significant changes in the orientation of the active site may result in a decrease in catalytic activity (Bower et al., 1998).

Moreover, the loss of catalytic activity of the T4L variants might also relate to the degree of interference caused by mutation in the hinge region of the enzyme. These amino acid substitutions are located directly opposite the active site in an area of the molecule that is responsible for the opening and closing movement of the active cleft. As such, an enzyme would retain full activity only if the substrate were able to force its way past the "hinged" cleft to reach the active site (Bower et al., 1998).

3 Lysozyme Applications

3.1 Diagnostic Applications

Lysozyme has been shown to be an indicator in some hematological malignancy; such as acute myeloid leukemia. In addition, an elevated serum lysozyme level could be helpful in differentiating between the types of cancer that affect blood, bone marrow and lymph nodes. However, lysozyme has been associated with other abnormalities since its activity increases in the gastric juice of patients with a peptic ulcer and in the colonic sections of patients with ulcerative colitis and regional enteritis (Mir, 1977). Elevated levels of lysozyme are also associated with certain abnormalities in humans. Deviations in lysozyme levels of blood and urine have been correlated to particular myeloid and renal abnormalities and lysozyme levels have subsequently been used to monitor the success of therapy. Although the lysozyme protein can be found in most tissues, studies at the cellular level have revealed that it is present in high concentrations, specifically in phagocytic cells and the Paneth cells (including the intestine and the proximal tubules of the kidney). While the kidney accumulates at least some of its lysozyme from the blood, endogenous lysozyme synthesis has been detected in mammalian myeloid cells (Cross et al., 1988). Additionally, lysozymuria has been reported in various renal tubular disorders as a manifestation of the disease. Furthermore, discrepancies could be caused by differences in methods, patient populations and stages of disease (Mir, 1977). It was also reported that lysozyme is an effective agent for killing HIV *in vitro* (Song & Hou, 2003)

3.2 Industrial Applications of Lysozyme

Lysozyme is gaining importance for its biotechnological applications; particularly in agriculture to control plant pathogens. It has also been used in food preservation due to its antibacterial activity. The success in employing lysozyme for different industrial applications depends on the supply of highly active preparations at a reasonable cost (Patil et al., 2000). Although egg white is the primary source for lysozyme production on an industrial scale, other sources such as mammals' milk contain lysozyme molecules with specific properties that are not present in the conventional egg white lysozyme (Benkerroum, 2008).

Chicken egg white lysozyme has also been recently used in some oenology processes (e.g. wine maturation) at a level of 100 mg/L to control contamination by lactic bacteria (Benkerroum, 2008; Ferreira et al., 2011). Furthermore, from a biochemical perspective, lysozyme has an optimum activity between 40°C and 45°C and can remain active up to 62°C, which is compatible with the temperature used for enzymatic hydrolysis. Also, enzymatic activity is effective within a pH value range between 3.5 and 7; however, it can remain active between pH 2 and 10 (Ferreira et al., 2011).

The use of lysozyme can reduce the use of sulfur dioxide (SO_2) to limit microbial contamination in many industrial processes. However, this usage is expensive and has health-related side effects (Ferreira et al., 2011).

3.2.1 Lysozyme in Food Packaging

Antimicrobial packaging is a form of active packaging because it has the ability to kill or inhibit spoilage by pathogenic microorganisms that contaminate foods unlike conventional packaging systems that only increase shelf-life (Han, 2000). Table 3 shows the application area when lysozyme is used as an antimicrobial agent and when packaging materials are used in antimicrobial food packaging. Several studies have shown that incorporating lysozyme as a preservative in packaging materials extends the shelf life of foods (Appendini & Hotchkiss, 1997; Buonocore et al., 2004; Mecitoglu et al., 2006).

Table 3. Application area, antimicrobial agents and packaging materials used in antimicrobial food packaging (Han 2000; Appendini & Hotchkiss, 2002)

Antimicrobial Agent	Packaging Material	Application area
Lysozyme	LDPE*	Bell pepper
	LDPE	Cheese

* LDPE: low-density polyethylene.

Food-borne microbial outbreaks have garnered much of the media's attention and as such have sparked a greater interest in antimicrobial food packaging. Antimicrobial agents may be mixed with the initial food formulations, or they may be applied to food surfaces by dusting, dipping or spraying (Min & Krochta, 2005). However, the antimicrobial agents may be neutralized in reactions within the food system, diminishing their protective abilities. Furthermore, antimicrobial agents within the food are unable to target food surfaces, where spoilage reactions take place most intensely. Antimicrobial packaging allows for a slow release of antimicrobial agent from the film onto the surface of the food, allowing for the maintenance of the concentration needed for the inhibition of microbial growth (De Roever, 1998; Devlieghere et al., 2004).

Typically, antimicrobial films are produced with organic or inorganic acids, metals, alcohols, ammonium compounds or amines (Appendini & Hotchkiss, 2002; Suppakul et al., 2003). However, natural bio-preservatives such as antimicrobial enzymes and bacteriocins have gained much interest due to a growing concern for healthy foods (Suppakul et al., 2003; Labuza & Breene, 1989). Naturally, lysozyme has become one of the most studied bio-preservatives for antimicrobial packaging applications (Han, 2000; Quintavalla & Vicini, 2002).

There is also interest in increasing the antimicrobial spectrum of lysozyme. For instance, the combination of lysozyme with EDTA makes lysozyme highly effective on Gram-negative bacteria (Mecitoglu et al., 2006), and conjugates of lysozyme with dextran, galactomannan or xyloglucan have good antimicrobial activity against both Gram-positive and Gram-negative bacteria when applied in combination with mild heating at 50°C (Nakamura et al., 1992).

3.2.2 Resistance to Lysozyme by Gram Positive/Negative Bacteria

While PG in Gram-positive bacteria cell walls is freely accessible to lysozyme, PG in Gram-negative bacteria is shielded by the lipopolysaccharidic (LPS) layer of the outer membrane. In addition to this protective effect against the hydrolytic action of lysozyme, Gram-negative bacteria have recently been shown to use another strategy that involves specific protein-inhibitors with high affinity to lysozyme (Benkerroum, 2008). Bacterial resistance to lysozyme is not exclusively related to the presence of the LPS layer because it does not provide absolute protection against the hydrolytic action of lysozyme. Sensitive Gram-negative bacteria have been described, along with mechanisms not hindered by LPS.

Specifically, Fleming and Allison (1927) were the first to report resistant strains of *Streptococcus*. They found that lysozyme was not effective against a range of bacteria that included *Haemophilus influenzae, Neisseria meningitidis, Escherichia coli, Streptococcus pneumonia* (Gram-positive), *Klebsiella pneumoniae* and group B *Streptococci* (Gram-positive). They then characterized lysozyme resistance as an inability to break down bacteria that are encapsulated with a gelatinous polysaccharide layer. This dramatically limits the effectiveness of lysozyme to bacteria that have no capsule. Lysozyme is therefore of little value as a therapeutic agent against bacterial growth, but the release of lysozyme as a product of phagocytosis and white blood cell turnover may be a useful marker for sepsis.

3.2.3 Mechanism of Resistance to Lysozyme

The complexity of the composition and structure of peptidoglycan is well known (Le Jeune et al., 2010). Its pathological effects are greatly enhanced by various modifications and substitutions to its basic composition and structure (Clarke & Dupont, 1992). While some bacterial species are sensitive to lysozyme, many bacteria (like some important human pathogens such as *Staphylococcus aureus, Neisseria gonorrhoeae, Enterococcus faecalis and Proteus mirabilis*) are resistant to lysozyme (Le Jeune et al., 2010; Herbert et al., 2007).

The exact mechanism of lysozyme resistance is not fully understood and may vary according to the bacterial strain or species (Benkerroum, 2008). One mechanism of resistance that has been suggested is interference of lysozyme action by surface attachment polymers (e.g., capsular polysaccharides and teichoic acids) (Benkerroum, 2008). For example, a high degree of peptide cross-linkage can add positively charged residues to teichoic and lipoteichoic acids, thereby modifying the net negative charge on the bacterial cell surface (Bera et al., 2005; Herbert et al., 2007; Benkerroum, 2008). This modification helps the bacteria to avoid the CAMP activity of lysozyme and disrupts many other antimicrobial peptides (Herbert et al., 2007; Le Jeune et al., 2010). In most cases, these peptide side chains are involved in the direct cross-linking between adjacent glycan strands; however, in some instances, a more complicated arrangement exists. For example, a pentaglycine peptide bridges the neighbouring peptide side chains within the peptidoglycan sacculus of *Staphylococcus aureus* (Clarke & Dupont, 1992); this is a modification of different sites of the PG structure by a type of enzyme (e.g., the peptidoglycan-specific *O*-acetyltransferase (*OatA*) of *S. aureus*) (Le Jeune et al., 2010). Similarly, this has been reported for 11 Gram-positive and Gram-negative species (Bera et al., 2005), which results in *O*-acetylation of hexosamine residues of the cell wall PG (Benkerroum, 2008) and the N-acetylglucosamine deacetylase (*PgdA*) of *Streptococcus pneumonia.* Both prevent the binding of lysozyme to its substrate and contribute to the muramidase resistance. *N-deacetylation* of the acetamido group of the hexosamine residues (Benkerroum, 2008), incorporation of D-Aspartic acid in the bacterial PG cross bridge, as was demonstrated in *Lactococcus lactis*, and the production of protein-inhibitors specific to lysozyme, *N*-non-substituted glucosamine residues in the peptidoglycan, as in *Bacillus cereus* and *Streptococcus pneumonia* (Bera et al., 2005). Lysozyme resistance may also consist of the production of lysozyme inhibitors such as the Streptococcal Inhibitor of Complement (SIC) in *streptococci*; the inhibitor of vertebrate lysozyme in *E. coli*; the Periplasmic Lysozyme Inhibitor of c-type lysozyme in *Salmonella enteritidis*; or the Membrane-bound Lysozyme Inhibitor of c-type lysozyme in *E. coli* (Le Jeune et al., 2010).

Over recent years, these mechanisms have been gaining attention. While there is a general agreement that surface attachment polymers and the degree of peptide cross-linking do not account for lysozyme-resistance, it is not clear whether there is only a single mechanism of resistance for all bacteria or if specific mechanisms may be used by specific strains or species. However, the above mechanisms may be modulated by other factors that are not directly involved in lysozyme resistance. For example, the presence of teichoic acid or a high degree of peptide cross-linking, though shown not to have an intrinsic effect on lysozyme resistance, has significantly enhanced the effect of *O*-acetylation on lysozyme resistance in *S. aureus* (Benkerroum, 2008). For example, some species of eubacteria known to possess *O*-acetylated peptidoglycan like *Neisseria gonorrhoeae, Proteus mirabilis* and *Staphylococcus aureus* have the presence of acetyl moieties at the C-6 hydroxyl group of *N-acetylmuramyl* residues (Clarke & Dupont, 1992). *O*-acetylation of peptidoglycan produces corresponding 2,6-diacetylmuramyl derivatives. The extent of naturally occurring *O*-acetylation in microorganisms ranges between < 10% and 70%, while a spontaneous mutant of *Micrococcus luteus*, cultured in the presence of HEWL, has been reported to have a molar ratio of 1:1 for *N-acetylmuramic* acid and *O*-acetyl. The role of *O*-acetylation in conferring resistance to the hydrolytic activity of HEWL was discerned soon after the initial discovery of the modification. Subsequent studies have confirmed the resistance of *O*-acetylated peptidoglycan to lysozyme; however, it should be noted that the efficacy of others, such as the N,O-diacetylmuramidases of *Chalaropsis* and *Streptomyces globisporus* is not affected by O-acetyl groups (Clarke & Dupont, 1992).

Moreover, as with other types of modifications to peptidoglycan, the chloramphenicol-induced increase in the degree of 6-*O*-acetylation of the *N. gonorrhoeae* PG was observed to immediately follow the cessation of protein synthesis; this subsequent acetylation was limited to newly incorporated PG. Therefore, while the process of *O*-acetylation occurs outside the cytoplasm, it would appear that it is regulated by agents that act directly on cytoplasmic targets. Of therapeutic concern, the increased 6-O-acetyl content of cell walls in bacteria that are treated with this antibiotic serve to increase its resistance to lysozyme (Clarke & Dupont, 1992).

To overcome the bactericidal action of lysozyme, bacteria have developed different mechanisms, among which some have been well dissected. Further, these are mainly based on the modification of the PG structure (Le Jeune et al., 2010). There is still a need for further investigation to explore mechanisms of lysozyme resistance to improve the use of lysozyme in therapeutics.

3.2.4 Inhibitory Activity of Lysozyme against Microorganisms Other than Bacteria

To date, no clear explanation has been provided for the sensitivity of some viruses to lysozyme. However, it was suggested that the inhibition of the herpes virus is due to interference with antiviral activity because of the basic nature of lysozyme, rather than the hydrolytic activity (Cisani et al., 1984). Moreover, *in vitro* inhibition of human immunodeficiency virus (HIV) by lysozyme was attributed to the hydrolysis of viral polysaccharides and genomic RNA (Benkerroum, 2008). Therefore, lysozyme may not always act by its hydrolytic activity, but it could inhibit the growth of some microorganisms either by permeabilizing the plasma membrane or acting on intracellular components by virtue of its cationic hydrophobic nature, as has been described for a variety of antimicrobial peptides. Sublethal concentrations (10 µg/ml) of lysozyme were shown to accumulate in the cytoplasm of *Candida albicans* and reduce the production and activity of aspartic proteinase, a putative virulence factor of yeast. This finding indicates that lysozyme acts at the transcriptional or translational level of DNA expression and, while at high concentrations, induces cell-swelling and invaginations near bud scars. This suggests that interference with the synthesis of cell-wall components may be alternative targets for this enzyme (Benkerroum, 2008).

Several factors that influence lytic efficiency have been discussed in other studies. Specifically, increases in the amount of lysozyme used were not particularly effective in increasing extent of lysis, but it was found that lysis improved with the lengthening of the incubation period. Another factor that is likely to produce a large change in lytic efficiency is the growth medium. It was found that supplementation of the growth medium with L-threonine, L-lysine or both, usually produced streptococci or lactobacilli cells that were lysed more easily. The action of threonine can be explained by its known interference in the establishment of cell wall cross-links. The mode of action of L-lysine is more difficult to explain, but it is obvious that the composition of the growth medium can affect the susceptibility of bacteria to lysozyme; therefore, it should be evaluated with strains that are difficult to lyse (Chassy & Giuffrida, 1980). In addition, the choice of buffers could be a major factor in determining the efficiency of lysozyme. Other studies have used high pH values, high buffer concentrations or phosphate buffers. Thus, the use of NaCl, MgCl$_2$, other salts or chelating anions should be avoided because these additives appear to interfere with lysis of Gram-positive, asporogenous bacteria. Finally, another factor that contributes to its effectiveness is the enzymatic stabilizer polyethylene glycol (PEG), which appears to act as a stimulant of lysis rather than an osmotic stabilizer (Chassy & Giuffrida, 1980).

4. Lysozyme Detection

Numerous methods have been reported for the detection of lysozyme and the determination of its concentration. A method was reported in 1946 in which an Ostwald or Uberlohde's viscometer was employed in the determination of lysozyme by monitoring the decrease in viscosity of various substrates (Song & Hou, 2003). Until now, the most widely used method was chromatography involving HPLC and affinity chromatography. Furthermore, a nephelometric immunoassay has also been developed and has a detection limit of 0.58 mg/ml for assay of lysozyme. Several other useful methods have been proposed, including response surface methodology and fluorimetry and acoustic wave viscosity sensing (Song & Hou, 2003).

4.1 Agar Plate Bioassays

The lysoplate method has been used as the basis for assay development because it showed that lysozyme was able to diffuse through the agar and prevent growth of the bacteria (Fleming & Allison, 1922). In another experiment, agar was mixed with heat inactivated bacteria so that it appeared turbid. The agar was allowed to solidify and then wells were bored into it. Lysozyme samples were added to these wells and allowed to diffuse over 12-18 hours. Then the zone of turbidity clearing was observed to determine the proportional concentration of lysozyme (Osserman & Lawlor, 1966). Similarly, the measurement of lysozyme activity by agar plate bioassay with *Micrococcus luteus* cells as a test strain is an indicator test. Enzyme-action produces a clear zone of inhibition of *M. luteus* with diameters that are proportional to the catalytic activity of the enzyme (Bower, Xuand and McGuire, 1998). Therefore, the use of agar plate bioassays is a quick, simple and efficient method for determining lysozyme activity.

4.2 Activity in Media Measurements (Turbidimetric)

Many of the turbidimetric methods used to measure the activity of lysozyme are similar and differ only with respect to the lysozyme activity and sample preparation. These differences might include buffer composition, pH, ionic strength, concentration of *M. luteus* substrate, temperature, duration of incubation and preparation of enzyme (Houser, 1983). An example of this method involves dissolving *M. luteus* cells in phosphate buffer (e.g. 0.01 M; pH 7.3) to give a stable suspension with an optical density of 0.9 at 450 nm. The activity of the lysozyme sample was determined by the decrease in turbidity of the *Micrococcus* suspension (at 450 nm) using a spectrophotometer. In this method, for each activity assay, 100 µl of T4L (0.5 mg/ml) was added to 3.0 ml of substrate and the decrease in optical density (absorbance) was recorded every 0.1 s for 60 s (Bower et al., 1998).

In a review, Klass et al. (1977) described the turbidimetric technique as having a high sensitivity and a rapid turnover rate; each sample took only one minute to process. They developed a method that has a detection limit of 1 µg/ml whereas a method by Ronan et al. (1975) has a detection limit of 1.5 mg/ml. Although the method by Klass et al. (1977) is the preferred method for routine testing of clinical samples because it is time effective, it is not the most reliable method as it produces a high inter-batch variability of results, with a precision of 2.5% as was obtained by Gorin et al. (1971). Bergmeyer (1965) however achieved a higher rate of precision. Regardless of its unreliability, this assay is easy to perform, requires small sample volumes, can be used with serum, urine and tears and is sensitive (Gorin et al., 1971; Bergmeyer, 1965).

An assay method that involved the use of two colorimeters, which correct the urine and serum discoloration problems experienced with controls (causing turbid solutions for blank samples), was developed. The bacterial suspension and clinical samples were continuously stirred and standards were run simultaneously. This automated assay was able to process 20 samples per hour and had an increased (yet undisclosed) sensitivity (Terry et al., 1971).

Furthermore, another method based on the principles of the turbidimetric assay was devised by Caballero et al. (1999). They described a micro-particle enhanced nephelometric immunoassay using serum and urine samples from patients. This assay involved the use of polystyrene particles covalently bound with anti-lysozyme antibodies. The polystyrene particles formed larger particles due to the binding with free lysozyme in the sample, which resulted in scattered light. The scattered light at the start and end of the reaction was compared and used to calculate the lysozyme concentration in the patient sample. The assay detection limit was 0.58 mg/L. However, this assay involves long incubation periods, requires biological fluid pre-treatment and has poor detection limit (Caballero et al., 1999).

Finally, a commercial assay currently on the market that employs the principle of turbidimetric assays is the EnzChek® Lysozyme by Molecular Probes (Leiden, Netherlands). This test measures lysozyme in solution at levels as low as 20 U/ml (equivalent to <0.5 µg/ml). The lysozyme activity is measured using *M. luteus* cell walls as substrate. The bacterial cell walls are specially labeled with a fluorophore in such a way that the fluorescence is quenched. Thus, activated lysozyme reduces the quenching while increasing the fluorescence. The fluorescence is therefore proportional to the lysozyme activity. The increase in fluorescence is measured using a spectro-fluorometer, mini fluorometer or a fluorescence microplate reader. Each assay takes around 30 minutes with less than an hour preparation time. The assay lays claim to being "simple and sensitive" (https://tools.invitrogen.com/content/sfs/manuals/mp22013.pdf).

1.4.3 Thermostability of Lysozyme on SDS-PAGE

To determine the thermostability of lysozyme enzymes, Akinalp et al. (2007) exposed the supernatant of various recombinant bacteria to various temperatures (from 37°C to 100°C) for 15 min followed by centrifugation at 15,000 rpm to remove denatured proteins. The supernatant was then mixed with an equal volume of trichloroacetic acid and the protein collected by centrifugation. Protein analysis was then performed using a denaturing polyacrylamide gel (SDS-PAGE, 12% (w/v)). After electrophoresis, protein bands were visualized by a coomassie blue staining. As a result, the thermostability of the lysozymes from the recombinant bacteria was found to be different. The lysozyme expressed by *S. salivarius* subsp. *thermophilus* cells has high thermoresistance and was not denatured at 70°C for 15 min. In contrast, the enzyme expressed by *L. lactis* and *E. coli* cells was easily denatured when exposed to the same temperature.

4.4 Immunoassays

The lack of sensitivity and long assay times of the turbidimetric and lysoplate techniques has led to the development of various immunoassay methods for the determination of lysozymes. Immunoassays rely on the reaction between the target analytic and a specific binding molecule of biological descent (the antibody). They can produce both quantitative and qualitative results and have shown considerable improvements in sensitivity (Ekins & Chu, 1997). For example, Porstmann et al. (1989) developed an enzyme immunoassay for the detection of lysozyme in patients with Crohn's disease and rheumatoid arthritis. Urine samples were taken from patients and were tested using three variations of the same method. The method showing highest sensitivity involved pre-coating a microtitre plate with anti-lysozyme IgG overnight at 4°C, followed by two hours incubation at room temperature with the clinical sample and IgG-HRP (Horseradish Peroxidase) conjugate. Then, the assay was stopped by addition of o-henylenediamine with 15 minutes incubation. The detection limit for this assay was 0.2 µg/L (Porstmann et al., 1989).

Francina et al. (1986) reported an immunoassay developed to test lysozyme secretion in serum of acute myeloid leukemia patients. A microtitre plate was incubated at 4°C overnight with anti-lysozyme IgG. After a wash, the plate was incubated for one hour at 37°C with the clinical sample then biotinylated antilysozyme was added and the plate was again incubated for a further hour. The plate was then washed and incubated with avidin peroxidase solution for 10 minutes at room temperature. The assay was stopped by the addition of enzyme substrate and incubated for five minutes. The total assay time was ~14.5 hours. To reduce assay time the two one-hour incubations were reduced to 20 minutes and the enzyme substrate incubation to five minutes. The detection limit for the standard assay was 0.1 ng/ml and for the rapid assays 1 ng/ml. In another example, Taylor et al. (1992) reported an immunoassay to measure lysozyme in a healthy adult's serum and urine samples. The method involved pre-coating a microtitre plate with rabbit anti-human lysozyme, followed by addition of the clinical sample and incubation for 90 minutes at room temperature. Additionally, a conjugated sheep anti-human lysozyme was added and incubated for a further 90 minutes at room temperature. Finally, enzyme substrate p-nitrophenyl sodium phosphate was added and incubated for 30 minutes at room temperature. The total assay time was 15 hours and the detection limit was 1 µg/L.

At present, there are some immunoassay kits available on the market and these include the Human Lysozyme EIA kit from Biomedical Technologies Inc. (Stoughton, USA). Referred to as sandwich ELISA, this method detects lysozyme from serum, plasma, urine, tears and saliva; the reference value for human lysozyme from serum ranges from 3-10 µg/ml. In this assay, specific lysozyme antibodies bound to polystyrene wells are incubated with a sample and then a second human lysozyme-specific antibody is added with a horseradish peroxidase conjugated secondary antibody. The total test time is 4 ¼ hours (http://www.funakoshi.co.jp/data/datasheet/BTI/BT-630.pdf).

Similarly, a company Orgentec (Mainz, Germany) supplies a kit known as Anti-Lysozyme kit for lysozyme in serum and plasma. This test can process 96 patient samples at a time, only requires 10 µl of sample and involves a plate pre-coated with antibody. The antigen from the patients' sample is added along with a horseradish peroxidase conjugate and then followed by TMB (a color substrate). The reaction is stopped by adding hydrochloric acid; the total time for this assay is two hours, due to the plates being purchased pre-coated (http://www.orgentec.com/products/pdfs/ELISA_en_IFU_ORG_526.pdf). All of these kits have the advantage of conducting analysis on several patient samples using minimal biological fluid.

4.5 Biosensor

Biosensor and related bioarray techniques are rapidly growing fields. A biosensor is an analytical device that converts the concentration of the target substances into an electrical signal via a combination of biological or biologically-derived recognition systems, either integrated within or intimately associated with a suitable physico-chemical transducer. Essentially, biosensors consist of three parts: recognition, transduction and signal output. A biological sensing element coordinated with a transducer will generate a signal that is proportional to the target analyses. An interaction between the biological receptor molecules and the target analyses causes a change in one or more physical or chemical parameters such as the generation of ions, gases, electrons, heat or mass. These changes are interpreted by the transducer and converted into electrical signals (Figure 5).

In addition, biological or biologically-derived elements are capable of recognizing the presence, activity, or concentration of a specific target analyte in a complex mixture of components. The recognition element may comprise one of three different types: affinity biosensors (based on ligand-receptor interactions such as those involving antibodies, nucleic acid, aptamers, peptides, protein or cell receptor); binding and catalysis (involves enzymes, microorganisms, organelles, plant or animal cells or tissue slices); or biomimetic receptors (based on various synthetic binding or catalytic systems). The interaction, or subsequent reaction, of the recognition element with the analyte in a sample matrix results in a reactant or formation of a product that is immediately proximal to the transducer. The latter converts the change in solution property into a quantifiable electrical signal. The transducer is a device, usually electronic, electroacoustic, electro-optical, electromagnetic, electrothermal or electromechanical that converts one type of energy (electricity, sound, light, magnetism, heat or mechanical) into another (usually electrical) for various purposes including measurement or information transfer (Marks et al., 2007).

More research is needed to develop an electrochemical approach for the detection of lysozyme; such a biosensor will offer an alternative, sensitive and versatile method for protein detection.

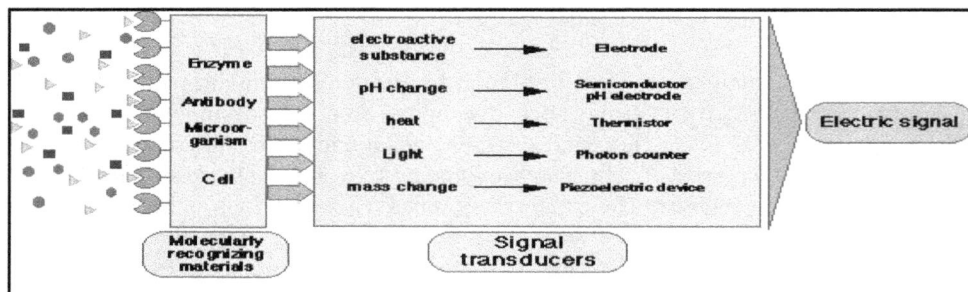

Figure 5. The principle of a biosensor (http://www.jaist.ac.jp/~yokoyama/images/biosensor.gif)

4.6 Flow Injection Chemiluminescence System

A new procedure using chemiluminescence (CL) detection has been developed to determine the levels of lysozyme at a ng/ml level. This procedure, termed Flow Injection Chemiluminescence (FICL) uses controlled reagent release technology within a flow injection system where the analytical reagents are immobilized on an anion-exchange resin. Upon injection with water, some of the analytical reagents, like luminol and periodate, are eluted from the column. Initially, CL is inhibited due to the presence of lysozyme, but upon release from the column, these reagents are free to luminesce (Song & Hou, 2003). The concentration of lysozyme can be calculated within a range of 30-1000 ng/ml based on the decline in CL intensity (Figure 6). Using this method, a typical analytical procedure could be performed in 0.5 minutes, including sampling and washing, when performed at a flow rate of 2.0 ml/min. This would provide a throughput of 120 ml/h with a relative standard deviation of less than 3.0%. Song and Hou (2003) successfully applied this method to determine the concentration of lysozyme in human tear and saliva samples with recovery rates from 92.0% to 105.7%.

Flow Injection Chemiluminescence (FICL) analysis is becoming increasingly important in various fields because of its high sensitivity, rapidity, simplicity and feasibility. The fast oxidation reaction between luminol and periodate in alkaline medium produces a strong CL signal. Furthermore, CL is greatly inhibited by lysozyme, so a sensitive CL inhibition assay for lysozyme combined with FI technology was described (Figure 7). This method offered simple and cheap instrumentation as well as rapid and reproducible means of detection; demonstrated in its application to determine lysozyme in human tear and saliva. Subsequently, a sensitive and convenient analytical method using control-reagent-release technology in a flow injection system was developed to determine lysozyme in human tears and saliva by its inhibitory effect on CL. In addition, incorporation of the immobilized reagents column into the flow injection analysis (FIA) manifold offers satisfactory stability, good reproducibility and precision in the analysis (Song & Hou, 2003).

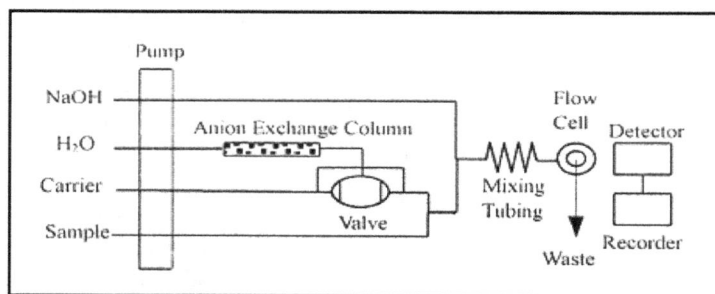

Figure 6. Schematic diagram of the flow-injection system for lysozyme determination (Song & Hou, 2003)

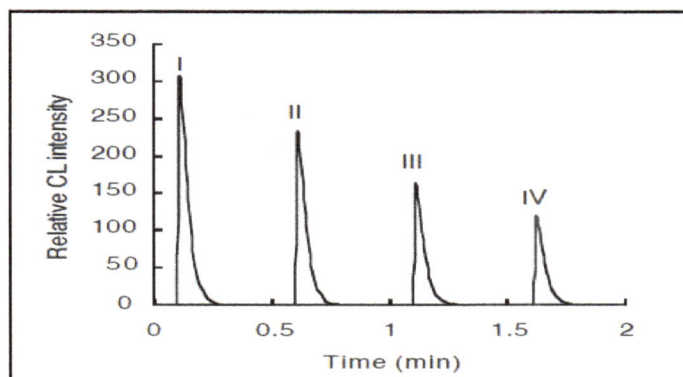

Figure 7. CL time profile in the batch system. I - CL intensity in the absence of lysozyme; II - CL intensity in the presence of lysozyme (50 ng ml^{-1}); III - CL intensity in the presence of lysozyme (150 ng ml^{-1}); IV - CL intensity in the presence of lysozyme (500 ng ml^{-1}) (Song & Hou, 2003)

5. Conclusion

This paper has covered multiple facets of lysozyme, including its history, functions and industrial purposes. Mechanisms of resistance were also explored, while protein structure and notable modifications were stressed. Lastly, various methods used in lysozyme detection were presented. Lysozyme is a crucial enzyme in the innate immune system and has many potential industrial uses such as in food processing, packaging and preservation. Studying it would also provide more knowledge about how the body fights infection and other intrusions. Although several sources for lysozyme exist, future studies should be directed to generating a modified structure suitable for large scale production or to produce a recombinant variation. It is also possible to explore other purposes for lysozyme such as biofilm management on varying surfaces, modified chitinase abilities or selective targeting for various specific bacterial peptidoglycans.

References

Akinalp, A. S., Asan, M., & Ozcan, N. (2007). Expression of T4 lysozyme gene (gene e) in Streptococcus salivarius subsp. thermophilus. *Afr. J. Biotechnol., 6*, 963-966.

Alberts, B., Bray, D., Hopkin, K., Johnson, A., Lewis, J., Raff, M., ... Walter, P. (2010). *Essential Cell Biology* (3rd ed.). London: Taylor and Francis Group.

Appendini, P., & Hotchkiss, J. H. (1997). Immobilization of lysozyme on food contact polymers as potential antimicrobial films. *Packag. Technol Sci., 10*, 271-279. http://dx.doi.org/10.1002/(SICI)1099-1522(199709/10)10:5<271::AID-PTS412>3.0.CO;2-R

Appendini, P., & Hotchkiss, J. H. (2002). Review of antimicrobial food packaging. *Innov. Food Set Emerg. Technol., 3*, 113-126. http://dx.doi.org/10.1016/S1466-8564(02)00012-7

Baase, W. A., Liu, L., Tronrud, D. E., & Matthews, B. W. (2010). Lessons from the lysozyme of phage T4. *Protein Sci, 19*(4), 631-641. http://dx.doi.org/10.1002/pro.344

Benkerroum, N. (2008). Antimicrobial activity of lysozyme with special relevance to milk. *Afr. J. Biotechnol., 7*, 4856–4867.

Bera, A., Herbert, S., Jakob, A., Vollmer, W., & Gotz, F. (2005). Why are pathogenic staphylococci so lysozyme resistant? The peptidoglycan O-acetyltransferase OatA is the major determinant for lysozyme resistance of Staphylococcus aureus. *Mol Microbiol, 55*(3), 778-787. http://dx.doi.org/10.1111/j.1365-2958.2004.04446.x

CCM international limited. (2011). Bio-products: BRI of CAAS awaiting investments in T4 lysozyme industrialization. *Industrial Biotechnologies China News, 3*(01), 4.

Blake, C. C., Koenig, D. F., Mair, G. A., North, A. C., Phillips, D. C., & Sarma, V. R. (1965). Structure of hen egg-white lysozyme. A three-dimensional Fourier synthesis at 2 A resolution. *Nature, 206*(4986), 757-61. http://dx.doi.org/10.1038/206757a0

Bower, C. K., Sananikone, S., Bothwell, M. K., & McGuire, J. (1999). Activity losses among T4 lysozyme charge variants after adsorption to colloidal silica. *Biotechnol Bioeng, 64*(3), 373-376. http://dx.doi.org/10.1002/(SICI)1097-0290(19990805)64:3<373::AID-BIT14>3.0.CO;2-J

Bower, C. K., Xu, Q., & McGuire, J. (1998). Activity losses among T4 lysozyme variants after adsorption to colloidal silica. *Biotechnol Bioeng, 58*(6), 658-662. http://dx.doi.org/10.1002/(SICI)1097-0290(19980620)58:6<658::AID-BIT13>3.0.CO;2-3

Buonocore, G. G., Sinigaglia, M., Corbo, M. R., Bevilacqua, A., La Notte, E., & Del Nobile, M. A. (2004). Controlled release of antimicrobial compounds from highly swellable polymers. *J Food Prot, 67*(6), 1190-1194.

Burgess, P. (1973). Lysozyme: History, Methods of Assay and Some Applications. *Industrial Year Report, Hatfield Polytechnic.*

Burgess, P., Appel, S. H., Wilson, C. A., & Polk, H. C., Jr. (1994). Detection of intraabdominal abscess by serum lysozyme estimation. *Surgery, 115*(1), 16-21.

Caballero, M., Ruiz, R., Marquez de Prado, M., Seco, M., Borque, L., & Escanero, J. F. (1999). Development of a microparticle-enhanced nephelometric immunoassay for quantitation of human lysozyme in pleural effusion and plasma. *J Clin Lab Anal, 13*(6), 301-307. http://dx.doi.org/10.1002/(SICI)1098-2825(1999)13:6<301::AID-JCLA9>3.0.CO;2-3

Chang, H. M., Yang, C. C., & Chang, Y. C. (2000). Rapid separation of lysozyme from chicken egg white by reductants and thermal treatment. *J Agric Food Chem, 48*(2), 161-164. http://dx.doi.org/10.1021/jf9902797

Chassy, B. M., & Giuffrida, A. (1980). Method for the lysis of Gram-positive, asporogenous bacteria with lysozyme. *Appl Environ Microbiol, 39*(1), 153-158.

Cisani, G., Varaldo, P.E., Ingianni, A., Pompei, R., & Satta, G., (1984). Inhibition of herpes simplex virus-induced cytopathic effect by modified hen egg-white lysozymes. *Current Microbiology, 10,* 35–40. http://dx.doi.org/10.1007/BF01576045

Clarke, A. J., & Dupont, C. (1992). O-acetylated peptidoglycan: its occurrence, pathobiological significance, and biosynthesis. *Can J Microbiol, 38*(2), 85-91. http://dx.doi.org/10.1139/m92-014

Cross, M., Mangelsdorf, I., Wedel, A., & Renkawitz, R. (1988). Mouse lysozyme M gene: isolation, characterization, and expression studies. *Proc Natl Acad Sci U S A, 85*(17), 6232-6236. http://dx.doi.org/10.1073/pnas.85.17.6232

Daly, S. (1938). Lysozyme of Nasal Mucus: Method of preparation and preliminary report on its effects on growth and virulence of common pathogens of paranasal sinuses. *Arch Otolaryngol, 27*(2), 189-196. http://dx.doi.org/10.1001/archotol.1938.00650030198007

Davis, C. S. (1971). Diagnostic value of muramidase. *Postgrad Med, 49*(4), 51-54.

De Roever, C. (1998). Microbiological Safety Evaluations and Recommendations on Fresh Produce. *Food Control., 9,* 321-347. http://dx.dci.org/10.1016/S0956-7135(98)00022-X

Devlieghere, F., Vermeiren, L., & Debevere, J. (2004). New preservation technologies: possibilities and limitations. Int. Dairy J. 14, 273–285. http://dx.doi.org/10.1016/j.idairyj.2003.07.002

Dong, S., Shew, H. D., Tredway, L. P., Lu, J., Sivamani, E., Miller, E. S., & Qu, R. (2008). Expression of the bacteriophage T4 lysozyme gene in tall fescue confers resistance to gray leaf spot & brown patch diseases. Transgenic Res, 17(1), 47-57. http://dx.doi.org/10.1007/s11248-007-9073-3

Ekins, R., & Chu, F. (1997). Immunoassay & other ligand assays: present status and future trends. *J Int Fed Clin Chem, 9*(3), 100-109.

Fett, J. W., Strydom, D. J., Lobb, R. R., Alderman, E. M., Vallee, B. L., Artymiuk, P. J., & Redfield, C. (1985). Lysozyme: a major secretory product of a human colon carcinoma cell line. *Biochemistry, 24*(4), 965-975. http://dx.doi.org/10.1021/bi00325a024

Fleming, A. (1922). On a remarkable bacteriolytic element found in tissues & secretions. *Proc. Roy. Soc. London, 93,* 306-317. http://dx.doi.org/10.1098/rspb.1922.0023

Fleming, A., & Allison, V. D. (1922). Further observations on a bacteriolytic element found in tissues & secretions. *Proc. Roy. Soc. London. Series B, 44,* 142-151. http://dx.doi.org/10.1098/rspb.1922.0051

Fleming, A., & Allison, V. D. (1927). On the Development of Strains of Bacteria Resistant to Lysozyme Action & the Relation of Lysozyme Action to Intracellular Digestion. *J. Brit. Exp. Path., 8,* 214-218.

Francina, A., Cloppet, H., Guinet, R., Rossi, M., Guyotat, D., Gentilhomme, O., & Richard, M. (1986). A rapid & sensitive non-competitive avidin-biotin immuno-enzymatic assay for lysozyme. *J Immunol Methods, 87*(2), 267-272. http://dx.doi.org/10.1016/0022-1759(86)90541-7

Gill, A., Scanlon, T. C., Osipovitch, D. C., Madden, D. R., & Griswold, K. E. (2011). Crystal structure of a charge engineered human lysozyme having enhanced bactericidal activity. *PLoS One, 6*(3), e16788. http://dx.doi.org/10.1371/journal.pone.0016788

Gorin, G., Wang, S. F., & Papapavlou, L. (1971). Assay of lysozyme by its lytic action on M. lysodeikticus cells. *Anal Biochem, 39*(1), 113-127. http://dx.doi.org/10.1016/0003-2697(71)90467-2

Grutter, M. G., & Matthews, B. W. (1982). Amino acid substitutions far from the active site of bacteriophage T4 lysozyme reduce catalytic activity and suggest that the C-terminal lobe of the enzyme participates in substrate binding. *J Mol Biol, 154*(3), 525-535. http://dx.doi.org/10.1016/S0022-2836(82)80011-9

Han, J. H. (2000). Antimicrobial food packaging. *Food Technol., 3*, 56–65.

Hansen, N. E., Karle, H., Andersen, V., & Olgaard, K. (1972). Lysozyme turnover in man. *J Clin Invest, 51*(5), 1146-1155. http://dx.doi.org/10.1172/JCI106907

Helal, R., & Melzig, M. F. (2010). In vitro Effects of Selected Saponins on the Production and Release of Lysozyme Activity of Human Monocytic and Epithelial Cell Lines. *Sci Pharm., 79*(2), 337-349. http://dx.doi.org/10.3797/scipharm.1012-15

Herbert, S., Bera, A., Nerz, C., Kraus, D., Peschel, A., Goerke, C., ... Gotz, F. (2007). Molecular basis of resistance to muramidase and cationic antimicrobial peptide activity of lysozyme in staphylococci. *PLoS Pathog, 3*(7), e102. http://dx.doi.org/10.1371/journal.ppat.0030102

Hinnrasky, J., Chevillard, M., & Puchelle, E. (1990). Immunocytochemical demonstration of quantitative differences in the distribution of lysozyme in human airway secretory granule phenotypes. *Biol Cell, 68*(3), 239-243. http://dx.doi.org/10.1016/0248-4900(90)90314-S

Houser, M. T. (1983). Improved turbidimetric assay for lysozyme in urine. *Clin Chem, 29*(8), 1488-1493.

Houser, M. T. (1983). Improved turbidimetric assay for lysozyme in urine. *Clin. Chem., 29*, 1488-1493

Ibrahim, H. R., Higashiguchi, S., Jugena, L. R., Kim, M., & Yamamoto, T. (1996). A Structural Phase of Heat-Denaturated Lysozyme with Novel Antimicrobial Action. *J. Agric. Food Chem., 44*, 1416-1423. http://dx.doi.org/10.1021/jf9507147

Ibrahim, H. R., Kato, A., Kobayashi, K. (1991). Antimicrobial Effects of Lysozyme Against Gram-Negative Bacteria due to Covalent Binding of Palmitic Acid. *J. Agric. Food Chem., 39*, 2077-2083. http://dx.doi.org/10.1021/jf00011a039

Jollès, P. (1969). Lysozymes: a chapter of molecular biology. *Angew. Chem. Int. Ed. Engl., 8*(4), 227–239. http://dx.doi.org/10.1002/anie.196902271

Kim, D. T., Blanch, H. W., & Radke, C. J. (2002). Direct imaging of lysozyme adsorption onto mica by atomic force microscopy. *Langmuir., 18*(15), 5841-5850. http://dx.doi.org/10.1021/la0256331

Klass, H. J., Hopkins, J., Neale, G., & Peters, T. J. (1977). The estimation of serum lysozyme: a comparison of four assay methods. *Biochem Med, 18*(1), 52-57. http://dx.doi.org/10.1016/0006-2944(77)90048-5

Kuroki, R., Weaver, L. H., & Matthews, B. W. (1999). Structural basis of the conversion of T4 lysozyme into a transglycosidase by reengineering the active site. *Proc Natl Acad Sci U S A, 96*(16), 8949-8954. http://dx.doi.org/10.1073/pnas.96.16.8949

Labuza, T. P., & Breene, W. (1989). Application of 'active packaging' technologies for the improvement of shelf-life and nutritional quality of fresh and extended shelf-life foods. *Bibl Nutr Dieta*, (43), 252-259.

Le Jeune, A., Torelli, R., Sanguinetti, M., Giard, J. C., Hartke, A., Auffray, Y., & Benachour, A. (2010). The extracytoplasmic function sigma factor SigV plays a key role in the original model of lysozyme resistance and virulence of Enterococcus faecalis. *PLoS One, 5*(3), e9658. http://dx.doi.org/10.1371/journal.pone.0009658

Lie, O., Evensen, O., Sorensen, A., & Froysedal, E. (1989). Study on lysozyme activity in some fish species. *Dis. Aquat. Org., 6*, 1-5. http://dx.doi.org/10.3354/dao006001

Lollike, K., Kjeldsen, L., Sengelov, H., & Borregaard, N. (1995). Purification of lysozyme from human neutrophils, and development of an ELISA for quantification in cells and plasma. *Leukemia, 9*(1), 206-209.

Malmsten, M. (Ed.) (2003). *Biopolymers at Interfaces* (2nd ed.). New York: Marcel Dekker. http://dx.doi.org/10.1201/9780824747343

Markart, P., Faust, N., Graf, T., Na, C. L., Weaver, T. E., & Akinbi, H. T. (2004). Comparison of the microbicidal & muramidase activities of mouse lysozyme M and P. *Biochem J, 380*(Pt 2), 385-392. http://dx.doi.org/10.1042/BJ20031810

Marks, R. S., Cullen, D., Lowe, C., Weetall, H. H., & Karube, I. (2007). *Handbook of Biosensors and Biochips.* Mississauga: John Wiley and Sons Ltd.

Marx, K. A. (2003). Quartz crystal microbalance: a useful tool for studying thin polymer films and complex biomolecular systems at the solution-surface interface. *Biomacromolecules, 4*(5), 1099-1120. http://dx.doi.org/10.1021/bm020116i

Matsumura, M., Becktel, W. J., Levitt, M., & Matthews, B. W. (1989). Stabilization of phage T4 lysozyme by engineered disulfide bonds. *Proc Natl Acad Sci U S A, 86*(17), 6562-6566. http://dx.doi.org/10.1073/pnas.86.17.6562

Matthews, B. W., & Remington, S. J. (1974). The three dimensional structure of the lysozyme from bacteriophage T4. *Proc Natl Acad Sci U S A, 71*(10), 4178-4182. http://dx.doi.org/10.1073/pnas.71.10.4178

Mecitoglu, Ç., Yemenicioglu, A., Arslanolu, A., Elmacı, Z.S., Korel, F., & Çetin, A.E. (2006). Incorporation of Partially Purified Hen Egg White Lysozyme into Zein Films for Antimicrobial Food Packaging. *Food Research International., 39*, 12-21. http://dx.doi.org/10.1016/j.foodres.2005.05.007

Meyer, K., Thompson, R., Palmer, J. W. & Khorazo, D. (1936b). On the mechanism of lysozyme action. *J. Biol. Chem., 113*, 479-486.

Meyer, K., Thompson, R., Palmer, J. W., & Khorazo, D. (1936a). The purification and properties of lysozyme. *J. Biol. Chem., 113*, 303-309.

Min, S., & Krochta, J. M. (2005). Inhibition of Penicillium Commune by Edible Whey Protein Films Incorporating Lactoferrin, Lactoferrin Hydrosylate, and Lactoperoxidase Systems. *J. Food Sci., 70*, 87-94. http://dx.doi.org/10.1111/j.1365-2621.2005.tb07108.x

Mir, M. A. (1977). Lysozyme: a brief review. *Postgrad Med J, 53*(619), 257-259. http://dx.doi.org/10.1136/pgmj.53.619.257

Nakamura, S., Kato, A., & Kobayashi, K. (1991) New antimicrobial characteristics of lysozyme-dextran conjugates. *J. Agric. Food Chem., 39*, 647-650. http://dx.doi.org/10.1021/jf00004a003

Nakamura, S., Kato, A., & Kobayashi, K. (1992). Bifunctional lysozyme-galactomannan conjugate having excellent emulsifying properties & bactericidal effect. *J. Agric. Food Chem., 40*, 735-739. http://dx.doi.org/10.1021/jf00017a005

Osserman, E. F., & Lawlor, D. P. (1966). Serum and urinary lysozyme (muramidase) in monocytic and monomyelocytic leukemia. *J Exp Med, 124*(5), 921-952. http://dx.doi.org/10.1084/jem.124.5.921

O'Sullivan, C. K., & Guilbgoldlt, G. G. (1999). Commercial quartz crystal microbalances-theory and applications. Biosens. *Bioelectron., 14*, 663-670. http://dx.doi.org/10.1016/S0956-5663(99)00040-8

Patil, R. S., Ghormade, V. V., & Deshpande, M. V. (2000). Chitinolytic enzymes: an exploration. *Enzyme Microb Technol, 26*(7), 473-483. http://dx.doi.org/10.1016/S0141-0229(00)00134-4

Porstmann, B., Jung, K., Schmechta, H., Evers, U., Pergande, M., Porstmann, T., & Krause, H. (1989). Measurement of lysozyme in human body fluids: comparison of various enzyme immunoassay techniques and their diagnostic application. *Clin Biochem, 22*(5), 349-355. http://dx.doi.org/10.1016/S0009-9120(89)80031-1

Poteete, A. R., Rennell, D., Bouvier, S. E., & Hardy, L. W. (1997). Alteration of T4 lysozyme structure by second-site reversion of deleterious mutations. *Protein Sci, 6*(11), 2418-2425. http://dx.doi.org/10.1002/pro.5560061115

Quintavalla, S., & Vicini, L. (2002). Antimicrobial food packaging in meat industry. *Meat Sci, 62*(3), 373-380. http://dx.doi.org/10.1016/S0309-1740(02)00121-3

Ramanathan, A., Savol, A. J., Langmead, C. J., Agarwal, P. K., & Chennubhotla, C. S. (2011). Discovering conformational sub-states relevant to protein function. *PLoS One, 6*(1), e15827. http://dx.doi.org/10.1371/journal.pone.0015827

Reitamo, S., Lalla, M., & Huipero, A. (1981). Serum lysozyme: evaluation of a nephelometric assay. *Scand J Clin Lab Invest, 41*(4), 329-332. http://dx.doi.org/10.3109/00365518109092053

Ronen, D., Eylan, E., Romano, A., Stein, R., & Modan, M. (1975). A spectrophotometric method for quantitative determination of lysozyme in human tears: description and evaluation of the method and screening of 60 healthy subjects. *Invest Ophthalmol, 14*(6), 479-484.

Smith, C. A., & Wood, E. J. (1991). *Molecular and Cell Biochemistry: Biological Molecules.* Chapman and Hall, London.

Song, Z. H., & Hou, S. (2003). A new analytical procedure for assay of lysozyme in human tear and saliva with immobilized reagents in flow injection chemiluminescence system. *Anal Sci, 19*(3), 347-352. http://dx.doi.org/10.2116/analsci.19.347

Sulakvelidze, A., Alavidze, Z., & Morris, J. G., Jr. (2001). Bacteriophage therapy. *Antimicrob Agents Chemother, 45*(3), 649-659. http://dx.doi.org/10.1128/AAC.45.3.649-659.2001

Suppakul, P., Miltz, J., Sonneveld, K., & Bigger, S. W. (2003). Active packaging technologies with an emphasis on antimicrobial packaging and its applications. *J. Food Sci., 68*(2), 408-420. http://dx.doi.org/10.1111/j.1365-2621.2003.tb05687.x

Takahashi, K., Lou, X. F., Ishii, Y., & Hattori, M. (2000). Lysozyme-glucose stearic acid monoester conjugate formed through the Maillard reaction as an antibacterial emulsifier. *J Agric Food Chem, 48*(6), 2044-2049. http://dx.doi.org/10.1021/jf990989c

Taylor, D. C., Cripps, A. W., & Clancy, R. L. (1992). Measurement of lysozyme by an enzyme-linked immunosorbent assay. *J Immunol Methods, 146*(1), 55-61. http://dx.doi.org/10.1016/0022-1759(92)90048-X

Terry, J. M., Blainey, J. D., & Swingler, M. C. (1971). An automated method for lysozyme assay. *Clin Chim Acta., 35*(2), 317-320. http://dx.doi.org/10.1016/0009-8981(71)90200-2

Trun, N., & Trempy, J (2004). *Fundamental Bacterial Genetics.* Malden: Blackwell Publishing.

Tsugita, A., & Inouye, M. (1968). Purification of bacteriophage T4 lysozyme. [Comparative Study]. *J Biol Chem, 243*(2), 391-397.

Van Dael, H. (1998). Chimeras of human lysozyme and alpha-lactalbumin: an interesting tool for studying partially folded states during protein folding. *Cell Mol Life Sci, 54*(11), 1217-1230. http://dx.doi.org/10.1007/s000180050249

Wohlkonig, A., Huet, J., Looze, Y., & Wintjens, R. (2010). Structural relationships in the lysozyme superfamily: significant evidence for glycoside hydrolase signature motifs. *PLoS One, 5*(11), e15388. http://dx.doi.org/10.1371/journal.pone.0015388

Zhang, X. J., Baase, W. A., Shoichet, B. K., Wilson, K. P., & Matthews, B. W. (1995). Enhancement of protein stability by the combination of point mutations in T4 lysozyme is additive. *Protein Eng, 8*(10), 1017-1022. http://dx.doi.org/10.1093/protein/8.10.1017

Zhang, X. J., Baase, W. A., Shoichet, B. K., Wilson, K. P., & Matthews, B. W. (1995). Enhancement of protein stability by the combination of point mutations in T4 lysozyme is additive. *Protein Eng, 8*, 1017-1022. http://dx.doi.org/10.1093/protein/8.10.1017

4

Gestational Diabetes Mellitus in Primi Gravida of Bangladesh in Different Trimesters

Ifat Ara Begum[1]

[1] Department of Biochemistry, Dhaka Medical College, Dhaka, Bangladesh

Correspondence: Ifat Ara Begum, Asset Avalon, Flat-B/2, House-23, Road-1, Sector-6, Uttara, Dhaka, Bangladesh. E-mail: ifat72@yahoo.com

Abstract

Background: Gestational Diabetes Mellitus is defined as any degree of glucose intolerance with onset or first recognition during pregnancy. It is an important dimension of the syndrome of Diabetes Mellitus. Similar to other members of the Asian race, Bangladeshi women are also considered to be at a high risk for developing gestational diabetes.

Materials and Methods: In order to better understand whether this heightened risk attributed to race really exists even in primi gravida of Bangladesh, what is the percentage of the disease among them and whether there is any association of classical risk factors of the disease, a hospital based observational study was performed and the glycemic status of the primi gravida women presenting to Chittagong Medical College Hospital were assessed.

Results: A total of 117 primi gravida women, 39 in each of the three trimesters of pregnancy were selected as study subjects during the study period on the basis of set criteria and a 2 h, 75 g OGTT was performed over them. The mean age of the subjects enrolled was 21.1 ± 2.29 years, mean BMI was 21.6 ± 2.72 kg/m^2, the mean fasting serum glucose values in mmol/l were 4.5 ± 0.63, 4.6 ± 0.62 and 4.7 ± 0.56 in 1st, 2nd and 3rd trimesters and the 2 h post 75 g glucose load serum glucose values in mmol/l were found 6.1 ± 1.08, 6.6 ± 1.76 and 7.0 ± 1.42 respectively. Using the cut-off value mentioned in the operational definition of GDM, 13.7% of the subjects were found to have the disease. Among them, 12.5% were in first trimester, 31.2% were in second trimester and 56.3 % were in third trimester of pregnancy. Maternal obesity/over-weight and a family history of diabetes, not the maternal older age, have shown statistically significant association with GDM. Prevalence of GDM has been noted in clinically non-risk group of subjects also.

Conclusion: From the above study, the inference could be drawn that, screening for GDM should be done as early as possible during pregnancy, even at first trimester, irrespective of presence or absence of established risk factors. But larger trials are needed to truly assess the disease burden of gestational diabetes among primi gravida women.

Keywords: Gestational Diabetes, hyperglycemia, trimester of pregnancy, Oral Glucose Tolerance Test

1. Introduction

Gestational Diabetes Mellitus is defined as any degree of glucose intolerance with onset or first recognition during pregnancy (The Expert Committee on the Diagnosis and Classification of Diabetes Mellitus, 2003; Buchanan & Xiang, 2005; American Diabetes Association, 2002; WHO Consultation, 1999; Schmidt et al., 2001). 1% to 14% of total pregnancies may be affected by it (American Diabetes Association, 2002). The prevalence of GDM in rural Bangladesh is comparable with any other population with higher prevalence of GDM (Sayeed et al., 2005).

Pregnancy is a complex endocrine-metabolic adaptation process. Placental secretion of hormones, such as – estrogen, progesterone, human chorionic somatotropin (hCS) or placental lactogen, prolactin and growth hormone is a major contributor to the insulin resistant state seen in pregnancy (Cianni et al., 2003; Setji et al., 2005) and this insulin resistance, which develops in 2nd trimester, makes pregnancy a diabetogenic condition.

As majority of women with GDM have no symptoms, a screening program is required (Jensen et al., 2003). Risk assessment for GDM should be done at the first prenatal visit (American Diabetes Association, 2002; Murthy et al., 2002). The peak effect of maternal hormones responsible for insulin resistance, which is seen in the 26th to the 33rd weeks of gestation, forms the basis for screening in the 24th to the 28th weeks of gestation (Mumtaz et al., 2000). But as diabetes should be diagnosed at the pre-pregnant state and as the patient does not come during this period, glucose screening should ideally be done at the 1st trimester of pregnancy and followed up at the 24th to the 28th weeks of gestation. But normal glucose tolerance in early pregnancy does not exclude the risk of development of GDM later. Screening earlier in pregnancy detects fewer women with GDM but identifies those at highest risk and allows for earlier intervention. Screening later in pregnancy detects a large number of women with GDM, many of whom are at lower risk but who would be treated for a shorter time (U.S. Preventive Services Task Force, 1996).

The ADA considers women to be at risk for GDM, unless they are younger than 25 years, have normal body weight (BMI < 25 kg/m^2), are not a member of high-risk ethnic group, have no first-degree relatives with diabetes and have no personal history of glucose intolerance or poor obstetrical outcome. But screening using only clinical and historic risk factors fails to diagnose the actual percentage of patient with GDM (Mumtaz et al., 2000; Khine, Winklestin, & Copel, 1999).

In a study of diabetes in pregnancy in Tianjin, China, gravidas with IGT were found to have poor pregnancy outcome, they were found to be at increased risk for premature rupture of membrane, pre-term birth, breech presentation and high birth weight, adjusting for maternal age, pre gravid BMI, hospital levels and other confounding factors (Yang et al., 2002).

There is significant association between increasing glucose intolerance and increased incidence of cesarean delivery, pre-eclampsia and length of maternal hospitalization (Mumtaz et al., 2000).

Gravidas with GDM generally demonstrates higher degrees of post-pregnancy insulin resistance, ß-cell dysfunction, higher BMI, central obesity and exaggerated hyperlipidaemia, which suggests that GDM is a transient manifestation of long standing metabolic dysfunction (Carpenter, 2007).

The prevalence of GDM is proportionate to the incidence of type-2 DM in a given population or ethnic group (Setji et al., 2005; U.S. Preventive Services Task Force, 1996; Senanayake et al., 2006). A study including 435 women with a singleton pregnancy done at Oulu University Hospital, Finland between 1984 and 1994, demonstrated occurrence of clinical type-1 diabetes in 4.6% and clinical type-2 diabetes in 5.3% of women with past history of GDM during their follow-up period in fertile age (Jarvela et al., 2006).

A study carried out for Japanese pregnant women at Kejo University Hospital from June'96 to December'00, demonstrated the occurrence of gestational hypertension more frequently in patients with mild glucose intolerance or GDM, than in those with normal glucose tolerance (Miyakoshi et al., 2004).

In the fetus or neonate, GDM is associated with higher rates of perinatal mortality, macrosomia, birth trauma, hyperbilirubinemia, polycythemia, hypocalcemia, shoulder's dystocia and neonatal hypoglycemia (Mumtaz et al., 2000; Turok, Ratcliffe, & Baxley, 2003). The children born to mother with GDM may have a higher incidence of obesity, IGT and DM in late adolescence and young adulthood.

Women with GDM need to be managed properly during pregnancy, as treatment of GDM reduces serious perinatal morbidity and may also improve the women's health related quality of life (Crowther et al., 2005). They should be screened for diabetes 6 weeks postpartum, should be re-classified into IFG, IGT, DM or normoglycemic group, and should be followed up with subsequent screening for the development of diabetes or pre-diabetes (esp. in person who had elevated fasting glucose level during pregnancy (Yang et al., 2002)), thus to reduce the risk for complications of diabetes and avoid conception of future pregnancy in the setting of uncontrolled hyperglycemia (Perkins, Dunn, & Jagasiasm, 2007; Ratner, 2007).

Present study was designed to determine the percentage of primi gravida, identifiable as GDM, by OGTT, in different trimesters of pregnancy, as there is lack of data relevant to it in Bangladesh. Significant association of well-known risk factors with GDM was also planned to be evaluated.

2. Materials and Methods

This study was carried out in the department of Biochemistry, CMC with cooperation of Obstetrics & Gynecology Out-Patient Department, CMCH, during the period of January 2007 to December 2007. A total of 117 primi gravida, 39 in each gestational trimester, who were not known to have previous history of glucose intolerance, were included in the study. Primi gravida with diagnosed pre gestational glucose intolerance, multiple pregnancy, complicated pregnancy (i.e. pregnancy with APH, PET etc.), diagnosed medical diseases (e.g. heart disease) and diseases requiring long term medical treatment that may affect glucose metabolism (e.g. thyroid disorder) were excluded from the study. Detail history was taken and relevant clinical examination was done for each subject. When necessary, a relevant clinical finding of index pregnancy was confirmed by investigation. Purpose and procedure of the study were elaborately explained to them. Only those subjects, willing to cooperate, underwent the investigation after giving informed consent in a prescribed form. They were motivated to come next day after overnight fasting condition and were referred for their usual antenatal advice after the procedure.

Data collection procedure included direct history taking, physical examination and laboratory procedure. Data were taken in a pre-designed data collection sheet. Detail history, clinical examination including anthropometry (height, weight and BMI) was recorded. A 2 h, 75 g OGTT was performed on each subject. Venous blood, taken at fasting condition and then 2 h after 75 g of oral glucose load, was collected in plain test tube and allowed clotting at room temperature. Serum was obtained by separation of the blood at 3000 rpm/min within 30 minutes of collection and was used for determination of glucose concentration by Glucose Oxidase Method (Randox Lab. UK). All blood samples were determined within one hour after sampling. Data were managed using the computer software Statistical Package for Social Sciences (SPSS) for Windows, version 12.00. The categorical variables were expressed in terms of percentage. Results were expressed as mean ± SD unless specified otherwise. The difference in the percentage was tested for significance by Chi-square test. A P value of < 0.05 was considered statistically significant.

Operational Definitions:

GDM was defined in this study on the basis of fasting and 2 h serum glucose values of ≥ 5.3 mmol/l and ≥ 7.8 mmol/l respectively following a 2 h, 75 g Oral Glucose Tolerance Test. If any one or both of the criteria were fulfilled, the diagnosis of GDM was made.

On the basis of family income per month, socio economic status was set, i.e. Poor: Tk. 1,000-5,000, Average: Tk. 5,001-15,000 and Affluent: Above Tk. 15,000.

Presence of any of the established risk factors for GDM like maternal overweight and obesity (BMI of 25 or more), maternal older age (age of 25 years or more) or a family history of diabetes was used to define the clinically risk group. Clinically non-risk group had none of these risk factors.

3. Results

The mean age of the total 117 subjects was found to be 21.1 ± 2.29 years and mean BMI, 21.6 ± 2.72 kg/m^2. Majority of subjects were from urban area, of average socio-economic status, were housewives and had the educational status, below SSC level. 25.6% of the subjects had at least one established risk factor, so considered as clinically risk group. 13.7% subjects were found to have GDM. 86.3% subjects had normal glucose tolerance. Of the subjects with GDM, 12.5% were found in 1st trimester, 31.2% were in 2nd trimester and 56.3% were in 3rd trimester of pregnancy (Figure 1).

In subjects with normal glucose tolerance (NGT), the mean fasting serum glucose concentration in 1st, 2nd and 3rd trimesters were 4.5, 4.6 and 4.6 mmol/l respectively and that of 2 h after 75 g of glucose load were 6.0, 6.1 and 6.3 mmol/l respectively (Figure 2). On the other hand, in subjects with gestational diabetes (GDM), the mean fasting serum glucose concentration in 1st, 2nd and 3rd trimesters were 4.2, 5.4 and 4.9 mmol/l respectively and that of 2 h after 75 g of glucose load were 8.3, 10.2 and 9.1 mmol/l respectively (Figure 3).

A higher prevalence of GDM was detected in clinically risk group of study subjects (Figure 4). A statistically significant association was found not between maternal older age with GDM but between a positive family history of DM as well as maternal overweight and obesity with GDM.

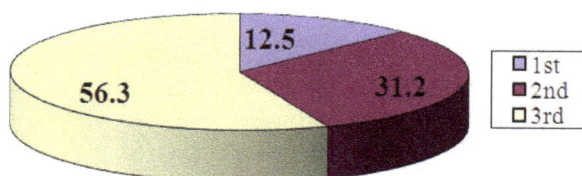

Figure 1. Distribution of GDM subjects in different trimesters of pregnancy

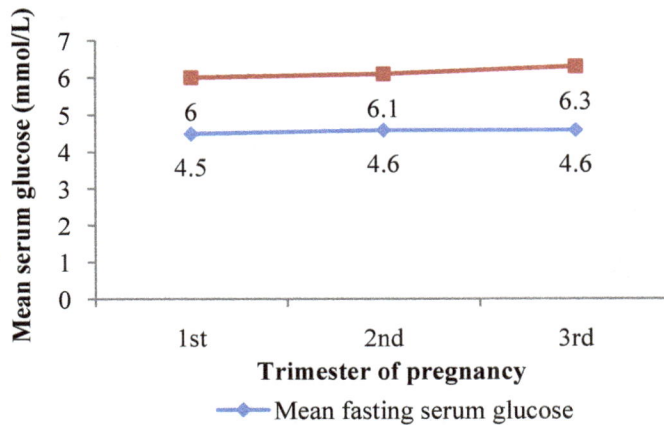

Figure 2. Serum glucose concentrations in subjects with NGT in different trimesters of pregnancy

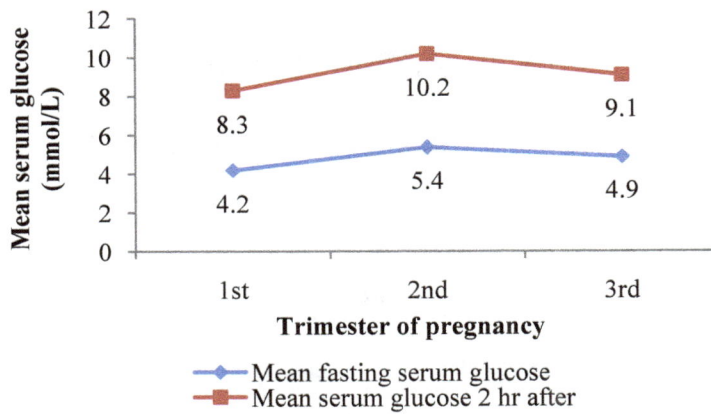

Figure 3. Serum glucose concentrations in subjects with GDM in different trimesters of pregnancy

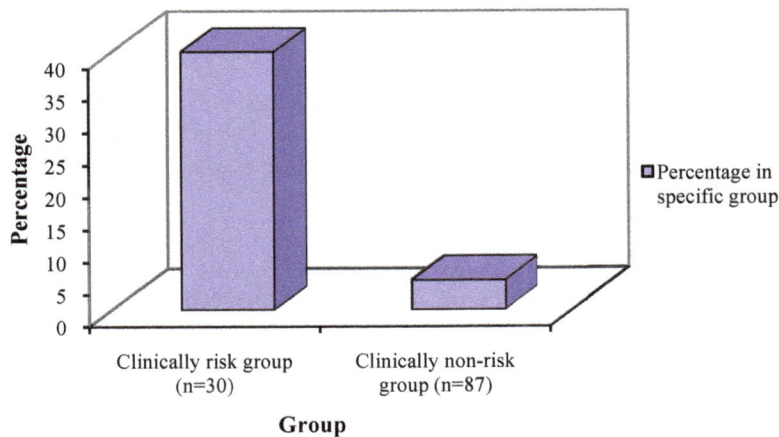

Figure 4. Distribution of GDM subjects in clinically risk group and clinically non-risk group of study subjects

4. Discussions

An overall GDM prevalence of 13.7% in primi gravida in different gestational trimesters on the basis of a 2 h, 75 g OGTT, as per revealed in the present study, is comparable with any other population with higher prevalence of GDM. The observed difference in GDM prevalence in different trimesters has been supported by different studies also. A study at Chennai revealed a prevalence of 16.3% at first trimester, 22.4% at second trimester and 65.3% at third trimester of pregnancy (Seshiah et al., 2007).

As gestational diabetes was diagnosed in clinically non-risk group as well as in first trimester group of study subjects in this study; the need of screening for GDM is strongly justified in all primi gravidas, even in first trimester, irrespective of presence or absence of clinical risk factors in them. Maternal over weight/obesity and family history of diabetes have shown an association with GDM and found to be statistically significant. The association of maternal older age and GDM has not found to be statistically significant here; but further study is required to come to a conclusion about it.

The present study had some limitations as follows:

1) The small size of sample- the result may not be representative of whole Bangladesh.

2) Subjects were selected from hospital in the city, so they do not represent the overall situation in our country.

3) OGTT was not well accepted by some subjects, esp. in early trimester, when morning sickness sets in. Some subjects started vomiting who were discarded from the test procedure and advised to come later. Many of them dropped out from the study.

4) The initial diagnosis of GDM could not be confirmed by a repeat OGTT on the consecutive day.

5. Recommendation

1) Screening for GDM in primi gravida should be done by OGTT as early as possible, even in first trimester of pregnancy.

2) All primi, irrespective of their clinically risk status, should be screened for GDM.

References

American Diabetes Association. (2002). Gestational Diabetes Mellitus. *Diabetes Care, 25*(Supplement 1), 94-96. http://dx.doi.org/10.2337/diacare.25.2007.S94

Buchanan, T. A., & Xiang, A. H. (2005). Gestational diabetes mellitus. *J. Clin. Invest., 115*, 485-491. http://dx.doi.org/10.1172/JCI24531

Carpenter, N. W. (2007). Gestational Diabetes Pregnancy Hypertension, and Late Vascular Disease. *Diabetes Care, 30*(Supplement 2), 246-250. http://dx.doi.org/10.2337/dc07-s224

Cianni, G. D., Miccoli, R., Volpe, L., Lencioni, C., & Prato, S. D. (2003). Intermediate metabolism in normal pregnancy and in gestational diabetes. *Diabetes Mtab Res. Rev., 19*, 259-270. http://dx.doi.org/10.1002/dmrr.390

Crowther, C. A., Hiller, J. E., Moss, J. R., McPhee, A. J., Jeffries, W. S., & Robinson, J. S. (2005). Effect of Treatment of Gestational Diabetes Mellitus on Pregnancy Outcomes. *The New England Journal of Medicine, 352*(24). http://dx.doi.org/10.1056/NEJMoa042973

Jarvela, I. Y., Juutinen, J., Koskela, P., Hartikainen, A. L., Kulmala, P., Knip, M., & Tapanainen, J. S. (2006). Gestational Diabetes Identifies Women at Risk for Permanent Type 1 and Type 2 Diabetes in Fertile Age: Predictive role of autoantibodies. *Diabetes Care, 29*(3), 607-612. http://dx.doi.org/10.2337/diacare.29.03.06.dc05-1118

Jensen, D. M., Pedersen, L. M., Nielsen, H. B., Westergaard, J. G., Ovesenp, & Damm, P. (2003). Screening for gestational diabetes mellitus by a model based on risk indicators: A prospective study. *Am. J. Obstet. Gynecol., 189*(5), 1383-1388. http://dx.doi.org/10.1067/S0002-9378(03)00601-X

Khine, M. L., Winklestin, A., & Copel, J. A. (1999). Selective screening for gestational diabetes mellitus in adolescent pregnancies. *Obstet Gynecol., 93*(5, Part 1), 738-742. http://dx.doi.org/10.1016/S0029-7844(98)00550-X

Miyakoshi, K., Tanaka, M., & Matsumoto, T. (2004). Hypertensive disorders in Japanese women with gestational glucose intolerance. *Diabetes Research and Clinical Practice, 64*, 201-205. http://dx.doi.org/10.1016/j.diabres.2003.11.002

Mumtaz, M. (2000). Gestational Diabetes Mellitus. *Malaysian Journal of Medical Sciences, 7*(1), 4-9.

Murthy, E. K., Renar, I. P., & Metelko, Z. (2002). Diabetes and Pregnancy. *Diabetologia Croatica, 31*(3).

Perkins, J. M., Dunn, J. P., & Jagasiasm. (2007). Perspectives in Gestational Diabetes Mellitus: A Review of Screening, Diagnosis and Treatment. *Clinical Diabetes, 25*(2). http://dx.doi.org/10.2337/diaclin.25.2.57

Ratner, R. E. (2007). Prevention of Type 2 Diabetes in Women with Previous Gestational Diabetes. *Diabetes Care, 30*(Supplement 2), 242-245. http://dx.doi.org/10.2337/dc07-s223

Sayeed, M. A., Mahtab, H., Khanam, P. A., Begum, R., Banu, A., & Khan, A. K. A. (2005). Diabetes and hypertension in pregnancy in a rural community of Bangladesh: a population-based study. *Diabetic Medicine, 22*(9), 1267-1271. http://dx.doi.org/10.1111/j.1464-5491.2005.01600.x

Schmidt, M. I., Duncan, B. B., Reichelt, A. J., Branchtein, L., & Yamashita, T. (2001). Gestational Diabetes Mellitus Diagnosed With a 2 h 75 g Oral Glucose Tolerance Test and Adverse Pregnancy Outcomes. *Diabetes Care, 24*(7), 151-1155. http://dx.doi.org/10.2337/diacare.24.7.1151

Senanayake, H., Seneviratne, S., Ariyaratne, H., & Wijeratne, S. (2006). Screening for gestational diabetes mellitus in Southern Asian women. *J. Obstet. Gynaecol. Res., 32*(3), 286-291. http://dx.doi.org/10.1111/j.1447-0756.2006.00400.x

Seshiah, V., Balaji, V., Balaji, M. S., Paneerselvam, A., Arthi, T., Thamizharasi, M., & Datta, M. (2007). Gestational diabetes mellitus manifests in all trimesters of pregnancy. *Diabetes Research and Clinical Practice, 77*, 482-484. http://dx.doi.org/10.1016/j.diabres.2007.01.001

Setji, T. L., Brown, A. J., & Peinglos, M. N. (2005). Gestational Diabetes Mellitus. *Clinical Diabetes, 23*(1). http://dx.doi.org/10.2337/diaclin.23.1.17

The Expert Committee on the Diagnosis and Classification of Diabetes Mellitus. (2003). Report of the Expert Committee on the Diagnosis and Classification of Diabetes Mellitus. *Diabetes Care, 26*(Supplement 1), 5-20. http://dx.doi.org/10.2337/diacare.26.2007.S5

Turok, D. K., Ratcliffe, S. D., & Baxley, E. G. (2003). Management of gestational diabetes mellitus. *American Family Physicians, 68*(9).

U.S. Preventive Services Task Force. (1996). Screening for gestational diabetes. *Guide to Clinical Preventive Services* (2nd Ed.). Washington, DC: Office of Disease Prevention and Health Promotion.

WHO Consultation. (1999). Definition, Diagnosis and Classification of Diabetes Mellitus and its Complication: report of a WHO Consultation. *Part 1: Diagnosis and Classification of Diabetes Mellitus*. Geneva, WHO/NCD/NCS/99.2, World Health Organization.

Yang, X., Hsuhage, B., Zhang, C., Zhang, H., Zhang, Y., & Zhang, C. (2002). Women With Imparted Glucose Tolerance During Pregnancy Have Significantly Poor Pregnancy Outcomes. *Diabetic Care, 25*, 1619-1624. http://dx.doi.org/10.2337/diacare.25.9.1619

5

Levels of Apigenin and Immunostimulatory Activity of Leaf Extracts of Bangunbangun (*Plectranthus Amboinicus* Lour)

Melva Silitonga[1], Syafruddin Ilyas[2], Salomo Hutahaean[2] & Herbert Sipahutar[1]

[1] Biology Education Department, Faculty of Mathematic and Science, State University of Medan, Medan, Indonesia

[2] Biology Department, Faculty of Mathematic and Science, University of North Sumatra, Medan, Indonesia

Correspondence: Melva Silitonga, Biology Education Department, Faculty of Mathematic and Science, State University of Medan, Medan, Indonesia. E-mail: Melvasil-2014@gmail.com

Abstract

Bangunbangun (*Plectranthus amboinicus Lour*) consumed by the mother, who just gave birth in North Sumatra, Indonesia in particular the Batak tribe, to increase the production of breast milk. This plant is known to have a high content of nutrients, especially iron and carotene. Also known to have many benefits, among others, as an antipyretic, analgesic, wound medicine, cough medicine, and thrush, antioxidant, antitumor, anticancer, and hypotensive. The study was conducted to determine levels of apigenin of Bangunbangun's leaves and evaluate its immunostimulatory activity in rats (*Rattus norvegicus*). Analysis of apigenin using High Performance Liquid Chramtography method (HPLC). Evaluation of immunostimulatory activity carried out by measuring the levels of imonoglobulin G (IgG), imonoglobulin M (IgM), Lysozyme and Monocytes. Analysis of IgG and IgM are using Elisa method (Sigma). Serum lysozyme activity was measured by the spectrophotometric method. Monocytes were analyzed by using ABX Micros 60. Organ histology preparations made by hematoxylin-eosin staining. Data were analyzed by ANOVA and showed that by giving the Ethanol Extract of Propolis (EEP) of 500 mg / kg bw in rats, with a significant increase of IgM and lysozyme activity with very significant. EEP give a very significant effect on levels of IgG. Monocytes were higher in mice given EEP, but did not differ significantly compared with mice not given EEP. Lymphoid organ weights are all under normal circumstances. Giving EEP 500 mg / kg bw mice, significantly increased the weight of the liver and spleen, but does not affect kidney weight.

Keyword: *Plectranthus amboinicus* Lour, IgG, IgM, Lysozyme, monocytes, spleen, liver, thymus, kidney

1. Introduction

Traditional medicine plays an important role in maintaining the health of humans and animals. By consuming medicinal plants, can boost the immune system and increase antioxidant activity in humans. The high level of use as a medicinal plant due to easily available, cheap, and relatively no side effects (Lie et al., 2010; Gabhe, Tathe, & Khan, 2006). Natural immunostimulants are biocompatible, biodegradatif, low prices and friendly environment. Therefore, its use has been widely used as a cure for several diseases, cosmetics, immunostimulant and a variety of other purposes (Ortuno et al., 2002). Plants have been used as medicine in India, Nigeria, China, Indonesia and some other countries (Uma et al., 2011). Immunomodulator is a substance that affects the immune system and how to restore and repair the impaired immune system and suppress excessive immune function (Małaczewska et al., 2010; Bellanti & Kadlec, 1993). Based on how it works can be divided into imunostimulant Immunomodulatory and imunosupresant.

Bangunbangun is a type of plant that is commonly consumed by mothers who gave birth in North Sumatra, in particular the Batak tribe. Bangunbangun's leaves is believed to increase the production of breast milk. These leaves have a high content of nutrients, especially iron and carotene. Consumption of leaf shapes significantly affect the elevated levels of some minerals such as iron, potassium, zinc and magnesium in milk and infant weight resulting in increased markedly. And is known to have many benefits, among others, as an antipyretic, analgesic, wound medicine, cough medicine, and thrush, antioxidant, antitumor, anticancer, and hypotensive ((Damanik et al., 2001; Duke, 2000). Usually drugs that have multiple properties have receptors on the target organ systems that perform the function of immune limforetikular. Apigenin is a flavonoid which is the active substance contained in the parsley and known to have antioxidant properties and effectively inhibiting the prooxidative activity of cadmium, the effects of anti-lung cancer Pru and inhibiting the growth of human colon carcinoma cells (Tong et al., 2007; Liu et al., 2005; Wang et al., 2000).

Bangunbangun's leaves contain various flavonoids apigenin, quercetin, luteolin, salvigenin, genkwanin and fat to fly, which is a phenolic compound that is widespread in plants and have been reported to have multiple biological effects, including antioxidant, eliminating free radicals have the ability of anti-inflammatory and anti-cancer activity . (Prasenjit, Hullanti, & Khumar, 2011; Preeja et al., 2011).

Immunostimulatory activity of a drug or medicinal plants observed by measuring several parameters, namely, *immunoglobulin G (IgG)* and *immunoglobulin M (IgM)* and some hematological parameters. Hematocrit, total leukocytes, total protein, total immunoglobulin and leukocrit are parameters immunostimulatory activity of the immune response and can be measured through serological and hematological tests (Dorucu et al., 2009; Khumar, Gupta, Sharma, & Khumar, 2011). Defense mechanisms of the simplest animals is phagocytosis. Although various cells in the body can perform phagocytosis, but the main cells involved in nonspecific defense is the mononuclear cells (monocytes and macrophages) and polymorphonuclear granulocytes or (Munasir, 2001).

Lysozyme is one of the parameters measured as the impact of the use of immunostimulant. It is found in many vertebrates including fish, and is an important factor to prevent the invasion of microorganisms (Ogier, Quentell, Fournierl, & Guovello, 1996). This issue became the basis for examining the role of bangunbangun's extract as an anti-bacterial that can increase lysozyme activity in the body.

2. Method and Material

2.1 Procurement and Extraction of Bangunbangun's Leaves

The fresh Bangunbangun's leaves was taken as much as 8 kg, and washed with water and then it dried naturally for 24 hours. To get the brittle bangunbangun's leaves, drying is done with an oven with a temperature of 40 °C. As much as 500 g of dried leaves were pulverized in a blender until powder form, which is placed in two container, ie each 250 grams of 95% ethanol and added to the already distilled some 2000 ml / container. Wheat leaves soaked for 5 days, and then stirred once a day. Immersion leaf powder was filtered using filter paper and add up to 3 liters of ethanol, and left for 2 days and filtered again. Extract obtained was concentrated by using a rotary evaporator using a water bath and then it dried to obtain the dried ethanol extract.

Bangunbangun ethanol extract dose to rats at 500 mg / kg (Jose, Ibrahim, & Janardhanan, 2005;   Patel et al., 2011). Ethanol extract of the bangunbangun's leaves dissolved in 1% CMC (Jose, Ibrahim, & Janardhanan, 2005). Bangunbangun ethanol extract solution was made 10% solution. Bangunbangun extract given orally every day by means of force-feeding, for 30 days.

2.2 Chemicals

In analyzing serum immunoglobulin was used Elisa Kit Rat IgG and IgM (Cat. No. E111-100), Rat IgM Elisa Kit (Sigma), Rat IgG Elisa Kit (Sigma), ELISA Coating Buffer (Cat. No. E107), ELISA wash Buffer Solution (Cat. No. E106), ELISA blocking buffer (Cat. No. E104) dan96- well plate (Cat. No. E105). Lysozyme kit is used to measure the levels of lysozyme in the serum. HPLC Apegenin Standards was used to determine levels of apigenin in bangunbangun's leaves.

2.3 Animal Test

Tests was conducted on a many of 24 white rats in aged 3 months with an average weight of 150-200 grams and adapted to laboratory conditions for 14 days. The health condition of the mice was controlled every day and there are no signs of illness such as no appetite, weight loss and decreased motor activity. Future provision of treatment carried out for 30 days. Rats were given food in the form pellets and drinking water. The temperature in the room ranged between (22 ± 2) °C with 12 hours dark cycle and 12 hours of light.

2.4 Sheep Red Blood Cells (SRBC)

SRBC antigen used is according to research conducted by Khoul and Khosa (2013). SRBC were made in the laboratory of the Veterinary Regional I Medan, North Sumatra, Indonesia. Making SRBC was done with fresh sheep blood that had been given an anticoagulant, centrifuged at 3000 rpm to separate plasma from red blood cells. The top layer is a pasteur pipette plasma discarded, and the bottom layer, which is a red blood cell sediment was added a solution of PBS pH 7.2 three times of the volume of SRBC remaining. Tube is then turned upside down slowly until SRBC suspended homogeneously, and then centrifuged again. This procedure is repeated until the top layer is really clear and colorless. Clear upper layer was discarded and the bottom layer is 100% SRBC suspension To get the suspension SRBC 50% is done by adding 0.5 ml of 100% SDMD suspension with 0.5 ml of PBS. And then to get the suspension SRBC 1%, ie by adding 1 ml of 50% suspension in PBS SRBC 50 ml.

2.5 Apigenin Measurement and Evaluation of Immunostimulatory Activity

Apigenin was measured by HPLC. Evaluation of immunostimulatory activity performed by Completely Randomized Design (CRD), with four treatments were each given six replicates. 24 white rats. There are four groups and each group consisted of 6 rats. In detail, these groupings are as in Table 1.

Table 1. Grouping of rats

Group	Treatment	
	Control	Experiment
1	CMC 1% orally every day	-
2	-	EEP 500mg/kg bw, no give SRBC
3	-	EEP 500mg/kg bw, give 0,1 ml SRBC
4	-	0,1 ml SRBC

EEP was given orally using a gastric tube every day for 30 days, and SRBC was given intraperitoneally on day 8 and day 15 of treatments. Weight recorded at the end of each week to determine the effect of EEP on body weight. Blood samples were obtained from all test animals at day 31 by decapitation neck. Serum was separated for measurement of IgG and IgM.

Serum lysozyme activity was measured following the factory procedure (Sigma Cat Number L7651). Measurement of lysozyme based lysis suspension bacteria Micrococcus process lysodeiktycus according to the method that developed by Ellis (1998). Details are described as follows. 0.15 mg / ml Micrococcus lysodeiktycus (Sigma) dissolved in 66 mM PBS (pH 6.2). 50 μL of serum was added to 1 ml of bacterial suspension. Decrease in absorbance was recorded at 0.5 and 4.5 min for 3 min on a spectrophotometer with a wavelength of 450 nM. One unit of lysozyme activity is defined as a decrease in absorbance of 0.001L / min.

2.6 Measurement of Monocytes

On day 31, the treatment were collected for analysis of blood monocytes. Blood was obtained by decapitation neck, then collected in a tube that has been coated with EDTA. The analysis is done using a monocyte ABX Micros 60.

2.7 Measurement of Organ Weight

The organs were measured liver, spleen, thymus and kidney. To get the weight of these organs, surgery on rats have blood drawn by means of decapitation neck. To the four above-mentioned organs removed and placed on strain paper, and then weighed. All ortikus was stored in a solution of 10 percent formalin for further-processed for histopathological observations.

3. Results

3.1 Evaluation of Immunostimulatory Activity

3.1.1 IgG dan IgM Avtivity

Giving EEP 500 mg / kg bw had no effect on the activity of rat's IgG (Table 2). IgG activity was highest in the control treatment, followed by treatment with 500 mg of EEP. Activity is the lowest IgG in mice that were given only SRBC.

Giving EEP 500 mg / kg bw rats that given antigen SRBC showed significant changes in the levels of IgM, compared with control mice. IgM of rats that given EEP 500 mg / kg bw and were not given SRBC , higher than IgM of control rat, and IgM higher than mice given only SRBC, but this difference was not significant. This data can be seen in Table 2. This shows the effect of ethanol extract of leaf shapes to increase levels of immunoglobulin M in white rats. Groups of rats were given only SRBC treatment, had higher levels of immunoglobulin M, higher than the control group who were given distilled water alone. This may be influenced antigen into the body where the proliferation of rat antibodies against these antigens.

Table 2. Effect of EEP against IgG, IgM and Monocyte Lysozyme of mice

No	Treatment	Parameter			
		IgM (ng/ml x 10^5)	IgG (ng/ml x10^7)	Lysozyme U/ml x 10	Monocytes (%)
1	CMC	3.19±1.26	3.18±0.66	0.191±0.0110	12.8±3.02
2	500 mg EEP/kg bw	4.33±2.26	5.73±1.12**	0.622±0.0051**	15.5±3.56*
3	500 mg EEP/kg bw+SRBC	8.01±4.52	3.68 ± 0.52	1.49±0.056**	15.15±2.45*
4	0.1 ml SRBC	3.93±1.86	3.79±0.85	0.442±0.0026*	15.12±4.21*

3.1.2 Monocyte

Highest levels of monocytes were found in mice given EEP 500 mg / kg bw, followed by monocyte levels in mice treated EEP 500 mg / kg bw + SRBC, and rats were given only SRBC (Table 2). Monocyte levels were lowest in the control mice, but was not statistically significantly different.

3.1.3 Lysoizime Activity

Lysozyme activity data can be seen in Table 1 Giving EEP 500 mg / kg bw to rats increases the activity of lysozyme is very significant. The highest lysozyme activity found in rats treated EEP 500 mg / kg bw + SRBC, followed by lysozyme treatment mice EEP 500 mg / kg bw. Lysozyme in mice both treatments was higher and significantly different compared with controls. Mice that were given only SRBC, the activity of its lysozyme was higher and significantly different compared with controls.

3.2 Effect of EEP on Organ Weights

The results of organ weight measurement was shown in Table 3, indicate that EEP administration of 500 mg / kg bw, significant effect on liver weight, and spleen. EEP effect of 500 mg / kg bw + SRBC to the weight of the spleen, and kidney and thymus weight in control rats significantly different compared to all treatments.

Table 3. Effect of EEP against organ weights of mice

No	Treatment	Organ Weight (g)			
		Liver	Spleen	Kidney	Thymus
1	CMC	4.68±1.14	0.28±0.05	1.62±0.09	0.80±0.08**
2	500 mg EEP/ kg bw	6.89±1.12*	0.51±0.04*	1.35±0.21	0.59±0.05
3	500 mg EEP/ kg bw + SRBC	5.73±0.60	0.68±0.19*	1.47±0.17	0.49±0.07
4	1 ml SRBC	6.38±1.50*	0.46±0.03	1.25±0.18	0.55±0.06

3.3 Mesurement of Apigenin Liver, Spleen, Kidney

Apigenin has been recognized in general medicine or in traditional medicine for its pharmacological activity. An important group of flavonoids are anthocyanins, flavonols, flavones, catechins, and flavanones. Apigenin is a group of flavonoids (Hertog et al., 1993). Apigenin was measured by HPLC technique. The measurement results showed that there were as many as apigenin 0.0236 ng / μl samples at bangunbangun's leaves. Figure 1a and 1b shows the retention times and measurements of leaf bangunbangun apigenin.

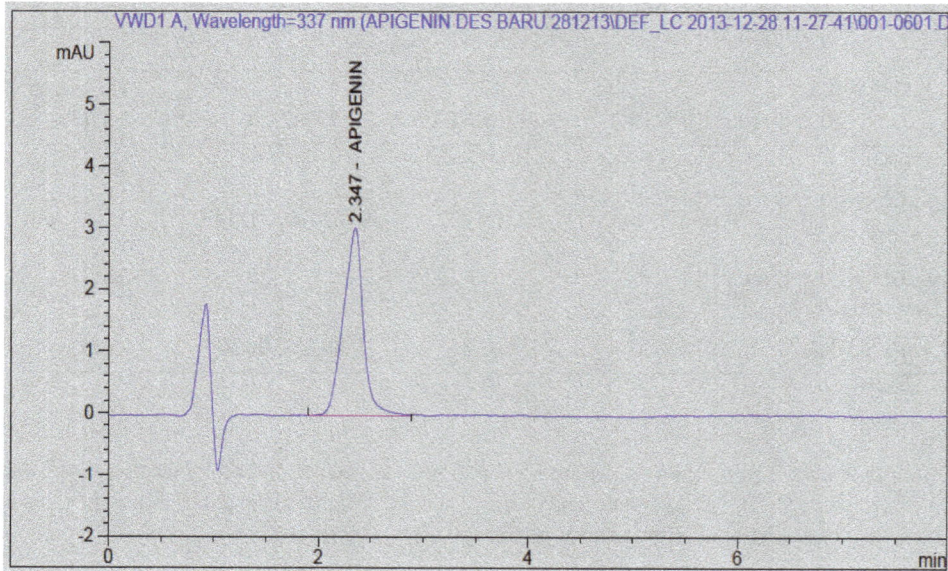

Figure 1a. Apigenin retention times

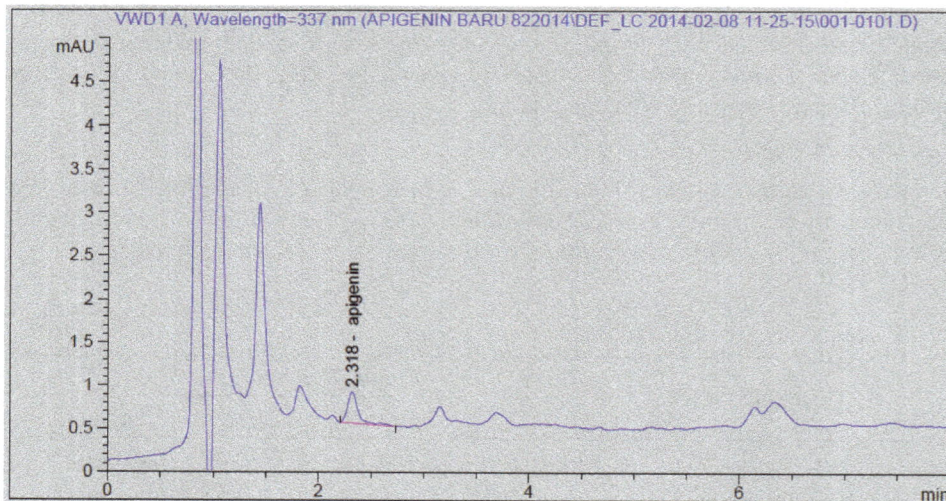

Figure 1. b. Apigenin content

4. Discussion

IgD and IgM is the main immunoglobulin found on the surface of B lymphocytes. There are two classes of immunoglobulin in the form and shape of a layered circular membrane that serve as receptors for specific antigens. The result of this interaction is further proliferation and differentiation of B lymphocytes, which produce antibody-secreting plasma cells. IgM circular also effectively activate the complement system, which is a group of plasma proteins made in the liver (Junqueira and Carneiro, 2003). Immunoglobulin M is the first immunoglobulin class formed on antigen stimulation, but IgM responses are generally short lasting only a few days and then decreased. Due to the high valency, IgM antibodies are very efficient as the cause clumping or cell destroyer. Because IgM are mostly found in the blood and rapidly increased in response to infection.

Giving EEP 500 mg / kg bw for 30 days and the mice were given antigen SRBC on day 8 and 15 significantly increased the levels of IgM, compared with controls. Increased IgM occurs in mice exposed to antigen SRBC. EEP IgM mice given 500 mg / kg bw higher than controls and were only given SRBC, but did not differ significantly in this regard EEP increase IgM production upon exposure to antigen. Giving EEP 500 mg / kg bw mice for 30 days with a very real increase of IgG compared with controls. This increase occurred without exposure to antigen SRBC. IgG levels was lower than IgM. This is consistent with the theory that exposure to foreign antigen will produce biphasik response. The first stage is associated with the production of IgM, followed by IgG production. The second stage is characterized by a decrease followed by an increase in IgG IgM just as shown in Figure 2 Antigen will select and expand clones of effector B cells that will develop into plasma cells and produce antibodies.

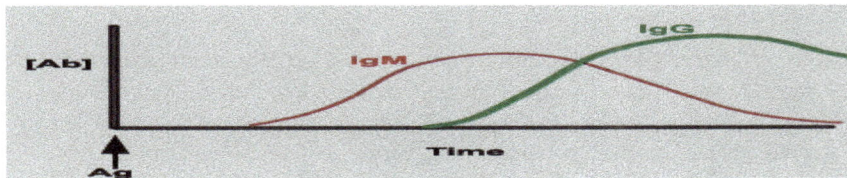

Figure 2 Production of IgG and IgM after antigen exposure

Lysozyme is an enzyme that has the power antiseptics, which can kill foreign organisms. This enzyme is present in the blood cells of granulocytes and monocytes, and also contained in saliva, sweat, and tears of the mammary gland (Millodot, 2009). Lysozyme protects several places in the body which is a potential place for food for bacterial growth. Lysozyme provides protection in the blood, with methods that are stronger than the one used by the immune system. Immunostimulatory bind specific receptors on the surface membrane of phagocytes. Lymphocytes were mengaktavasi these cells, produce several enzymes including lysozyme to destroy pathogens (Fazlolahzadeh et al., 2011). EEP is as an antibacterial as reported by Shiney, Ganesh, and Kumar (2012). Thus the antibacterial properties of the EEP increases with increasing serum Lysozyme. Increased activity of lysozyme in the serum of mice along with increased levels of monocytes. This is in accordance with that stated by Millodot (2009), that was formed by the lysozyme enzyme neutrophill or monocytes.

Cells monosit`merupakan one of the components of the white blood cells (leukocytes), which function to phagocytosis. Monocites role in phagocytic cells when the host first meeting with a foreign object (Bellanti, 1993). These cells will be stimulated sums in the event of infection or chronic inflammation, have a short shelf-life in the blood circulation then into the network and turn into macrophages (Guyton, 1995). Monocytes are motile and move with amuboid movement to areas experiencing chronic infection for the phagocytic response (Ganong, 1999). Monocytes and macrophages are the second most frequent type of leukocytes. Macrophages play an important role in the immune response caused by fagositiknya properties, its ability to secrete cytokines and kemokinin, and his ability as a pointer cell antigens.

The measurement results showed that there apigenin apigenin as 0.0236 ng / ml sample. This is in accordance with the opinion of Prasenjit, Hullanti, and Khumar (2011) and Preeja et al. (2011) which says that bangunbangun leaves contain various flavonoids apigenin, quercetin, luteolin, salvigenin, genkwanin and fat to fly. However the measurement of quercetin, luteolin, salvigenin, genkwanin and fat still needs to be done.

The organs involved in the body's immune system are the spleen, liver, kidneys, thymus and lung. There are two main functions of the spleen is to produce a specific immune response and destroy the red blood cells are abnormal. Increased functional activity of the spleen, which is accompanied by hyperplasia limphoid cells or macrophages are usually caused by the stimulation of antigen or the presence of foreign bodies in the blood. Just as shown in Table 3, that the spleen of mice given EEP and SRBC heavier in comparison with other treatments. In this case give SRBC as antigen stimulation of spleen enlargement. Spleen mice given EEP has a greater weight than that given only SRBC. It also can stimulate the EEP shows an enlarged spleen. Magnification is caused by activities in producing immunoglobulins. The absence of stimulants in the control group of mice spleen menyebapkan can not optimally capture antigen and its size was smaller than the other treatments. Despite the significant changes in spleen weight with EEP administration, but all the organs of the spleen is still in the normal range is 0.58 ± 0.10 - 0.72 ± 0.14g. Spleen size bigger makes it easier spleen in preparation to capture antigen. In addition, a large spleen size also because the number of lymphocytes that formed in the spleen, this EEP showed good immunostimulatory effect on spleen is in line with the statement Piao, Liu, and Xie. (2013).

Giving EEP 500 mg / kg bw significantly affect liver weight. The highest liver weight was 6.89 ± 1.12. This size is still within the normal range, because according to Piao, Liu, and Xie (2013) for rats aged 13 months with an average weight of 6.70 ± 0.96 g liver. The liver is an organ with predominant innate immunity, plays an important role not only in the body's defense against invading microorganisms and tumor transformation but also in liver injury and recovery (Gao, Jeong, & Zhigang, 2008).

The thymus is the first lymphoid organ to form and grow slowly after birth in response to antigen stimulation posnatal and to ask for a large number of mature T cells. Genetic factors also affect, age, the rate and degree of immunologic function needs. In rats and mice, along with the maximum size is reached sexual maturity and then slowly experiencing involution (Pearse, 2006). In this study, the thymus weight decreased by administration of EEP 500 mg / kg bw. In the control rat thymus weight was higher than all other treatments. Weight of the thymus in mice is 12:50 ± 0.004 to 0.62 ± 0.001.

In newborn animals thymus is relatively large and mature sex before gradually shrinking (having involution) and replaced by fatty tissue (Hartono, 1989). But the rest of the thymus were found in the thoracic space remain on the animal until some old. Besides involution was associated with age, thymic atrophy was also faster in response to stress, so that the animals that died after a long illness that has a very small thymus (Tizard, 2004). The results of measurements performed thymus weight all showed still within the normal range, but the thymus weight control treatment was significantly different compared to the other treatments. Based on this can be called that thymus organ weight have anything to do with giving bangunbangun that spur these organs in improving the body's immune system.

Kidney is most often targeted by the immune response against pathogens kidney autoantigen or local manifestations of systemic autoimmunity (Kurts et al., 2013). Kidney failure affects immunity in general, cause intestinal barrier dysfunction, systemic inflammation and immunodeficiency that contribute to morbidity and mortality of patients with kidney disease. Giving EEP 500 mg / kg bw did not affect kidney weight. This is shown in Table 3 that all the kidney in the normal range.

Acknowledgements

The author would like to thank Mr. Hamonangan Tambunan, Lecturer of the Faculty of Engineering, State University of Medan for helping the author.

References

Alvin Jose, M., & Janardhanan, S. (2005). Modulatory effect of Plectranthus amboinicus Lour. on ethylene glycol-induced nephrolithiasis in rats. *Indian journal of pharmacology, 37*(1), 43. http://dx.doi.org/10.4103/0253-7613.13857

Bellanti, J. A. (1993). *Immunologi III*. Terjemahan dari Immunology III oleh A. Samik Wahab. Yogyakarta: Gadjah Mada University Press.

Damanik, R., Damanik, N., Daulay, Z., Saragih, S., Premier, R., Wattanapenpaiboon, N., … Wahlqvist, L. (2001). *Consumption of Bangun-Bangun Leaves (Coleus amboinicus Lour) to Increase Breast Milk Production Among Bataknesse Women in North Sumatra Island, Indonesia.* Retrieved January 2011 from www.healthyeatingclub.com/APJCNutSOO1/Damanik67.pdf

Dorucu, M., Colak, O. S., Ispir, U., Altinterin, B., & Celayir, Y. (2009). The Effect of Black Cumin *Nigella sativa* on the immune response of Rainbow Trout, *Oncorhynchus mykiss. Mediterranian Aquaculture Journal, 2*(1), 27-33. http://dx.doi.org/10.1016/j.aquaculture.2013.01.008

Duke. (2000). *Dr. Duke's Constituens and Ethnobotanical Databases.* Phytochemical database, USDA - ARS-NGRL. http://dx.doi.org/10.5860/choice.38-3317

Ellis, A. (1988). *Fish faccination* (pp. 1-16). San Diego: Academic Press.

Gabhe, S. Y. Tatke, P. A., & Khan, T. A. (2008). Evaluation of the immunomodullatory activity of the methanol extract of *Ficus benghalensis* roots in rats. *Indian J. Pharmacol, 38*(4), 271-275. http://dx.doi.org/10.4103/0253-7613.27024

Ganong, W. F. (1999). *Review of Medical Physiology.* Translater: Adji Dharma. Fisiologi Kedokteran. Edisi 9. Jakarta: Penerbit Buku Kedokteran.

Gao, B., Jeong, W., & Zhigang, T. (2008). Liver: An Organ with Predominant Innate Immunity. *Hepatology, 47,* 729-736. http://dx.doi.org/10.1002/hep.22034

Guyton, A. C. (1995). *Buku Ajar Fisiologi Kedokteran.* Ed ke-7. Jakarta: Penerbit Buku Kedokteran EGC.

Hartono. (1989). *Bahan Pengajaran Histologi Veteriner.* Departemen Pendidikan dan Kebudayaan Direktorat Jendral Pendidikan Tinggi Pusat Antar Universitas Ilmu Hayat, Institut Pertanian Bogor.

Junqueira, L., & Carnero, J. (2007). *Histologi Dasar.* Jakarta: Buku kedokteran EGC.

Khoul, S., & Khosa, R. L. (2013). *Immunomodullatory activity of phytoconstituent of Melissa officinalis. Der Pharmacia Lettre, 5*(1), 141 -145.

Khumar, S., Gupta, P., Sharma, S., & Khumar, D. (2011). A Review on Immunostimulatory plant. *Journal of Chinese Integrative Medicine, 9*(2), 117 -128. http://dx.doi.org/10.3736/jcim20110201

Kurts, Ch., Panzer, U., Anders, H., & Andrew, R. J. (2013). The immune system and kidney disease: Basic concepts and clinical implications. *Nature ReviewsImmunology, 13*, 738-753. http://dx.doi.org/10.1038/ nri3523

Kurts, Ch., Panzer, U., Hans-Joachim, A., & Andrew, J. R. (2013). The immune system and kidney disease: basic concepts and clinical implications. *Nature Reviews Immunology, 13, 738-753.* http://dx.doi.org/10.1038/nri3523

Lee, T. T., Huang, C. C., Shieh, X. H., Chen, C. L., Chen, L. J., & Yu, B. I. (2010). Flavonoid, phenol and polysaccharide contents of Echinacea purpurea L. and its immunostimulant capacity in vitro. *Int J Environ Sci Dev, 1*, 5-9. http://dx.doi.org/10.7763/ijesd.2010.v1.2

Liu, L. Z., Fang, J., Zhou, Q., Hu, X., Shi, X., & Jiang, B. H. (2005). Apigenin inhibits expression of vascular endothelial growth factor and angiogenesis in human lung cancer cells implication of chemoprevention of lung cancer. *Mol Pharmacol., 68*, 635-43. http://dx.doi.org/10.1124/mol.105.011254

Małaczewska, J., Wójcik, R., Jungi, L., & Siwicki, A. K. (2010). Effect of Biolex β-HP on Selected Parameters of Specific and Non-Specific Humoral. and Cellular Immunity in Rats. *Bull Vet Ins Pulawi., 54*, 75-80

Millodot. (2009). *Dictionary of Optometry and Visual Science* (7th ed.). Butterworth-Heinemann.

Munazir, Z. (2001). Respon Imun Terhadap Infeksi Bakteri. *Sari Pediatri, 2*(4), 193-197.

Ogier de Baulny, M., Quentel, C., Fournier, V., Lamour, F., & Le Gouvello, R. (1996). Effect of long-term oral administration of β-glucan as an immunostimulant or an adjuvant on some non-specific parameters of the immune response of turbot Scophthalmus maximus. *Diseases of Aquatic Organisms, 26*(2), 139-147. http://dx.doi.org/10.3354/dao026139

Ortuno, J., Cuesta, A., Rodriguez, A., Esteban, M. A., & Meseguer, J. (2002). Oral administration of yeast, *Saccharomyces cerevisiae*, enchances the cellular innate immune respon of glithead seabream (*Sparus aurata* L). *Veterinary Immunology and Immunopathology, 85*, 41-50. http://dx.doi.org/10.1016/s0165-2427 (01)00406-8

Pearse, G. (2006). Normal Structure, Function and Histology of the Thymus. *Toxicologic Pathology, 34*(5), 504-514. http://dx.doi.org/10.1080/01926230600865549

Piao, Y., Liu, Y., & Xie, X. (2013). Change Trends of Organ Weight Background Data In Sprague Dawley Rats at Different Ages. *J Toxicol Pathol., 26*(1), 29-34. http://dx.doi.org/10.1293/tox.26.29

Prasenjit, B. H., Hullanti, K. K., & Kumar Vijay, M. L. (2011). *Antithelmintic and antioxidant activity of alcoholic extracts of diffrent parts of Coleus amboinicus Lour.*

Preeja, G., Pillai, S., Mishra, G., & Annapura, M. (2011). Evaluation of the acute and sub acute toxicity of the methanolic leaf extract of *Plectranthus amboinicus* (Lour) Spreng in balb c mice. *Euro. J. Exp. Bio., 1*(3), 236-245

Shiney, R. B., Ganesh, P., & Khumar, R. S. (2012). Phytochemical Screening of Coleus aromaticus and Leucas aspera and Their Antibacterial Activity against Entheric Phatogens. *International Journal of Pharmaceutical and Biological Archives, 3*(1), 162-166.

Tizard, I. R. (2004). *Veterinary Immunology an Introduction.* Edisi ke-7. USA: Saunders.

Tong, X, Van Dross, R. T., Abu-Yousif, A., Morrison, A. R., Pelling, J. C. (2007). Apigenin prevents UVB-induced cyclooxygenase 2 expression: Coupled mRNA stabilization and translational inhibition. *Mol Cell Biol., 27*, 283-96. http://dx.doi.org/10.1128/mcb.01282-06

Uma, M., Jothinayaki, S., Kumaravel, S., & Kalaiselvi, P. (2011). Determination of bioactive components of Plectranthus amboinicus Lour by GC-MS analysis. *New York Science Journal, 4*, 66-69. Retrieved from http://www.Sciencepub.net/newyork

Wang, W., Heideman, L., Chung, C. S., Pelling, J. C., Koehler, K. J., & Birt, D. F. (2000). Cell-cycle arrest at G2/M and growth inhibition by apigenin in human colon carcinoma cell lines. *Mol Carcinog, 28*, 102-10. http://dx.doi.org/10.1002/1098-2744

6

Inferring Human Phylogenies Using Three CODIS STR Markers (CSF1PO, TPOX and TH01)

Nuzhat A. Akram[1] & Shakeel R. Farooqi[1]

[1] Department of Genetics, University of Karachi, University Road, Karachi 75270, Pakistan

Correspondence: Shakeel R. Farooqi, Department of Genetics, University of Karachi, University Road, Karachi 75270, Pakistan. E-mail: farooqis@uok.edu.pk, smr.akram@gmail.com

Abstract

Over the past several decades polymorphic genetic loci have been discussed for their utility in human phylogenetic inferences. Short Tandem Repeat (STR) loci have shown promising results for this purpose. Unfortunately, allele frequency data of polymorphic loci are largely confined to few populations. Therefore, the number of shared loci declines as the number of population increases. We hypothesize that even a smaller number of STR loci can be used efficiently for phylogenetic purposes if an appropriate theoretical and statistical strategy is employed. This strategy provides a feasible and cost effective method to choose appropriate STR loci for phylogenetic studies. For this purpose, an empirical study was conducted using allele frequency data of three STR loci CSF1PO, TPOX, and TH01 across 98 human populations from the literature (references are available at http://dnaa.bravehost.com/ index.html and http://www.cstl.nist.gov/strbase/population/Omnipop). The choice of markers was based on locus polymorphism, high heterozygosity, low mutation rate, less artifacts and independence between the loci. Three methods were used to measure genetic distances between the populations; Cavalli Sforza's chord distance (D_C), Nei's genetic (D_A) and Nei's standard genetic distances (D_{ST}). Coefficient of variation (CV) was calculated across hundred (100) datasets obtained by re-sampling of the original dataset for each of the genetic distance methods. CV was in order of $D_{ST} > D_A > D_C$. Therefore, a consensus tree based on D_C was constructed using Neighbour Joining (NJ), Unweighted Pair Group Method with Arithmatic mean (UPGMA) and Maximum Likelihood (ML) methods. NJ and UPGMA methods got more statistical support that is higher bootstrap values than ML (NJ> UPGMA> ML). Validation study was performed using (A) Principal Component Analysis (B) Comparison with trees reported for other molecular markers (C) STR genotyping of five Pakistani subpopulations. Results strongly supported our hypothesis that the three STR markers CSF1PO, TPOX, and TH01 are successful in delineating ethnic, geographic and linguistic differentiation between the populations.

Keywords: human phylogenetic inferences, STR loci *CSF1PO TPOX TH01*, world population data, Pakistani subpopulations.

1. Introduction

Phylogenetic inferences are premised on the inheritance of ancestral characteristics and on the existence of an evolutionary history defined by changes in these characteristics (Li, Pearl, & Doss, 2000). Indeed, many human populations carry distinct genetic markers, and by tracing these markers through the generations their origin can be traced out (Adams, 2008). Since many decades allele frequency data have been used to reconstruct evolutionary histories of human populations (Ayub et al., 2003; Agrawal & Khan, 2005). A number of statistical problems related to the number of loci, sample size of the populations, degree of locus/loci polymorphism, distance methods, methods of reconstructing phylogenetic trees and the methods to ensure the reliability of the trees…etc have been addressed in the literature (Zhivotovsky & Feldman, 1995; Takezaki & Nei, 1996; Nei & Takezaki, 1996; Felsenstein, 2003; Holder & Lewis, 2003; Takezaki & Nei, 2008). However, polymorphic loci for which the allele frequency data are available are largely confined to European, North American and East Asian populations (Nei & Roychoudhury, 1993). For this reason the number of shared loci declines as the number of population increases. Therefore if one wants to use a large number of loci, the number of populations that can be used becomes very small. Moreover, the missing elements in locus × population matrix often introduce unreasonable branching patterns in phylogenetic trees. In the present study we hypothesize that the minimum number of markers can perform efficiently for phylogenetic inferences if the theoretical and statistical strategy applied is correct. For this purpose three microsatellite loci, also called Short Tandem Repeats (STRs), were chosen from Combined DNA Index System (CODIS) of American Federal Bureau of Investigation (Budowle, Moretti, Niezgoda, & Brown, 1998; Budowle, Moretti, Baumstark, Defenbaugh, & Keys, 1999; Butler, 2006). STRs are regions of tandemly repeated DNA segments found throughout the human genome that show length polymorphism with a core repeated DNA sequence (Butler & Hill, 2012). These markers are used for human identification purposes in forensic caseworks.

In the present work, an empirical study was conducted using STR allele frequency data of 98 human populations (references available at http://dnaa.bravehost.com/index.html and http://www.cstl.nist.gov/strbase/population/Omnipop). For the validation of the phylogenetic inferences five Pakistani subpopulations were genotyped for the three STR loci and the allele frequency data was incorporated into world population data. Statistical analyses performed on the datasets of empirical and validation studies are explained in section 'Theory and Calculations'.

2. Materials and Methods

2.1 Choosing the STR Loci

Three STR loci (CSF1PO, TPOX, and TH01) were chosen from Combined DNA Index System (CODIS) of American Federal Bureau of Investigation (FBI). Different chromosomal locations of these loci minimize the chances of linkage disequilibrium between them (Table 1). Moreover the variance of the number of heterozygous loci was found to be within 95% confidence interval which means there is no association between the loci under study. Two of the loci namely TPOX and TH01 showed the lowest mutation rate among all the CODIS STR loci, hence making them suitable for phylogenetic purposes (Keim et al., 2004). Frequency of biological "artifacts" associated with theses loci such as null alleles, stutter products, non-template nucleotide addition is low.

Table 1. Chromosomal locations and other information of the three STR loci (CSF1PO, TPOX and TH01)

S. No.	Locus Name	Locus Definition	Chromosomal Location	Repeat Sequence	Mutation Rate (%)
1	CSF1PO	Human c-fms proto-oncogene for CSF-1 receptor gene	5q33.3-34	AGAT	0.16
2	TPOX	Human thyroid peroxidase gene	2p25.1-pter	AATG	0.01
3	TH01	Human tyrosine hydroxylase gene	11p15.5	AATG	0.01

2.2 World Population Data

Allele frequency data of 62 world populations from the literature (http://dnaa.bravehost.com/index.html) and 36 world populations from the Omnipop excel file (http://www.cstl.nist.gov/strbase/population/Omnipop) were available to reconstruct the phylogenetic trees. These populations encompass major geographical areas of the world and show different ethnic and linguistic affiliations (Table 2 and Table 3). All three loci were reported to be in Hardy Weinberg equilibrium across all the populations under study.

Table 2. Ethnic, Linguistic and Geographic classification of the populations (http://dnaa.bravehost.com/index.html)

S. No.	Ref. No.[a]	Population/Subpopulation	Abbreviation used	Ethnic group	Language
1	37	African Jordanian	AfricanJ	Negroid	Niger-Congo
2	2	African Mozambique	AfricanM	Negroid	Niger-Congo
3	25	Andhra Pradesh, Golla caste1	APGolla1	Australoid	Indo-European
4	25	Andhra Pradesh, Golla caste2	APGolla2	Australoid	Indo- European
5	25	Andhra Pradesh, Golla caste3	APGolla3	Australoid	Indo- European
6	25	Andhra Pradesh, Golla caste4	APGolla4	Australoid	Indo-European
7	25	Andhra Pradesh, Golla caste5	APGolla5	Australoid	Indo- European
8	25	Andhra Pradesh, Golla caste6	APGolla6	Australoid	Indo-European
9	25	Andhra Pradesh, Golla caste7	APGolla7	Australoid	Indo-European
10	8	Bangladeshis	BNDeshis	Caucasoid	Indo-European
11	31	Bengal Tribe1	BengalT1	Mongoloid	Indo-European
12	31	Bengal Tribe2	BengalT2	Mongoloid	Indo- European

13	31	Bengal Tribe3	BengalT3	Mongoloid	Indo- European
14	31	Bengal Tribe4	BengalT4	Mongoloid	Indo- European
15	17	Bhutan	Bhutan	Mongoloid	Sino-Tibetan
16	35	Bohemian	Bohemian	Caucasoid	Indo-European
17	7	Bolivian	Bolivian	Caucasoid	Indo-European
18	11	Brazilian	Brazilian	Caucasoid	Indo-European
19	36	Brazilian2	Brazilian2	Caucasoid	Indo-European
20	27	Central India,Agharia	C.IndiaA	Caucasoid	Indo-European
21	27	Central India,Dheria Gond	C.IndiaD	Australoid	Indo-European
22	27	Central India,Satmani	C.IndiaS	Caucasoid	Indo-European
23	27	Central India,Teli	C.IndiaT	Caucasoid	Indo-European
24	15	Chinese	Chinese	Mongoloid	Austronesian
25	19	Chinese Hong Kong	ChnHkong	Mongoloid	Austronesian
26	38	Chinese Korean	ChnKorean	Mongoloid	Austronesian
27	1	Dubai, Bangladeshis	DubaiBang	Caucasoid	Indo- European
28	1	Dubai, Iranians	DubaiIran	Caucasoid	Afro-Asiatic
29	1	Dubai, Omanis	DubaiOman	Caucasoid	Afro-Asiatic
30	1	Dubai, Saudi Arabians	DubaiSArab	Caucasoid	Afro-Asiatic
31	1	Dubai, Yemenites	DubaiYemen	Caucasoid	Afro-Asiatic
32	9	Eastern India, Garo	E.IndiaG	Mongoloid	Indo-European
33	9	Eastern India,Brahmin	E.IndiaBr	Caucasoid	Indo-European
34	9	Eastern India,Kayastha	E.IndiaKa	Caucasoid	Indo-European
35	26	Ecuadorian	Ecuadorian	Caucasoid	Indo- European
36	6	Greek Cyprus	GrkCyprus	Caucasoid	Afro-Asiatic
37	34	Gurkha, Malaysia	GurkhaMLY	Caucasoid	Indo-European
38	3	India, Bihar Baniya	BiharBaniy	Caucasoid	Indo-European
39	3	India, Bihar Kurmi	BiharKurmi	Australoid	Indo-European
40	3	India, Bihar Yadav	BiharYadav	Caucasoid	Indo-European
41	4	India,Andhra Pradesh, Dravidian1	APDravidn1	Caucasoid	Dravidian
42	4	India,Andhra Pradesh, Dravidian2	APDravidn2	Caucasoid	Dravidian
43	4	India,Andhra Pradesh, Dravidian3	APDravidn3	Caucasoid	Dravidian
44	29	Iran	Iranian	Caucasoid	Afro-Asiatic
45	33	Japanese	Japanese	Mongoloid	Afro-Asiatic
46	24	Kashmiris	Kashmiris	Caucasoid	Indo-European
47	30	Kurd	Kurdish	Caucasoid	Afro-Asiatic
48	28	Malaysian Chinese	MLYchinese	Mongoloid	Afro-Asiatic
49	28	Malaysian Indians	MLYindians	Caucasoid	Indo-European
50	28	Malaysian Malays	MLYmalays	Mongoloid	Afro-Asiatic
51	10	Muslim Tamil Bohra	TamilBohra	Australoid	Dravidian
52	10	Muslim Tamil Sunni	TamilSunni	Australoid	Dravidian
53	16	Northern Greece	NGreece	Caucasoid	Indo-European
54	21	Singapore Indians	SPIndian	Caucasoid	Indo-European
55	12	South African Blacks	S.AfBlack	Negroid	Niger-Congo
56	12	South African Whites	S.AfWhite	Caucasoid	Indo-European
57	32	Thailand	Thailand	Mongoloid	Austronesian
58	20	Tibet Lassa	TibetLassa	Mongoloid	Austronesian
59	5	Zimbabwe(Black African)	Zimb. Bl. Af	Negroid	Niger-Congo
60	18	Nepal	Nepal	Mongoloid	Sino-Tibetan
61	23	South India Tamil	SIndTamil	Australoid	Dravidian
62	14	Bavarian Caucasians	Caucasians	Caucasoid	Indo-Europeans

[a] It refers to the number of reference provided at the website http://dnaa.bravehost.com/index.html

Table 3. Ethnic, Linguistic and Geographic classification of the populations (http://www.cstl.nist.gov/strbase/population/Omnipop)

S.No.	Ref. No.[a]	Population/Subpopulation	Abbreviation used	Ethnic group	Language
1	1	FBI African American	AfrAmer	Negroid	Niger-Congo
2	2	Bahama African American	AfrBahm	Negroid	Niger-Congo
3	2	Jamaica African American	AfrAmJm	Negroid	Niger-Congo
4	2	Trinidad African American	AfrAmTr	Negroid	Niger-Congo
5	2	California African American	AfrAmCa	Negroid	Niger-Congo
6	2	Alabama African American	AfrAmAL	Negroid	Niger-Congo
7	2	Florida African American	AfrAmFL	Negroid	Niger-Congo
8	2	Virginia African American	AfrAmVr	Negroid	Niger-Congo
9	2	New York African American	AfrAmNY	Negroid	Niger-Congo
10	2	Illinois African American	AfrAmIL	Negroid	Niger-Congo
11	2	Alabama Caucasians	CaucaAL	Caucasoid	Indo-European
12	2	Virginia Caucasians	CaucaVr	Caucasoid	Indo-European
13	2	Michigan Caucasians	CaucaMi	Caucasoid	Indo-European
14	2	Florida Hispanics	HispaFL	Caucasoid	Indo-European
15	2	Arizona Hispanics	HispaAr	Caucasoid	Indo-European
16	2	Chinese	Chinese	Mongoloid	Austronesian
17	2	Japanese	Japanes	Mongoloid	Afro-Asiatic
18	2	Korean	Koreans	Mongoloid	Austronesian
19	2	General Asians	GAsians	Caucasoid	Indo-European
20	3	Swiss Caucasians	CaucaSw	Caucasoid	Indo-European
21	5	Connecticut African American	AfrAmCo	Negroid	Niger-Congo
22	5	Connecticut Caucasians	CaucaCo	Caucasoid	Indo-European
23	5	Connecticut Hispanics	HispaCo	Caucasoid	Indo-European
24	7	Turkish	Turkish	Caucasoid	Indo-European
25	9	Southern Spain (Andalusia)	SSpaini	Caucasoid	Indo-European
26	10	Brazilian	Brazils	Caucasoid	Indo-European
27	13	Tamil (India)	TamilsI	Australoid	Dravidian
28	2	New York Caucasians	CaucaNY	Caucasoid	Indo-European
29	6	Basques	Basques	Caucasoid	Language Isolate
30	56	CFS Asian Canada	AsianCa	Caucasoid	Indo-European
31	56	CFS East Indian Canada	EIndian	Caucasoid	Indo-European
32	58	Desasthbrahmin (India)	DBrahmn	Caucasoid	Indo-European
33	58	Chitpavanbrahmin (India)	CBrahmn	Caucasoid	Indo-European
34	59	Tunisian	Tunisia	-	Afro-Asiatic
35	61	Oriya Brahmin (India)	OBrahmn	Caucasoid	Indo-European
36	61	Khandayat (India)	KIndian	Caucasoid	Indo-European

[a] It refers to the number of reference provided at the website (http://www.cstl.nist.gov/strbase/population/Omnipop).

2.3 Genotyping of the Three STR Loci (CSF1PO, TPOX and TH01) across Five Pakistani Subpopulations

One hundred and seventy five unrelated individuals (2n) were chosen from five Pakistani subpopulations residing in Karachi. Individuals within each subpopulation were selected through randomization. These subpopulations were Baloch (n = 64), Muhajir (Urdu speaking Indian immigrants) (n = 94), Pathan (n = 60), Punjabi (n = 74) and

Sindhi (n = 58). Each individual was genotyped for the three STR loci after taking informed consent. All the three loci were co amplified in a single PCR reaction using CTT (CSF1PO, TPOX and TH01) primers in 2400 thermal cycler. The protocols provided by the Promega Geneprint STR System Technical Manual (tm#004) were followed. Allele frequencies were estimated using maximum likelihood method (Li, 1976; Hedrick, 2011).

2.4 Theory and Calculations

2.4.1 STR Polymorphism

STR polymorphism was estimated by (i) *Heterozygosity (h).* Heterozygosity of each of the three loci was estimated by the Nei's unbiased formula

$$h = (2n/(2n\text{-}1))(1 - \sum pi^2) \tag{1}$$

where pi is the frequency of ith allele in the sample, and n is the number of diploid individuals examined at the locus. The average Heterozygosity *(H)* of each of the three loci was estimated by

$$H = (\sum h)/s \tag{2}$$

where s is the total number of populations under study. (ii) Average number of alleles per locus (n_a). n_a was computed by the formula

$$n_a = (\sum N_a)/s \tag{3}$$

where N_a is the number of alleles of a locus in a population and s is the total number of populations under study. (iii) Polymorphism Information Content (PIC) was calculated using the formula

$$PIC = 1 - \sum pi^2 - 2\sum pi^2 pj^2, \tag{4}$$

where pi and pj stands for the frequencies of ith and jth alleles of a locus (Shete, Tiwari, & Elston, 2000; Kobilinsky, Liotti, & Oeser Sweat, 2005). (iv) Probability of Identity (PI) and Power of Discrimination (PD): PI is derived by the formula

$$PI = \sum (xi)^2 + \sum (xij)^2 \tag{5}$$

where xi stands for the frequency of homozygotes and is equal to pi^2. While xij stands for the frequency of heterozygotes and is equal to *2 pi pj*, where pi and pj stands for the frequencies of ith and jth alleles of a locus. PD is defined as,

$$PD = 1 - \sum (xi)^2 + \sum (xij)^2 \quad \text{or} \quad 1\text{- PI} \tag{6}$$

Different measures/statistics of locus polymorphism across the world populations are shown in Table 4 and Table 5.

Table 4. Measures of locus polymorphism for the three STR loci (CSF1PO, TPOX and TH01) averaged over 62 world populations (http://dnaa.bravehost.com/index.html)

Measures of locus polymorphism	Abbreviation used	CSF1PO	TPOX	TH01
Average Heterozygosity	H	0.713	0.690	0.756
Average number of alleles	n_a	6.7	5.9	6.1
Polymorphism Information Content	PIC	0.690	0.619	0.708
Power of Discrimination	PD	0.837	0.798	0.849
Power of Identity	PI	0.163	0.202	0.151

Table 5. Measures of locus polymorphism for the three STR loci (CSF1PO, TPOX and TH01) averaged over 36 world populations (http://www.cstl.nist.gov/strbase/population/Omnipop)

Measures of locus polymorphism	Abbreviation used	CSF1PO	TPOX	TH01
Average Heterozygosity	H	0.739	0.689	0.763
Average number of alleles	n_a	10	10	10
Polymorphism Information Content	PIC	0.694	0.641	0.727
Power of Discrimination	PD	0.886	0.852	0.906
Power of Identity	PI	0.113	0.147	0.093

Heterozygosities of the three loci across each subpopulation of Pakistan are shown in Table 6.

Table 6. Observed heterozygosities of the three STR loci (CSF1PO, TPOX and TH01) across the five Pakistani subpopulations. These subpopulations were Baloch, Muhajir, Pathan, Punjabi and Sindhi

Pakistani Subpopulations	CSF1PO	TPOX	TH01
Baloch	0.531	0.625	0.875
Muhajir	0.532	0.851	0.787
Pathan	0.6	0.7	0.7
Punjbi	0.486	0.594	0.702
Sindhi	0.482	0.689	0.875

2.4.2 Genetic Distance Measure

Three distance measure were used to infer genetic distances between the populations under study. (1) D_A (Nei's genetic distance) is formulated for Infinite Allele Model (IAM) in which there is a rate of neutral mutation and each mutation give rise to a distinguishable allele (Nei, Tajima, & Tateno, 1983). D_A is calculated as

$$D_A = 1 - 1/r \sum_j^r \sum_i^{mj} \sqrt{X_{ij} Y_{ij}} \qquad (7)$$

where X_{ij} and Y_{ij} are the frequencies of the *ith* allele at the *jth* locus in populations X and Y, respectively, and mj is the number of alleles at the *jth* locus. D_A was computed through a statistical program *Poptree*. (2) D_{ST} is the Nei's standard genetic distance (Nei, 1972) given by

$$D_{ST} = -In\, J_{XY} / \sqrt{J_X\, J_Y} \qquad (8)$$

where $J_X = \sum_j^r \sum_i^{mj} x_{ij}^2 / r$ and $J_Y = \sum_j^r \sum_i^{mj} y_{ij}^2 / r$ are the average heterozygosities over the loci for populations X and Y, respectively, and $J_{XY} = \sum_j^r \sum_i^{mj} x_{ij} y_{ij}/r$. D_{ST} was computed through *Phylip version 3.68*. (3) D_C is the chord distance proposed by Cavalli Sforza and Edward (1967). D_C is defined by

$$D_C = 2/\pi\, r \sum_j^r [\, 2\, (\, 1 - \sum_i^{mj} \sqrt{X_{ij} Y_{ij}})\,]^{1/2} \qquad (9)$$

D_C was computed through *Phylip version 3.68*.

2.4.3 Coefficient of Variation (*CV*) of D_A, D_{ST} and D_C

Nei's distance and Cavalli Sforza's distance measures are different estimators of the same quantity under the same model. Therefore a measurement of relative variance (*CV*) was used for each distance measure. *CV* was calculated across 100 replicates of the original dataset obtained through re-sampling of the original dataset for each of the three distance measures. A random sample of fourteen populations (14 X 14 populations distance matrix) was used for this purpose.

2.4.4 Construction of Phylogenetic Trees

Three methods were used to reconstruct phylogenetic trees. (1) Neighbor Joining (NJ) Method. NJ method constructs a tree by successive clustering of lineages, setting branch lengths as the lineages join (Saitou, & Nei, 1987). (2) Unweighted Pair Group Method using Arithmetic Mean (UPGMA). UPGMA merge closest pair of taxa (by distance) and then recomputes distances to merged nodes via arithmetic mean of pairwise distances to leaves of the tree. (3) Continuous Character Maximum Likelihood Method (CONTML). This is a program in PHYLIP which estimates phylogenies by the restricted maximum likelihood method based on the Brownian motion model. It assumes that each locus evolves independently by pure genetic drift.

2.4.5 Consensus Tree

Consensus trees were generated by bootstrapping (100 to 1000 replications) of the original data taken from http://dnaa.bravehost.com/index.html and http://www.cstl.nist.gov/strbase/population/Omnipop (Figure 1 and Figure 2). Consensus trees were also constructed between the populations who either have a strong ethnic (Figure 3) or linguistic affiliation (Figure 4). Allele frequency data of Pakistani subpopulations were incorporated into the 62 world populations' data (*http://dnaa.bravehost.com/index.html*) as well as 36 world populations' data from Omnipop file (http://www.cstl.nist.gov/strbase/population/Omnipop). Two measures of genetic distance were used i.e. Nei's genetic distance (Figure 5 and Figure 6) and Cavalli Sforza chord distance (Figure 7 and Figure 8). Trees were constructed using NJ method.

2.4.6 Comparison with other Phylogenetic Trees

Consensus tree was then compared with the trees obtained from other molecular markers such as Alu and RFLP (Nei & Roychoudhury, 1993; Nei & Takezaki, 1996).

2.4.7 Principal Component Analysis (PCA)

PCA was performed for the allele frequency data of 62 world populations (http://dnaa.bravehost.com/index.html) (Figure 9). Pakistani subpopulation allele frequency data was incorporated into 62 world population data and PCA was performed again (Figure 10).

3. Results

- Coefficient of variation (CV) was in order of $D_{ST} > D_A > D_C$. Therefore, a consensus phylogenetic tree based on DC was constructed using NJ, UPGMA and Maximum Likelihood (ML) methods. NJ method showed higher bootstrap values. Comparison of the resultant trees showed that the tree topology was consistent with the trees reported for other molecular markers (Nei & Roychoudhury, 1993; Nei & Takezaki, 1996). Geographic, ethnic and linguistic demarcation between the populations was appreciable (Figure 1 through Figure 4).

- Tree topology was consistent with 'out of Africa' theory of human origin. African populations formed a distinct cluster with high bootstrap value (>950) and the remaining populations branched off from the African cluster.

- Geographical and ethnic demarcations between the populations were more obvious than linguistic demarcation i.e. populations who are in close geographical proximity to each other or who have a common ethnic origin showed tendency to form a separate cluster. For example, Chinese, Tibet, Bhutan, Thai, Malays and Japanese formed a separate cluster though they have diverse linguistic affiliations. China, Thailand and Malaysia belong to Austronesian class of languages while Bhutan, Tibet and Nepal belong to Sino Tibetan class. All these populations belong to the Mongoloid ethnic group. It showed that the STRs were more successful in delineating ethnic rather than linguistic partitioning.

- Phylogenetic efficiency of the three STRs for the populations and subpopulations of the Indian subcontinent was remarkable. Central Indian, south Indian and eastern Indian populations were well differentiated according to their ethnic and linguistic backgrounds. All the Dravidian speaking Australoid Golla subpopulations of Andhra Pradesh (seven in number) consistently formed a separate cluster in phylogenetic trees. Similarly, Eastern India castes Brahmin and Kayasth consistently showed a single cluster while another Eastern India caste Garo was closer to European Caucasians rather than their neighboring populations. Likewise Tamil Bohra muslims did not cluster with their neighboring Tamil sunni muslims, instead they were closer to the Mongoloid populations.

- PCA showed a distinct position of all the African populations in the of score plot of PC1 and PC2 (Figure 9). Indian populations and subpopulations were lying in the left upper quadrant while Mongoloids were in the right lower quadrant. Caucasoids were dispersed in the right half around the median axis.

- Phylogenetic tree (Figure 5 through Figure 8) showed a distinct cluster of the five Pakistani subpopulations with high bootstrap values (≥ 840). It also showed the close affinity of Pakistani subpopulations to the Caucasoid and Mongoloid populations.

- PCA showed all the Pakistani subpopulations in the left lower quadrant except Muhajir that was in the left upper quadrant (Figure 10).

Figure 1. Phylogenetic tree of 62 world populations (*http://dnaa.bravehost.com/index.html*) based on Cavalli Sforza's Chord distance and NJ method (consensus tree= 1000 bootstrapping)

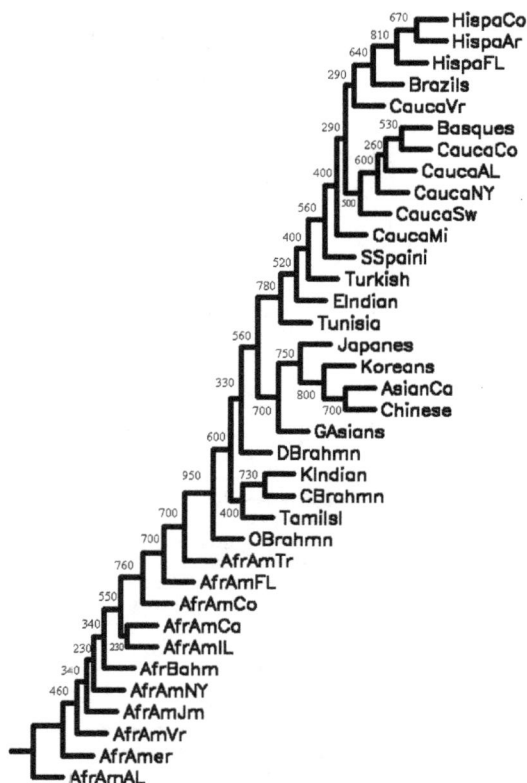

Figure 2. Phylogenetic tree of 36 world populations from Omnipop file
(http://www.cstl.nist.gov/strbase/population/Omnipop) based on Cavalli Sforza's Chord distance and NJ method
(consensus tree= 1000 bootstrapping)

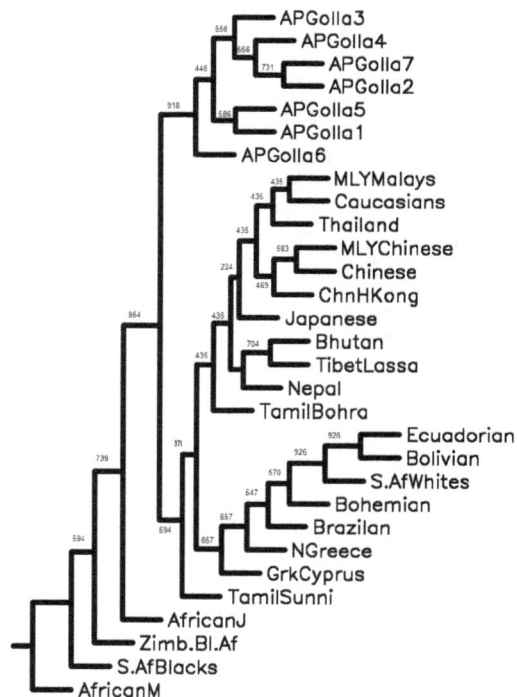

Figure 3. Phylogenetic tree of 30 world populations (*http://dnaa.bravehost.com/index.html*) with strong ethnic
affiliations and less degree of genetic admixture based on Cavalli Sforza's Chord distance and NJ method
(consensus tree = 1000 bootstrapping)

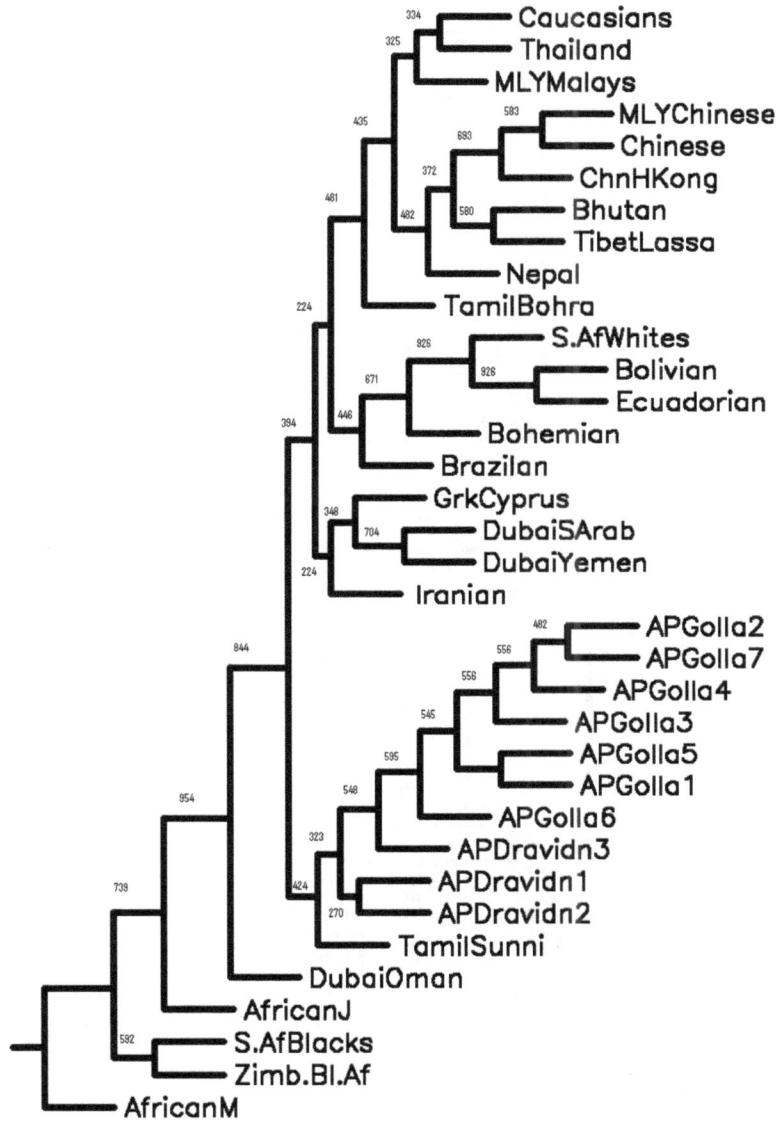

Figure 4. Phylogenetic tree of 35 world populations(*http://dnaa.bravehost.com/index.html*) with strong linguistic affiliations and less degree of genetic admixture based on Cavalli Sforza's Chord distance and NJ method (consensus tree = 1000 bootstrapping)

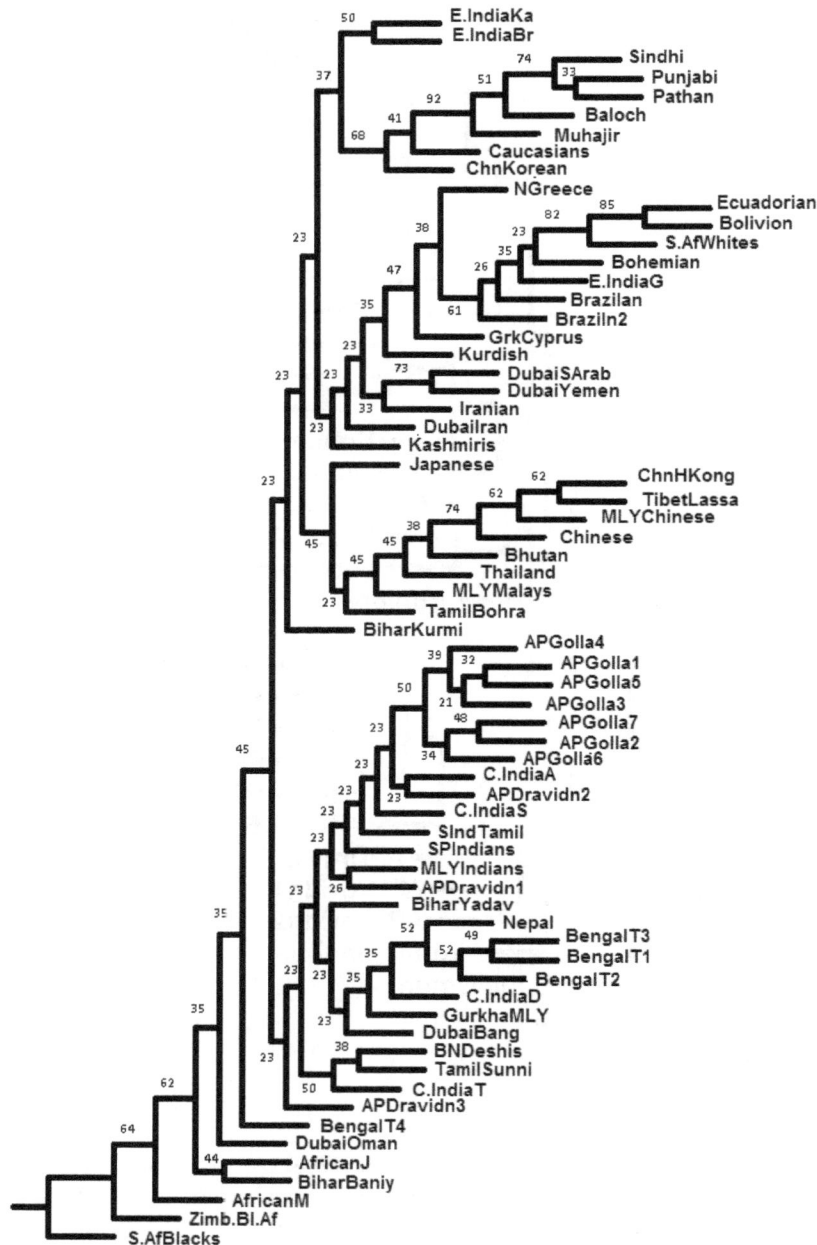

Figure 5. Phylogenetic tree of 67 world populations (*http://dnaa.bravehost.com/index.html*) including five Pakistani subpopulations of the present study. The phylogenetic tree is based on Nei's genetic distance and NJ method (consensus tree= 100 bootstrapping)

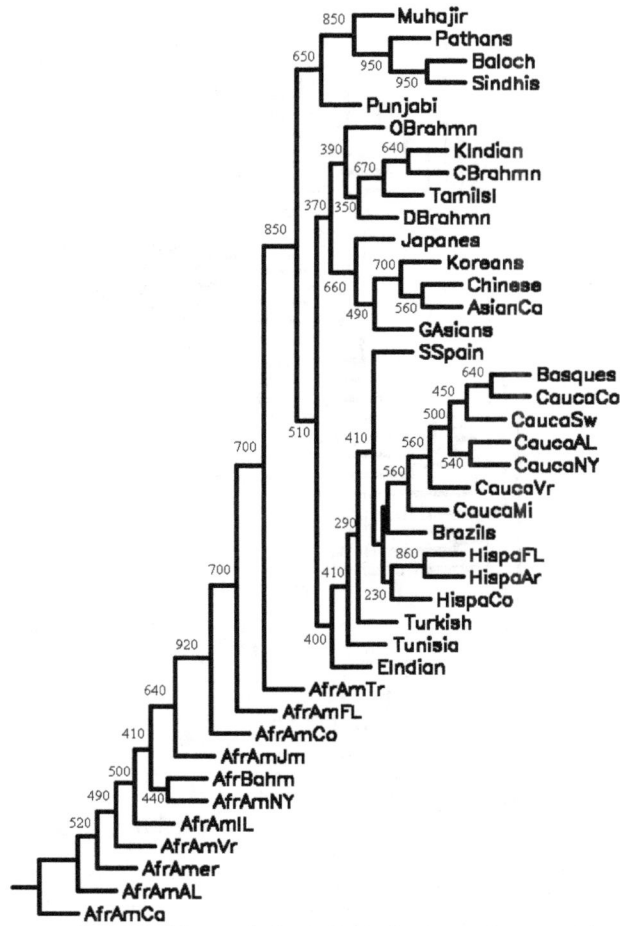

Figure 6. Phylogenetic tree of 41 world populations (*http://www.cstl.nist.gov/strbase/population/*Omnipop) including five Pakistani subpopulations of the present study. The phylogenetic tree is based on Nei's genetic distance and NJ method (consensus tree= 1000 bootstrapping)

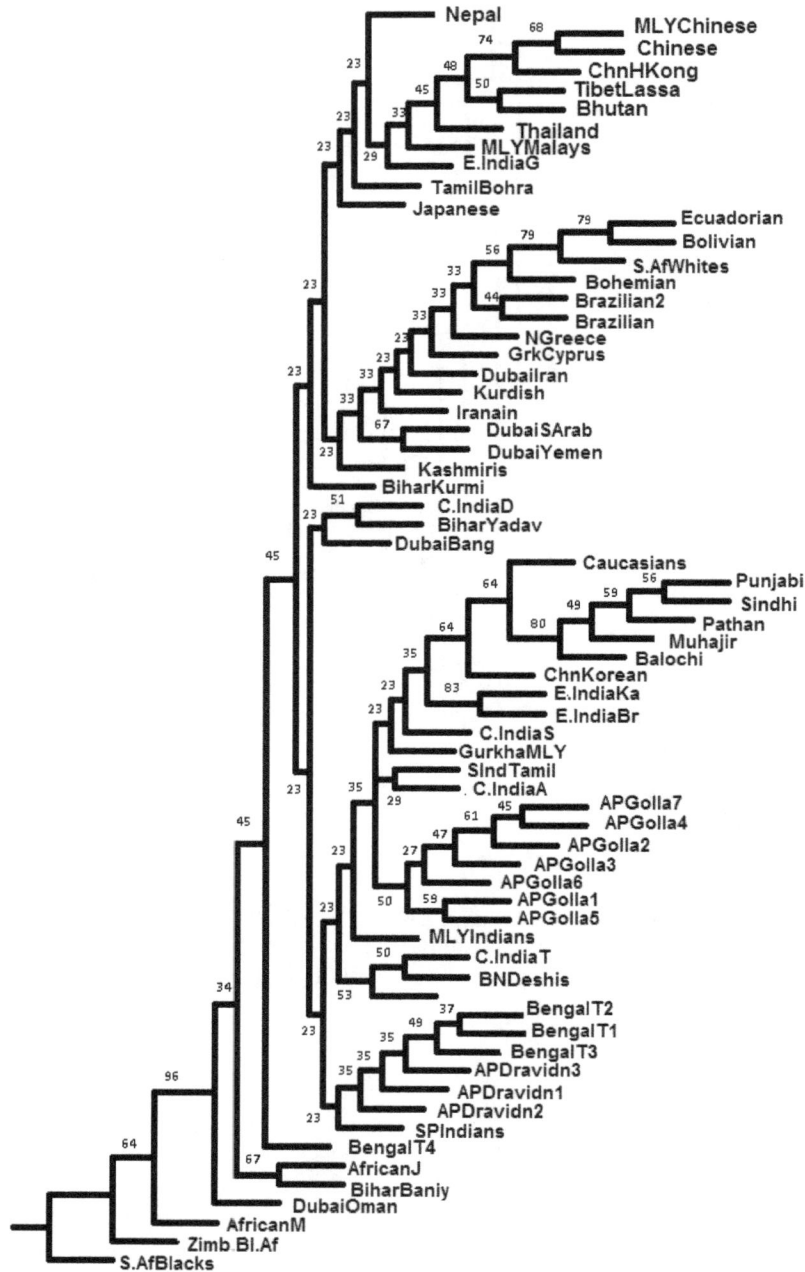

Figure 7. Phylogenetic tree of 67 world populations (*http://dnaa.bravehost.com/index.html*) including five Pakistani subpopulations of the present study. The phylogenetic tree is based on Cavalli Sforza's Chord distance and NJ method (consensus tree= 100 bootstrapping)

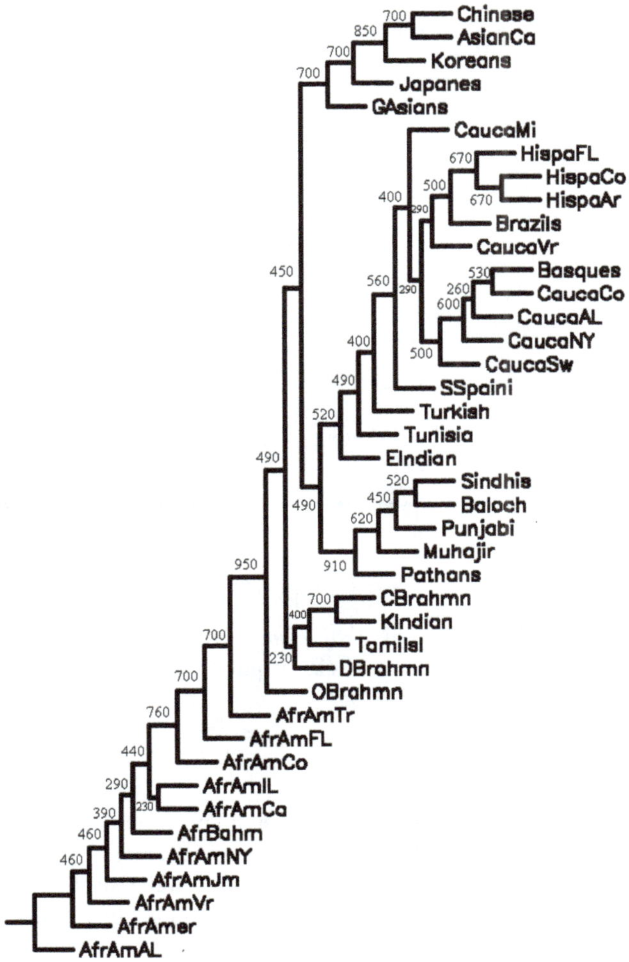

Figure 8. Phylogenetic tree of 41 world populations (*http://www.cstl.nist.gov/strbase/population/*Omnipop) including five Pakistani subpopulations of the present study. The phylogenetic tree is based on Cavalli Sforza's Chord distance and NJ method (consensus tree= 1000 bootstrapping)

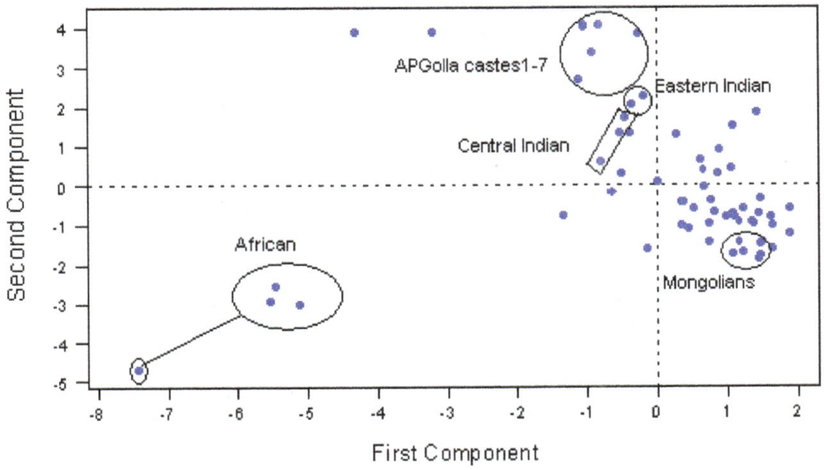

Figure 9. Score plot of PCA of 62 populations (*http://dnaa.bravehost.com/index.html*) based on allele frequencies of three STRs, CSF1PO, TPOX, and TH01

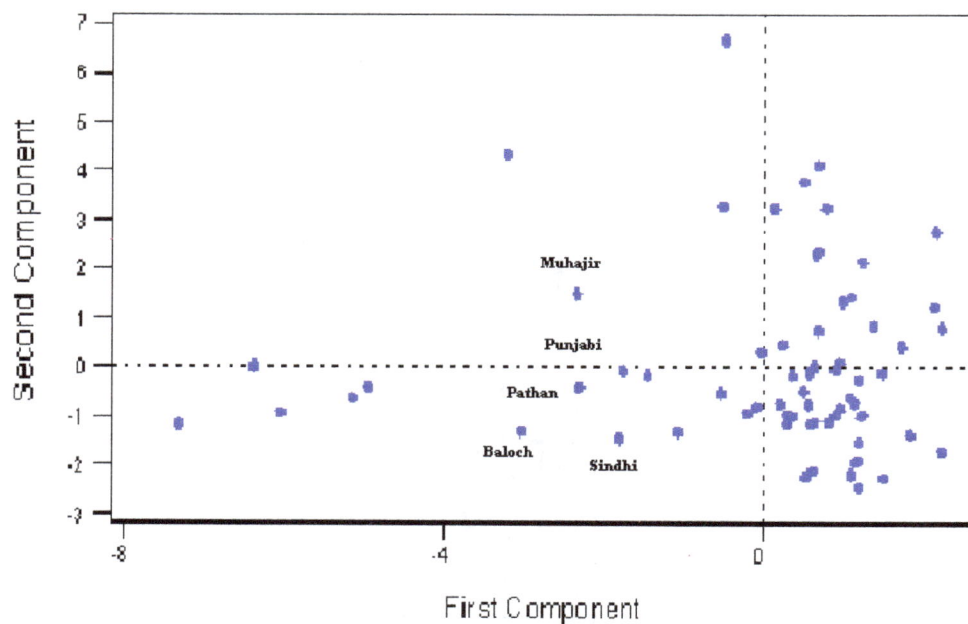

Figure 10. Score plot of PCA of 67 world populations (*http://dnaa.bravehost.com/index.html*) including five Pakistani subpopulations of the present study based on allele frequencies of three STRs, CSF1PO, TPOX, and TH01

4. Discussion

Evolutionary histories and phylogenetic relationship of many extant human populations have been explored using microsatellite loci (Bowcock et al., 1994; Deka et al., 1995; Gonser, Donnelly, Nicholson, & Rienzo, 2000; Rowold & Herrera, 2003). Rowold and Herrera (2003) and Agrawal and Faisal (2005) used five STR loci including CSF1PO, TPOX and TH01 for phylogenetic analyses and concluded that these STR loci are successful in reconstructing recent human evolutionary histories. However, they used only ten (10) and twenty one (21) population groups respectively in comparison to ninety eight (98) world populations used in the present study. Moreover the strategy employed in the present study was more comprehensive and each decision making step was explained logically. Findings were also supported by the validation studies.

Understanding the pattern and rate of mutations is very relevant to the applications of these hypervariable genetic markers in evolutionary studies as well as in gene mapping studies (Goldstein, Linares, Cavalli-Sforza, & Feldman, 1995; Shriver et al., 1995). The utility of a genetic marker for determining phylogenetic relationships within a given population is a function of the mutation rate of the marker and the overall genetic diversity of the examined population (Keim et al., 2004). When population genetic diversity is high, only markers with low mutation rates will yield accurate phylogenetic patterns. TPOX and TH01 showed the lowest mutation rates among all thirteen CODIS STR loci; hence likely to be suitable for phylogenetic purposes.

Topology of the phylogenetic trees (Figure 1 through Figure 4) was consistent with those obtained from other molecular markers. For example a phylogenetic tree for 26 human populations based on D_A using 29 polymorphic loci (Nei & Roychoudhury, 1993) showed the same partitioning of human populations as shown by the trees reconstructed in the present study. The trees were also compared with the trees based on RFLP data and Alu insertion polymorphism data using D_A distance measure (Nei & Takezaki, 1996). Tree topology and partitioning of the populations into ethnic groups were consistent with those of RFLP and Alu insertions. It should be mentioned that the performance of D_A distance measure in obtaining the correct tree topology is considered to be the same as that of D_C (Takezaki & Nei, 1996). Major ethnic groups identified in the present study were more or less similar to those recognized by classical anthropologist (Nei & Rouchaudhry, 1993). They were four in number namely, Negroid (Africans), Caucasoids (European and their related populations), Mongoloids (East Asians) and Australoid (Andhra Pradesh Golla castes). Tree topology was also supportive of 'out of Africa theory' which has gained popularity among geneticist and anthropologist during the last two decades (For example Nei, 1995; Templeton, 2002; Adams, 2008; Hanihara, 2008; Sun, Mullikin, Patterson, & Reich, 2009).

Stability of tree topology and the adequacy of the data to validate the topology are assessed by bootstrap values (Berry & Gascuel, 1996). Tree topology showed higher bootstrap values when applied to the dataset of populations who have lesser degree of admixture and a strong affiliation with a single ethnic or linguistic group (Figure 3 and Figure 4). It indicates that apart from the number of markers used, there are certain other factors which affect the phylogenetic efficiency of a marker. These factors include STR locus polymorphism, distance measures and methods to reconstruct phylogenetic trees (Nei & Roychoudhury, 1974; Nei, Kumar, & Takahashi, 1998; Goldstein & Pollock, 1994; Tajima & Takezaki, 1994; Takezaki & Nei, 1996; Takezaki & Nei, 2008). Ethnic demarcation showed more statistical support than linguistic demarcation. Another study using 182 autosomal microsatellites could not reveal any phylogenetic relationship between the two language isolate populations namely Hunza Burusho and Basques (Ayub et al., 2003). It was argued then that the microsatellites are best suited for the study of more recent population separations.

Phylogenetic efficiency of the three STR markers was worth noticing for the populations and subpopulations of the Indian subcontinent. Eastern India castes Brahmin and Kayasth consistently formed a single cluster. These two population groups are considered upper classes of Hindu caste system where intermarriages are not prohibited, while Garo which was closer to European Caucasians, is the middle class of Hindu caste system. Likewise Tamil Bohra Muslims were closer to the Mongoloid populations rather than their neighboring Tamil sunni Muslims. Bohra and Sunni are the two religious sects of Muslims between which marriages are generally prohibited. Dravidian speaking Golla castes of Andhra Pradesh remained separated from their neighboring subpopulations and branched off as a single cluster. Even within the Golla castes, western Golla castes (APGolla1 and APGolla5) were closer to each other than other Golla castes. The results established the efficiency of the three STRs (CSF1PO, TPOX, TH01) in delineating genetic relationships of the subpopulations of Indian subcontinent. The finding was validated for the subpopulations of Pakistan which is geographically a part of Indian subcontinent. All five Pakistani subpopulations namely Baloch, Muhajir, Pathan, Punjabi and Sindhi were united in a single cluster with a high bootstrap value that possibly suggests their common origin. Most of the Indian subcontinent populations are thought to be Caucasoid in origin (Cavalli Sforza, Menozzi, & Piazza, 1994). Pakistani subpopulations also showed their affiliation with other Caucasoid and Mongoloid populations. According to a hypothesis of populations' evolution and migration Indian subcontinent has been invaded by both the Caucasoid as well as Mongoloid populations (Nei & Roychoudhury, 1993). Mongoloids are also believed to originate from later splitting in Caucasoid race. Theories of gene flow and varying degrees of admixture between south Asian Indian populations and Mongoloid populations have also been proposed (Bamshad et al., 2003; Watkins, 2003; Shriver et al., 2005). Close affiliation of Pakistani subpopulations may be the results of gene admixture between the two populations. Another study based on HLA-A, -B, -C and –DRB, -DQB1 loci have also shown admixture of Pakistani ethnic groups with Caucasoids and Oriental populations (Mohyuddin, 2000).

In April 2011 American FBI recommended to remove few STR loci from the CODIS list due to their lower polymorphism observed across the world populations (Butler & Hill, 2012; Hares, 2012). TPOX was the least polymorphic of all the CODIS STR loci. In the present study TPOX showed heterozygosity values higher than those for CSF1PO (Table 5). Tri allelic pattern frequently observed for TPOX (Butler, 2005, Lane, 2008; Diaz, Rivas, & Carracedo, 2009) was not observed across the five subpopulations of Pakistan. Results emphasized that the STR loci should be investigated extensively for their efficiency as human identification markers across the populations and subpopulations of the Indian subcontinent. Heterogeneity of extant populations of the Indian subcontinent may deserve a separate standard set of STR loci. It is worth mentioning that European standard set of STR loci are/is different from American core set of STR loci (Butler & Hill, 2012).

It can be concluded that the three STRs successfully exhibited ethnic and linguistic as well as the(omit it) intra-ethnic differentiation across the populations and the subpopulations. Results also suggest that minimum number of markers can be used for reconstructing phylogenetic trees with high bootstrap values provided the markers are efficient for this purpose and a correct statistical strategy is employed. This study may help to identify the STR loci that can be used for forensic as well as phylogenetic purposes.

References

Adams, J. (2008). Human Evolutionary Tree. *Nature Education, 1(1)*. Retrieved from http://www.nature.com/scitable/topicpage/human-evolutionary-tree-417

Agrawal, S., & Khan, F. (2005). Reconstructing recent human phylogenies with forensic STR loci: A statistical approach. *BMC Genetics, 6,* 47. Retrieved from http://www.biomedcentral.com/1471-2156/6/47

Astolfi, P., Kidd, K. K., & Cavalli Sforza, L. L. (1981). A comparison of methods for reconstructing evolutionary trees. *Syst . Zool., 30,* 156-169. Retrieved from http://www.jstor.org/discover/10.2307/2992414

Ayub, Q., Mansoor, A., Ismail, M., Khaliq, S., Mohyuddin, A., Hameed, A., ... & Mehdi, S. Q. (2003). Reconstruction of human evolutionary tree using polymorphic autosomal microsatellites. *Am. J. Phys. Anthrop.*, *122*, 259-268. Retrieved from http://www.ncbi.nlm.nih.gov/pubmed/14533184

Bamshad, M. J., Wooding, S., Watkins, W. S., Ostler, C. T., Batzer, M. A., & Jorde, L. B. (2003). Human population genetic structure and inference of group membership. *Am. J. Hum. Genet.*, *75*, 578-589. Retrieved from http://www.ncbi.nlm.nih.gov/pmc/articles/PMC1180234/

Berry, V., & Gascuel, O. (1996). On the interpretation of bootstrap trees: Appropriate threshold of clade selection and induced gain. *Mol. Biol. Evol.*, *13*, 999-1011. http://mbe.oxfordjournals.org/cgi/doi/10.1093/molbev/13.7.999

Bowcock, A. M., Ruiz – Linares, A., Tomfohrde, J., Minch, E., Kidd, J. R., & Cavalli Sforza, L. L. (1994). High resolution of human evolutionary trees with polymorphic microsatellites. *Nature*, *368*, 455-457. Retrieved from http://www.nature.com/nature/journal/v368/n6470/abs/368455a0.html

Budowle, B., Moretti, T. R., Baumstark, A. L., Defenbaugh, D. A., & Keys, K. M. (1999). Population data on the thirteen CODIS core short tandem repeat loci in African Americans, U.S. Caucasians, Hispanics, Bahmians, Jamaicans and Trinidadians. *J. Forensic. Sci.*, *44*, 1277-1286. Retrieved from http://www.ncbi.nlm.nih.gov/pubmed/ 10582369

Budowle, B., Moretti, T. R., Niezgoda, N. R., & Brown, B. L. (1998). CODIS and PCR based short tandem repeat loci: Law enforcement tools. *Second European symposium on human identification. Promega corporation*, 73-88.

Butler, J. M. (2005). *Forensic DNA typing: Biology, technology and genetics of STR markers*. USA: Elsevier.

Butler, J. M. (2006). Genetics and genomics of core short tandem repeat loci used in human identity testing. *J. Forensic. Sci.*, *51*, 253-265. http://www.cstl.nist.gov/strbase/pub_pres/Butler2006JFS

Butler, J. M., & Hill, C. R. (2012). Biology and genetics of new autosomal STR loci useful for forensic DNA analysis. *Forensic Sci. Rev.*, *24*, 15-26. Retrieved from http://www.cstl.nist.gov/strbase/pub_pres/Butler-Hill-FSR2012-newSTRloci.

Cavalli Sforza, L. L., & Edwards, W. (1967). Phylogenetic analysis: Models and estimation procedures. *Am. J. Hum. Genet.*, *19*, 233-257. http://www.ncbi.nlm.nih.gov/pmc/articles/PMC1706274/

Cavalli Sforza, L. L., Menozzi, P., & Piazza, A. (1994). *The history and geography of human genes*. New Jersey: Princeton University Press.

Deka, R., Jin, L., Shriver, M. D., Yu, L. M., Decroo, S., Hundrieser, J., ... & Chakraborty, R. (1995). Population genetics of dinucleotide (dCdA);(dG-dT), polymorphisms in world populations. *Am. J. Hum. Genet.*, *56*, 461-474. http://www.ncbi.nlm.nih.gov/pmc/articles/PMC1801145/

Diaz, V., Rivas, P., & Carracedo, A. (2009). The presence of tri-allelic TPOX genotypes in Dominican Population. *Forensic Sci. Int. Genetics Supplement Series.*, *2*, 371-372. http://dx.doi.org/10.1016/j.fsigss. 2009.09.021

Felsenstein, J. (2003). *Inferring phylogenies*. Sunderland, MA: Sinauer Associates.

Goldstein, D., Linares, A., Cavalli-Sforza, L., & Feldman, M. (1995). An evaluation of genetic distances for use with microsatellite loci. *Genetics*, *139*, 463–471. http://www.ncbi.nlm.nih.gov/pmc/articles/PMC1206344/

Goldstein, D. B., & Pollock, D. D. (1994). Least square estimation of molecular distance- noise abatement in phylogenetic reconstruction. *Theor. Popul. Biol.*, *44*, 219-226. Retrieved from http://www.ncbi.nlm.nih.gov/pubmed/8066551

Gonser, R., Donnelly, P., Nicholson, G., & Rienzo, A. D. (2000). Microsatellite mutations and inferences about human demography. *Genetics*, *154*, 1793-1807. Retrieved from http://www.ncbi.nlm.nih.gov/pmc/articles/PMC1461043

Hanihara, T. (2008). Morphological variation of major human populations based on nonmetric dental traits. *Am. J. Phys. Anthrop*, *136*, 169-182. Retrieved from http://onlinelibrary.wiley.com/doi/10.1002/ajpa.20792/abstract

Hares, D. R. (2012). Expanding the CODIS core loci in United States. *Forensic Sci. Int. Genetics*, *6*, e52-e54. http://dx.doi.org/10.1016/j.fsigen.2011.04.012

Hedrick, P. W. (2011). *Genetics of populations* (4th ed.). Sudbury, Massachusetts: Jones and Bartlett publishers.

Holder, M., & Lewis, P. O. (2003). Phylogeny estimation: Traditional and Bayesian approaches. *Nature Genetics,* *4,* 275-284. Retrieved from http://www.nature.com/nrg/journal/v4/n4/full/nrg1044.html

Keim, P., Matthew, N. V. E., Talima, P., Amy, J. V., Lynn, Y. H., & David, M. W. (2004). Anthrax molecular epidemiology and forensics: Using the appropriate marker for different evolutionary scales. *Infection, Genetics and Evolution., 4,* 205–213. Retrieved from http://jan.ucc.nau.edu/aa238/Kenefic%20reference1

Kobilinsky, L., Liotti, T. F., & Oeser Sweat, J. (2005). *DNA: Forensic and legal applications.* Hoboken, New Jersey: Wiley International.

Lane, A. B. (2008). The nature of tri allelic TPOX genotypes in African populations. *Forensic Sci. Int. Genetics, 2,* 134-137. http://www.ncbi.nlm.nih.gov/pubmed/19083808.

Li, C. C. (1976). *First course in population genetics.* Boxwoods. Pacific Grove, CA. USA.

Li, S., Pearl, D. K., & Doss, H. (2000). Phylogenetic tree construction using Markov chain Monte Carlo. *J. Am. Stat. Assoc., 95,* 493-508. Retrieved from http://www.stat.ufl.edu/~doss/Research/mc-trees.pdf

Mohyuddin, A. (2000). *Genetic diversity of Pakistani subpopulations.* PhD Dissertation, Quaid-e-Azam University, Islamabad. Retrieved from http://eprints.hec.gov.pk/2110/1/2028.htm

Nei, M. (1972). Genetic distance between populations. *The American Naturalist., 106,* 283-292. Retrieved from http://www.jstor.org/discover/10.2307/2459777.

Nei, M. (1995). Genetic support for the out of Africa theory of human evolution. *Proc. Natl. Acad. Sci., 92,* 6720-6722. http://www.ncbi.nlm.nih.gov/pmc/articles/PMC41400/

Nei, M., & Roychoudhury, A. K. (1974). Genic variation within and between the three major races of man, caucasoids, negroids, and mongoloids. *Am. J. Hum. Genet., 26,* 421-443. Retrieved from http://www.ncbi.nlm.nih.gov/pmc/articles/PMC1762596/

Nei, M., & Roychoudhury, A. K. (1993). Evolutionary relationship of human populations on global scale. *Mol. Biol. Evol., 10,* 927-943. Retrieved from http://mbe.oxfordjournals.org/content/10/5/927

Nei, M., & Takezaki, N. (1996). The root of the phylogenetic tree of human populations. *Mol. Biol. Evol., 13,* 170-177. http://mbe.oxfordjournals.org/content/13/1/170.

Nei, M., Kumar, S., & Takahashi, K. (1998). The optimization principle in phylogenetic analysis tends to give incorrect topologies when the number of nucleotides or amino acids used is small. *Proc. Natl. Acad. Sci. USA, 95,* 1239-12397. Retrieved from http://www.ncbi.nlm.nih.gov/pubmed/9770497

Nei, M., Tajima, F., & Tateno, Y. (1983). Accuracy of estimated phylogenetic trees from molecular data, II gene frequency data. *J. Mol. Evol., 19,* 153-170. Retrieved from http://link.springer.com/article/10.1007%2FBF02300753

Rowold, D. J., & Herrera, R. J. (2003). Inferring recent human phylogenies using forensic STR technology. *Forensic Sci. Int., 133,* 260-265. Retrieved from http://www.fsijournal.org/article/S0379-0738(03)00073-2/fulltext

Saitou, N., & Nei, M. (1987). The neighbor joining method: A new method for reconstructing phylogenetic trees. *Mol. Biol. Evol., 4,* 406-425. Retrieved from http://mbe.oxfordjournals.org/content/4/4/406

Shete, S., Tiwari, H., & Elston, R. C. (2000). On estimating the heterozygosity and polymorphism information content value. *Theor. Popul. Biol., 57,* 265-271. Retrieved from http://www.ncbi.nlm.nih.gov/pubmed/10828218

Shriver, M., Jin, L., Boerwinkle, E., Deka, R., Ferrell, E., & Chakraborty, R. (1995). A novel measure of genetic distance for highly polymorphic tandem repeat loci. *Mol. Biol. Evol., 12,* 914–920. Retrieved from http://mbe.oxfordjournals.org/content/12/5/914.short

Shriver, M., Mei, R., Parra, E. J., Sonpar, V., Halder, I., Tishkoff, S. A., ... Jones, K. W. (2005). Large scale SNP analysis reveals clustered and continuous patterns of human genetic variation. *Human Genomics, 2,* 81-89. Retrieved from http://www.ncbi.nlm.nih.gov/pubmed/16004724.

Sun, J. X., Mullikin, J. C., Patterson, N., & Reich, D. E. (2009). Microsatellites are molecular clocks that support accurate inferences about history. *Mol. Biol. Evol., 26,* 1017-1027. Retrieved from http://www.ncbi.nlm.nih.gov/pmc/articles/PMC2734136/

Tajima, F., & Takezaki, N. (1994). Estimation of evolutionary distance for reconstructing molecular phylogenetic trees. *Mol. Biol. Evol., 11,* 278-286. Retrieved from http://www.ncbi.nlm.nih.gov/pubmed/8170368

Takezaki, N., & Nei, M. (1996). Genetic distances and reconstruction of phylogenetic trees from microsatellite DNA. *Genetics, 144,* 389-399. Retrieved from http://www.genetics.org/content/144/1/389

Takezaki, N., & Nei, M. (2008). Empirical tests of the reliability of phylogenetic trees constructed with microsatellite DNA. *Genetics, 178,* 385-392. Retrieved from http://www.ncbi.nlm.nih.gov/pmc/articles/PMC2206087/

Templeton, A. (2002). Out of Africa again and again. *Nature, 416,* 45-51. Retrieved from http://www.nature.com/nature/journal/v416/n6876/abs/416045a.html

Watkins, W. S., Rogers, A. R., Ostler, C. T., Wooding, S., Bamshad, M. J., Brassington, A. E., ... Jorde, L. B. (2003). Genetic variation among world populations: Inferences from 100 Alu insertion polymorphisms. *Genome Research, 13,* 1607-1618. Retrieved from http://www.ncbi.nlm.nih.gov/pubmed/12805277

Zhivotovsky, L. A., & Feldman, M. W. (1995). Microsatellite variability and genetic distances. *Proc. Natl. Acad. Sci. USA, 92,* 11549-11552. Retrieved from http://www.ncbi.nlm.nih.gov/pmc/articles/PMC40439/

7

Antimicrobial and Antioxidant Activity of Crude Extracts of *Rauvolfia caffra var. caffra* (Apocynaceae) From Tanzania

Efrem-Fred A. Njau[1], Jane Alcorn[2], Patrick Ndakidemi[1], Manuel Chirino-Trejo[3] & Joram Buza[1]

[1] School of Life Science and Bioengineering, The Nelson Mandela African Institution of Science and Technology, P.O.Box 447, Arusha, Tanzania

[2] College of Pharmacy and Nutrition, University of Saskatchewan, 105 Clinic Place, Saskatoon, SK, Canada

[3] Department of Microbiology, Western College of Veterinary Medicine, University of Saskatchewan, SK, Canada

Correspondence: Efrem A. Njau, School of Life Sciences and Bioengineering, The Nelson Mandela African Institutions of Science and Technology, P.O.Box 447, Arusha, Tanzania. E-mail: efrednjau@gmail.com

Abstract

As part of an effort to search for extracts and compounds with new antimicrobial efficacy to fight against bacterial resistance, the antibacterial activity of *Rauvolfia caffra var. caffra* (Sond.), a plant of family Apocynaceae used in Traditional Medicine in Tanzania, was investigated. Ethanol, methanol and water extracts from leaf, stem and root barks were tested against three species of bacteria namely *Escherichia coli* (ATCC 25922) (Gram -ve), *Staphylococcus aureus* (ATCC 25923) (Gram +ve) and *Enterococcus faecalis* (ATCC 51299) (Gram +ve) using Agar-well diffusion assay method and minimum inhibitory concentration on Mueller-Hinton Agar plates. The extracting solvents were removed by vacuo evaporator to obtain gummy-like extracts. This was then dissolved in dimethylsulfoxide (10% DMSO). The DMSO without plant extracts was used as a negative control whereas Gentamicin® as the standard antibiotic was used as a positive control. The Zone of Inhibition (ZOI) measured in mm and Relative Inhibitory Zone Diameter (RIZD) was calculated. Results showed that *R. caffra* exhibited antimicrobial inhibitory activity at a range of 1.25 to 5.0 mg/ml with activity most prominent with methanol extract (ZOI of 28.33 ± 0.33 mm and RIZD of 95% for *S. aureus*; and ZOI of 26.66 ± 0.33 mm for *E.coli* and 19.0 ± 0.57 mm for *E. faecalis* at $P < 0.05$). To characterize further, the alkaloid from the root bark was extracted according to the standard procedure. The antioxidant activity of the alkaloids and ethanolic extracts was determined using 2, 2-diphenyl-1-picrylhydrazyl (DPPH) and reducing capacity assays. The results indicated that alkaloid fraction of the root and 80% ethanolic extracts of stem bark exhibited high antioxidant activity. The phytochemical analysis indicated that *R. caffra* is rich in alkaloids, anthraquinones, anthocyanoides, flavonoids, saponins, tannis and reducing sugars. This study provides supportive evidence that methanol and ethanol extracts of *R. caffra* can be used as herbal medicine in control of *E.coli, S. aureus* and *E. faecalis*.

Keywords: antimicrobial, *E. faecalis, E. coli, S. aureus, Rauvolfia caffra*, ethnopharmacology, phytochemicals

1. Introduction

The increasing resistance of bacteria and fungi to currently marketed antimicrobial agents is becoming a world-wide medical problem (World Health Organization [WHO], 2004). In recent years many bacteria have developed antimicrobial drug resistance, these include but not limited to *Staphylococcus aureus* and most of the *Enterobacteriaceae*, such as *Klebsiella pneumonia* (WHO, 2004). Statistics indicate that more than 70% of the bacteria causing infections are resistant to at least one of the drugs most commonly used to treat them (WHO, 2004). For underdeveloped countries like Tanzania, HIV/AIDs pandemic, poverty and an upsurge of new and re_emerging infectious diseases, and high cost and side effects of available drugs (Humber, 2002) further aggravate this situation. Antimicrobial resistance has increased both severity of infectious disease and mortality rates from certain infections. This has necessitated studies on potential source of additional effective, safe and cheap antimicrobial alternatives, and plants are one of these sources that have not been exhaustively utilized. Plants have ability to synthesize a wide variety of chemical compounds such as alkaloids, glycoside, saponins, resins, lactose and essential oils (Soraya, 2011). Many of these phytochemicals have beneficial effects on a long term human health and may be used to effectively treat human disease (Lai & Roy, 2004).

Rauvolfia caffra var. caffra is a plant species belonging to family Apocynaceae. It is commonly known as "quinine tree" and is widely used in Sub-sahara Africa by natives as a medicine (Nkunya, 1992). In Tanzania, *R. caffra* is widely distributed in riverine *Brachystegia* woodlands, lowlands, in dry montane rainforests and in swamps (Food and Agricultural Organization [FAO], 1986). It is commonly found in highlands of Arusha, Kilimanjaro, Morogoro, Mbeya and Tanga regions in Tanzania. It is known as **Mwembemwitu, or mkufi** (Kiswahili), **Msesewe** (Kichagga), **Olchapukalyan** or **Oljabokaryan** (Maasai) (Njau, 2001; Mbuya, Msanga, Ruffo, Birnie, & Tengnäs, 1994). Ethnobotanical information indicates that *R. caffra* has been used widely for treatment of various diseases. Decoctions derived from stem bark are taken as an astringent, purgative or emetic to treat fever, swellings, abscesses, hepatitis and pneumonia (Tshikalange, Meyer, & Hussein, 2005). The pounded stem bark is applied against measles skin lessions or itching rashes (Schmelzer & Gumb-Fakin, 2008). A stem bark is chewed to cure cough and toothaches and also for treatment of venereal diseases (Njau, 2001). Root decoction is taken to treat fever and swollen legs (Tshikalenge et al., 2005). The bitter bark is strongly purgative and said to produce severe abdominal pains; nevertheless the Pondos of South Africa use the bark for abdominal disorders (Bryant, 1995). The Vendas people of South Africa regard the plant as insecticidal and use the powdered bark to kill maggots in wounds. Bryant (1995) reported that Zulus use *R. caffra* bark along with other plants in a decoction for scrofula and also employ the powdered bark as an application to skin rashes caused by measles, urticaria and other forms of rashes. Chagga and Meru of North Eastern Tanzania use stem bark, which yields an astringent latex and mix with banana and millet to increase the potency of local brew commonly known as (***pombe ya mbege***) (Njau, 2001). Furthermore, *Rauvolfia* species are commonly used in the treatment of malaria, diabetes, and both parasitic and microbial infections (Amole, Onabanjo, & Agbaje, 1998; Campbell, Mortensen, Mølgaard, 2006). Although different species of *Rauvolfia* grow in Tanzania, *R. caffra var. caffra* is the most used species in North Eastern Tanzania in the treatment of various diseases. It is used in the treatment of coughs, gastrointestinal disturbances, skin infections, hypertension, diarrhea, dysentery, scabies, worm infections and malaria (Pesewu, Cutler, & Humber, 2008; Oyedeji, 2007). The ethnomedicinal uses make it one of the most important medicinal plants used in the suppression of skin diseases and opportunistic infections in HIV/ AIDS patients in Tanzania (McMillen, 2004). *R. caffra var. caffra* (Apocynaceae) are rich in indole alkaloids most of which have been isolated and identified (Malik & Siddiqui, 1979; Nasser & Court, 1984). These alkaloids have various pharmacological properties including antimalarial, antitumor and antidiabetes efficacy (Katic, Kusan, Prosek, & Bano, 1980; Dewick, 2002). Furthermore, the extracts and alkaloids derived thereof may have high antimicrobial activity which is due to inhibition of some redox pathways and other biochemical processes in the bacterium cell (Mazza, Fukumoto, Delaquis, Girard & Ewert, 1999). As part of our efforts to search for extracts and compounds with high antibacterial efficacy, the stem bark, root barks and leaf extracts of *R. caffra* were screened for antibacterial activity. The most effective and active extracts was screened for antioxidant activity and evaluated further for pharmacological effects for its use in ethnomedicines.

Although studies on *Rauvolfia caffra* have been done elsewhere little information exist about the Tanzania species. In this study, the antimicrobial efficacy of *R. caffra* var. *caffra* was examined using three species of bacteria namely *Escherichia coli*, *Staphyllococcus aureus* and *Enterococcus faecalis;* The antioxidant properties and phytochemical potential of the plant was also examined. The results will be useful and may contribute in the development of pharmaceutical industry.

2. Materials and Methods

2.1 Collection of Plant Materials

Plant material of *R. caffra var. caffra* (Figure 1) was collected from a coffee plantation at Njari village, in Uru North, Moshi, Tanzania, near a Catholic church (grid 3°16' 60'' S and 37°22'0''E) about 15 km north of Moshi Town in February, 2013. The village where *R. caffra* plant parts were collected has typical volcanic soils and receives biannual rainfall with short rain in October to December and long rains in March to June. The plant was authenticated by Mr. Emmanuel Mboya, a Botanist from National Herbarium of Tanzania (NHT) from Tropical Pesticides Research Institute (TPRI)-Arusha. A voucher specimen was kept at the herbarium with reference no. EN 2981. Leaves stem barks and root barks were collected, washed with tap water to remove soil debris followed with distilled water. They were then allowed to dry under shade at The Nelson Mandela African Institute of Science and Technology for 2 weeks. The plant material was pounded to fine powder using motor and pestle, packed and sealed in cellophane papers and transported by DHL to University of Saskatchewan-Canada, College of Pharmacy and Nutrition laboratories for analysis. All research ethical issues were adhered to, including material transfer agreement, sanitary and phytosanitary, plant export permit and biosafety regulation for plant material handling.

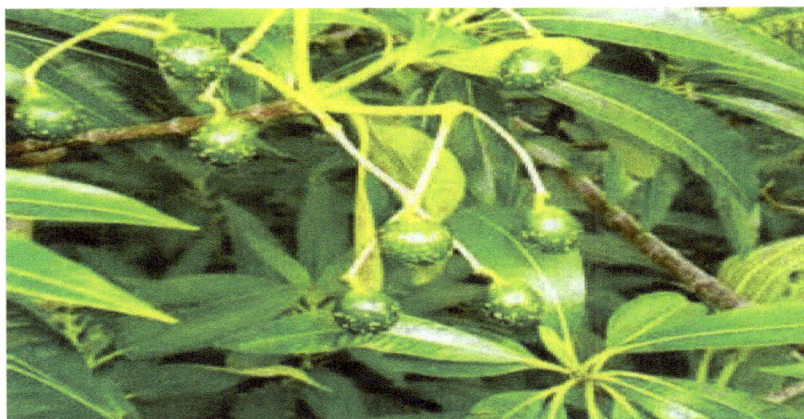

Figure 1. *Rauvolfia caffra* as found in the field at Uru North-Moshi

Source: field photo.

2.2 Extraction of the Plant

Powdered (100 g) *R. caffra* plant parts were weighed into three different conical flasks. To each of these flasks, 500 ml of one of the following solvents was added (95 % methanol, ethanol, or distilled water) and then stirred by means of magnetic stirrer for 30 minutes. The mixtures were macerated under sonicator (Bransonic Ultrasonic 5510 OR-DTH, Alberta, Canada). Sonication was carried in water bath, at room temperature (25°C) for 30 minutes. These mixtures contained in the three conical flasks were thereafter transferred to a shaker. The shaker was set at 60 rpm for 48 hours.

2.3 Filtration and Evaporation of the Sample

The extracts were filtered using Whatman® no.1 filter paper in Büchi funnel under vacuum pump. To evaporate, the samples were poured in a 250 ml round bottomed flask (Pyrex® USA). Concentration by evaporation was achieved using vacuum evaporation (Büchi Vacuum V-850 Switzeland). The bath temperature was set at 45°C. Evaporation was run until a gummy like material was formed. The concentrate extracts were stored in a refrigerator at 4°C until used.

2.4 Phytochemical Analysis

Powdered (20 g) *R. caffra* plant parts namely root barks; stem barks and leaf were weighed into separate conical flasks. To each of these flasks, 250 ml of one of the following solvents was added (95% methanol, ethanol or distilled water) and boiled. Thereafter, the solution was filtered in Whatman® no.1 filter paper in Büchi funnel using vacuum pump. The resulting plant filtrates were used for phytochemical screening for secondary metabolites. The different phytochemicals such as alkaloids, flavonoids, saponins, phenolics, anthraquinones, anthocyanosides and reducing sugars were tested using established methods.

2.4.1 Alkaloid Test

The presence of alkaloids in extracts was tested by using Wagner reagent prepared by dissolving 2 g of iodine and 6 g of potassium iodide in 100 ml of water as previously described by (Mamta & Jyoti, 2012). Two milliliters of Wagner reagents was added to 2 ml of extracts. The formation of reddish brown precipitate indicates the presence of alkaloids.

2.4.2 Test for Steroids

Test for steroids was done according to the method described by (Mohammad, Khulood, Salim, & Zawan, 2013) with modification. The plant extracts (1 mg) was taken in a test tube and dissolved with chloroform (10 ml) then added equal volume of concentrated H_2SO_4 to the test tube by sides. The upper layer in the test tube appears red and Sulphuric acid layer showed yellow with green fluorescence, which indicates the presence of steroids.

2.4.3 Flavonoids Test

A stock solution (2 ml) was taken in a test tube and 2-3 drops of dilute NaOH were added as per (Mohammad et al., 2013). An intense yellow colour appeared in the test tube. This solution becomes colourless when few drops of dilute H_2SO_4 are added confirming the presence of flavonoids.

2.4.4 Test for Saponins

Two grams (2 g) of the powdered sample was boiled in 20 ml of distilled water in a water bath and filtered as previously described by (Doherty, Olaniran, & Kanife, 2010). To the filtered sample (10 ml), about (5 ml distilled water) was added, shaken vigorously and observed for a stable persistent frothing for 25 minutes.

2.4.5 Test for Tannins

Test for tannins was done according to (Doherty et al., 2010) with some modification. Dried powdered sample (0.5 g) was boiled in water (20 ml) in a test tube and then filtered. One milliliters of 0.1% ferric chloride (0.01 Mol/dm^3) was added to 2 ml of each extract sample. Brownish green colourations indicate the presence of tannins.

2.4.6 Anthraquinones

Approximately 1 ml of the plant extract to be tested was shaken with 10 ml of benzene and then filtered as previously described (Doughari, Ndakidemi, Human, & Benade, 2012; Doherty et al., 2010). Five millilitres of the 10 % ammonia solution was then added to the filtrate and shaken. Appearance of a pink, red or violet color in the ammoniacal (lower) phase is an indication of presence of free anthraquinones.

2.4.7 Anthocyanosides

Test for anthocyanosides was done according to method previous descibed (Mamta & Jyoti, 2012). One millilitre of the plant filtrate was mixed with 5 ml of dilute HCI; a pale pink color indicates positive test.

2.4.8 Reducing Sugars

One millilitre of the plant filtrate was mixed with Fehling A and Fehling B separately; a brown color with Fehling B and a green color with Fehling A indicate the presence of reducing sugars as previously described (Doughari et al., 2012).

2.5 Chemicals

Gallic acid, 2, 2-diphenyl-1-picrylhhydrazyl (DPPH), potassium ferricynadine [$K_3Fe (CN)_6$] trichloroacetic acid and ferric chloride were purchased from Sigma Aldrich (Alberta, Canada). All solvents were of HPLC grade and purchased from (Alberta, Canada). Middlebrook 7H9 broth base was obtained from HIMEDIA. Gentamicin® was purchased from Sigma (UK), 96 wells microtitre plates supplied by KAS Medics.

2.6 Test organisms

The testing microorganisms were obtained from the Western College of Veterinary Medicine, Department of Microbiology, University of Saskatchewan-Canada. These bacterial samples were pure and imported from the American Type Culture Collection (ATCC) as indicated in Table 1.

Table 1. Microorganisms used for activity tests on *R. caffra*

Name of bacteria	Stain type	ATCC number
Escherichia coli	Gram-	ATCC 25922
Enterococcus faecalis	Gram +	ATCC 51299
Staphylococcus aureus	Gram +	ATCC 25923

Key: Gram+ = Gram positive bacteria; Gram- = Gram negative bacteria.

2.7 Inoculum Preparation

Eighteen-hour broth culture of the test organism was suspended into sterile Mueller Hinton broth (MHB). It was standardized according to National Committee for Clinical Laboratory Standards (NCCLS, 2002) by gradually adding 9% normal saline to compare its turbidity to McFarland standard of 0.5, which is approximately 1.0×10^8 CFU/mL

2.8 Susceptibility Tests

Overnight broth culture was diluted to approximately 0.1 ml using McFarland scale (0.5 McFarland which is about 1×10^8 CFU/ml). The molten sterile Mueller-Hinton agar (20ml) was poured into sterile Petri plates and allowed to settle. The sterile Mueller-Hinton agar plates were swabbed (sterile cotton swabs) with the 18 hour old-broth culture of respective bacteria by careful striking and rotating the plates about 360 degrees. Wells (6 mm in diameter and about 4 cm apart) were made in each of these plates using sterile borer no. 4. The extracts of the *R. caffra* plant were tested as follows: - An extract (0.2 g) was weighted and dissolved in 10 ml of dimethyl sulphoxide (DMSO) to obtain a concentration of 20 mg/ml. Extractions of 10 mg/ml of methanol, ethanol and water extracts of *R. caffra* stem bark, root bark and leaves for the same solvents were tested in turn. 100 µl of the different concentrations (20 mg/ml - 2.5mg/ml) of the extract was added to fill the bore holes. The negative control was prepared by putting 100 µl of (10 % DMSO) in one of the bored holes. Gentamicin® 10 µg discs were placed in each of the testing Petri plates as positive controls.   One hour of pre-diffusion time was allowed, after which the plates were incubated at 37 °C for 18 to 24 hours. The diameters of zone of inhibition were then measured in millimeters. The above method was carried out in triplicate and the mean of the triplicate results were taken.

2.9 Minimum Inhibitory Concentrations (M.I.Cs) and Minimum Bactericidal Concentrations (M.B.Cs)

Graded concentrations of the extracts ranging from 20 mg/ml to 0.625 mg/ml were used. Extract solution (concentration 20 mg/ml) was serially diluted two fold in Mueller-Hinton broth (MHB) to give decreasing concentration of 10 mg/ml, 5 mg/ml, 2.5 mg/ml, 1.25 mg/ml and 0.625 mg/ml. An aliquot (0.1 ml) of overnight broth culture of test microorganism (concentration 1.5×10^8 CFU/ml) in a sterile normal saline was introduced into each extract dilution. The mixture in test tubes were incubated at 37°C for 24 hrs and observed for turbidity (signifying growth) or absence of it (signifying inhibition). Gentamicin®, a standard antibacterial drug was used as a positive control and sterile normal saline without extract nor drug was used as a negative control. The minimum inhibitory concentration was the lowest concentration of extract solution that inhibited microbial growth.

2.10 Percentage Relative Inhibitory Zone Diameter for E. coli, S. aureus and E. faecalis for Different Extracts

Antibacterial activity was determined by measuring the inhibition zone diameter (mm) against each test organism. The antimicrobial activity expressed as percentage relative inhibition zone diameter (RIZD) was calculated according to (Rojas, Veronica, Saul, & John , 2006) as follows:

$$\%RIZD = \left(\frac{\text{IZD sample} - \text{IZD negative control}}{\text{IZD standard antibiotic}}\right) X\ 100 \tag{1}$$

RIZD is the percentage of relative inhibition zone diameter and IZD is the inhibition zone diameter (mm).

2.11 Total Alkaloid Extraction

A total alkaloid extraction from *R. caffra* root was done according to (Rujjanawate, Kanjanapothi, & Pathong, 2003) with some modification. Approximately 15.58 g of the extracts was dissolved in 350 ml distilled water in a 500 ml flask. The flask and its contents were shaken in the water bath set at 40°C to further dissolve the extracts and obtain a homogenous solution. Thereafter the mixture was acidified by adding 10 ml of 5% sulphuric acid in water. Subsequently the acid solution was repeatedly washed with 100 ml Chloroform to remove neutral substances. The aqueous acidic solution was then made basic with 15 ml of 20% Ammonium hydroxide and extracted again with Chloroform until the aqueous layer was free of alkaloids. The combined total chloroform extracts were evaporated in vacuo evaporator (Büchi vacuum V-850, Switzeland) to yield 2.91 g of alkaloid as shiny brown powder. The crude alkaloids were spotted on thin layer chromatographic plates and developed using dichloromethane / methanol (10: 1). In order to confirm the presence of alkaloids, the TLC plates were sprayed with Dragendorf reagent to give three major orange spots and some other minor compounds.

2.12 Determination of DPPH Radical Scavenging Activity

In order to determine the DPPH radical scavenging activity of *Rauvolfia caffra* stem bark extracts and alkaloid extracts from the root the method described by (Liyana-Pathirana & Shahid, 2005; Cuendet, Hostettmann, & Potterat, 1997) was employed with some modifications. A volume of 0.5 ml of 0.12mM DPPH solution in methanol was separately mixed with 2 ml of 0.01, 0.025, 0.05 and 0.075 mg/ml of the extracts/ alkaloids in methanol and vortexed thoroughly. The absorbance of the mixture at ambient temperature (25°C) was recorded for 30 minutes at 10 minute intervals. Gallic acid (GA) was used as reference antioxidant compound. The absorbance of remaining DPPH radical was read at 519 nm using a Jenway 6505 UV/Vis spectrophotometer (Cole-Parmer, Canada). The analysis of each assay solution was replicated thrice. The scavenging of DPPH radical was calculated according to the following equation:

$$\text{DPPH radical scavenging activity } (\%) = \left(\frac{\text{A control} - \text{A sample}}{\text{A control}}\right) X\ 100 \tag{2}$$

Where A control is the absorbance of DPPH radical in methanol, A sample is the absorbance of DPPH radical + sample extract or / standard.

2.13 Reducing Capacity

For reducing capacity determination, the method of Oyaizu (1986) was adapted with some modification.Stem bark extracts at 0.01, 0.025, 0.05 and 0.075 mg/ ml of were mixed with 2.5 ml of 0.02 M phosphate buffer ($_p$H 6.6) and 2.5 ml of 1% potassium ferricyanide [$K_3Fe(CN)_6$]. The mixture was then incubated at 50°C for 20 min. Aliquots (2.5ml) of 10 % trichloroacetic acid were added to the mixture, which was then centrifuged for 10 minutes at 1000 rpm. The upper layer of the solution (2.5 ml) was mixed with 2.5 ml of distilled water and 0.5 ml of 0.1 % $FeCl_3$, and absorbance was measured at 700 nm in a Jenway 6505 UV / Vis spectrophotometer (Cole-Parmer, Canada). The same procedure was done using alkaloid extract of the root part of *R. caffra*. Gallic acid was used as a standard antioxidant compound. The analysis of each assay was done in triplicates.

2.14 Statistical Analysis

The statistical analysis was performed according to (Steel, Torrie, & Dickey, 1980) using the one way analysis of variance (ANOVA) with the computations being performed with STATISTICA software program. The Fisher's Least Significance Difference (L.S.D) was used to compare treatment means at $P= 0.05$ level of significance. Results are expressed as mean ± standard error (Mean ± S.E).

3. Results

3.1 The Determination of Diameter of Zone of Inhibition (mm) using Agar wells Assay

The inhibitory effects of different parts of *R. caffra* namely stem and roots barks as well as leaf using different extracting solvents against the three pathogens *E. coli*, *S. aureus* and *E. faecalis* were determined. The results of zone of inhibition (ZOI) against tested pathogens are presented in Table 2. A general trend shows that methanolic root bark extracts (RBMe) was more effective against all the three tested pathogens with values of 28.33± 0.33 mm, 26.66 ± 0.33 mm and 19.0 ± 0.57 mm against *S. aureus*, *E. coli and E. faecalis*, respectively.This was followed by stem bark ethanolic extracts (SBEt). Root bark (RBWa) and stem bark (SBWa) aqueous extracts demonstrated moderate effectiveness against the tested pathogens. The lowest value was given by leaf aqueous extracts with value of 8.3± 0.67 mm, 6.67 ± 0.33 mm and zero against *S. aureus*, *E. coli* and *E. faecalis*, respectively. However, Gentamicin®,the standard antibiotic demonstrated the highest zone of inhibition diameter with values ranging from zero to 30.0± 0.57 mm (*E. coli*), 29.6 ± 0.33 mm (*S. aureus*) and 27.0± 1.2 (*E. faecalis*) at ($P \leq 0.05$) level of significance. The higher ZOI values for methanol and ethanol extracts may be explained in terms of solvent polarity. It is likely that *R. caffra* contains active ingredients that are more soluble in polar solvent and has been the reason for higher activity.

Table 2. Activity of plant extracts indicating zone of inhibition (mm) on selected bacterial species

Zone of inhibition in (mm) for different bacteria species		
Extract/ drug	*E. coli*	*S. aureus*	*E. faecalis*
SBEt	20.66 ± 0.33e	18.66 ± 0.33e	11.0 ± 0.57e
RBEt	22.66 ± 0.33d	20. 66 ± 0.66d	12.66 ± 0.88d
SBMe	24.66 ± 0.33c	24.33 ± 0.33c	16.67 ± 0.33c
RBMe	26.66 ± 0.33b	28.33± 0.33b	19.0 ± 0.57b
SBWa	17.66 ± 0.33g	14.33 ± 0.33g	7.0 ± 0.33g
RBWa	19.33 ± 0.33f	17.0 ± 0.0f	9.0 ± 0.0f
LEt	11.33 ± 0.33i	13.0 ± 0.0h	0 ± 0.0h
LMe	13.33 ± 0.33h	15.0 ±0.0g	0 ±0.0h
LWa	6.67 ± 0.33j	8.3 ± 0.67i	0 ±0.0h
DMSO	0 ± 0.0k	0 ± 0.0j	0 ±0.0h
Gentam	30.0 ± 0.57a	29.6 ± 0.33a	27.0 ± 1.2a
One way ANOVA F –statistics value		
Extract/drug	66.55***	57.08***	30.04**

Values presented are Mean ± SE; **, *** significant at P ≤ 0.01, P ≤ 0.001 respectively, ns = not significant; SE = standard error; Means followed by dissimilar letter(s) in a column are significantly different from each other at P= 0.05 according to Fischer Least Significance Difference (LSD); **Key:** SBEt-Stem Bark Ethanolic extract; RBEt-Root Bark Ethanolic extract; SBMe-Stem Bark Methanolic extract; RBMe-Root Bark Methanolic extract; SBWa-Stem Bark Water extract; RBWa-Root Bark Water extract; LEt-Leaf Ethanolic extract;LMe- Leaf Methanolic extract; LWa- Leaf Water extract; DMSO- Dimethylsulfoxide; Gentam-Gentamicin, standard antibiotic.

3.2. Percentage RIZD for E. coli, S. aureus and E. faecalis for Different Extracts

The results of percentage relative inhibition zone diameter (RIZD) against the tested pathogens are presented in Table 3. The general trend revealed that methanolic root barks extracts of *R. caffra* (RBMe) gave higher percentage relative inhibition zone diameter with values ranging from zero to 96.0 ± 1.2%, 89.0 ± 2.0 % and 69.0 ± 0.88% against *S. aureus, E. coli and E. faecalis,* respectively. The ethanolic root barks (RBEt) and stem bark (SBEt) had RIZD values ranging between 60-70%. Lowest values were given by aqueous extracts of leaf parts. However, it was observed further that the ethanolic, methanolic and aqueous extracts of leaf part of *R. caffra* had zero effect against *E. faecalis*. This indicates that *E. faecalis* has higher resistance against these extractions and / or drugs.

Table 3. Calculated percentage RIZD for *E. coli, S. aureus* and E. *faecalis* for different extracts

Extract/ drug	E. coli	S. aureus	E. faecalis
SBEt	69.0 ± 1.5d	63.0 ± 1.73d	40.0 ± 3.0d
RBEt	75.0 ± 1.8c	69.0 ± 1.76c	46.0 ± 1.85c
SBMe	82.0 ± 1.8b	82.0 ±1.0b	60.0 ± 2.40b
RBMe	89.0 ± 2.0a	96.0 ±1.2a	69.0 ± 0.88a
SBWa	59.0 ± 0.57f	48.0 ±0.88f	27.0 ± 2.33f
RBWa	64.0 ± 0.88e	57.0 ±0.66e	33.0 ±1.45e
LEt	39.0 ± 1.2h	43.0 ±0.66g	0.0 ± 0.0g
LMe	44.0 ± 0.66g	50.0 ±0.66f	0.0 ± 0.0g
LWa	22.0 ± 0.66i	28.0 ±2.0h	0.0 ± 0.0g
DMSO	0.0 ± 0.0j	0.0 ±0.0i	0.0 ±0.0g
One way ANOVA F-statistics value			
Extract/drug	64.9***	69.1***	28.9**

Values presented are Mean ± SE; **, *** significant at $P \leq 0.01$, $P \leq 0.001$ respectively; ns = not significant; SE = standard error; Means followed by dissimilar letter(s) in a column are significantly different from each other at P= 0.05 according to Fischer Least Significance Difference (LSD). **Key:** SBEt-Stem Bark Ethanolic extract; RBEt-Root Bark Ethanolic extract; SBMe-Stem Bark Methanolic extract; RBMe-Root Bark Methanolic extract; SBWa-Stem Bark Water extract; RBWa-Root Bark Water extract; LEt- Leaf Ethanolic extract; LMe- Leaf Methanolic extract; LWa- Leaf Water extract; DMSO- Dimethylsulfoxide; Gentam-Gentamicin, standard antibiotic.

3.3 Qualitative Phytochemicals Analysis of R.caffra

The phytochemical screening of *R.caffra* for secondary metabolites indicated the presence of the following major compounds namely alkaloids, anthracene, flavonoids, glycoside, reducing sugars, saponins and tannis.

Table 4: Results of qualitative phytochemicals analysis of *R. caffra*

Phytochemical	SBEt	RBEt	SBMe	RBMe	SBWa	RBWa	LEt	LMe	LWa
Alkaloid	+	+	+	+	+	+	+	+	+
Anthracene	–	–	+	+	+	+	–	–	–
Flavonoids	+	+	+	+	+	+	+	+	+
Glycoside	+	+	+	+	+	+	+	+	+
Saponin	+	+	+	+	+	+	+	+	+
Reducing sugars	+	+	+	+	+	+	+	+	+
Tannins	+	+	+	+	+	+	+	+	+

Key: SBEt-Stem Bark Ethanolic extract; RBEt-Root Bark Ethanolic extract; SBMe-Stem Bark Methanolic extract; RBMe-Root Bark Methanolic extract; SBWa-Stem Bark Water extract; RBWa-Root Bark Water extract;LEt- Leaf Ethanolic extract; LMe-Leaf Methanolic extract; LWa-Leaf Water extract.

3.4 Minimum Inhibitory Concentration

The minimum inhibitory concentrations (MIC) for different extracts of *R. caffra* are presented in Table 5. The results showed that root extracts generally had more promising activity when compared with the leaf extracts at a given concentration. The best extracts that demonstrated inhibitory effects at lower concentrations include both methanol and ethanol root extracts with a value of 1.25 mg/ml against *E. coli and S. aureus*. The ethanol and methanol leaf extracts followed with the MIC value of 2.50 mg/ml against *E. coli and S. aureus*. In contrast the minimum inhibitory concentration that had effect against *E. faecalis* had a value of 5.00 mg/ml for both root and leaves methanol and ethanol extracts. However, *E. faecalis* showed resistance against extracts of leaves and using ethanol as well as methanol.

Table 5. Minimum inhibitory concentration of *R. caffra* extracts with antibacterial activity

Bacteria spp	Minimum inhibitory concentration of *R. caffra* extracts mg/ml									
	SBEt	RBEt	SBMe	RBMe	SBWa	RBWa	LEt	LMe	ALk	Gent
E.coli	1.25	1.25	1.25	1.25	2.50	2.50	5.0	5.0	1.25	1.25
S.aureus	1.25	1.25	1.25	1.25	2.50	2.50	5.0	5.0	0.625	1.25
E. faecalis	5.00	5.00	5.00	5.00	5.00	5.00	N/A	N/A	2.50	5.00

Key: SBEt-Stem Bark Ethanolic extract; RBEt-Root Bark Ethanolic extract; SBMe-Stem Bark Methanolic extract; RBMe-Root Bark Methanolic extract; SBWa-Stem Bark Water extract; RBWa-Root Bark Water extract; Leaf Ethanolic extract; Leaf Methanolic extract; Leaf Water extract; ALk-Alkaloid; DMSO- Dimethylsulfoxide; Gentam-Gentamicin, standard antibiotic; N/A-no activity observed.

3.5 DPPH Radical Scavenging Activity

The free radical scavenging activity of 80% ethanolic aqueous extract of the stem bark (SBEt) and alkaloids from the root of *Rauvolfia caffra* was determined from a reduction of absorbencies of DPPH radical at 519 nm. In this assay, both 80% ethanolic stem bark aqueous and alkaloid extracts exhibited higher antioxidant activity than gallic acid, a standard and natural antioxidant compound (Figure 2). The order of activity was alkaloids >80% ethanolic (SBEt) aqueous > Gallic acid. At 0.03 mg /ml the alkaloid extract scavenged 98 ± 0.88 % of the DPPH radical while 80% ethanolic aqueous extract had 86 ± 0.88 % and gallic acid had 82 ± 1.2 % (Figure 2). The observed trend was such that, the activity of each test sample increased with concentration and time.

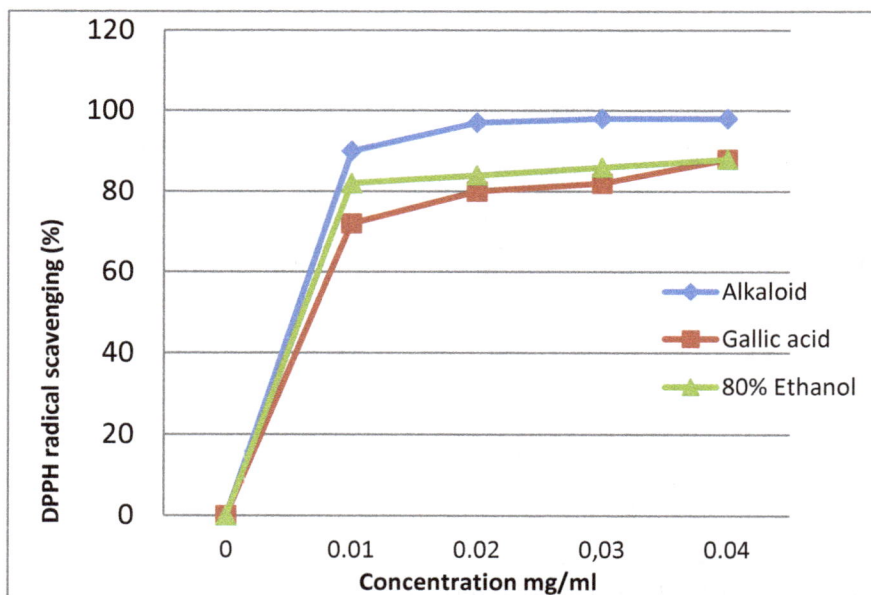

Figure 2. The DPPH radical scavenging activity of ethanolic aqueous and alkaloid extract from the stem bark and root bark of R. caffra compared with gallic acid after 60 minutes of reaction. Each value is expressed as mean ± S.E (n=3)

3.6 Reducing Capacity

The antioxidant activity of the stem bark of R. caffra was further manifested through their reducing power as shown in (Figure 3). In this assay, the Fe^{3+}----$\rightarrow Fe^{2+}$ transformation was established as reducing capacity. Again alkaloid had superior reducing power than the other assayed samples, followed by 80% ethanolic aqueous (SBEt) extract and then gallic acid. At 0.03 mg/ml the absorbancies of alkaloids, 80% ethanolic aqueous extract and gallic acid (at 700nm) were 0.75 ± 0.057, 0.72 ± 0.057 and 0.65 ± 0.09, respectively. However, at 0.075 mg /ml the absorbencies of all samples were of the same order (Figure 3). This trend shows that reducing capacity increased with increasing concentration of the sample. This implies that the strength of donation of electrons is directly related to the concentration of extracts.

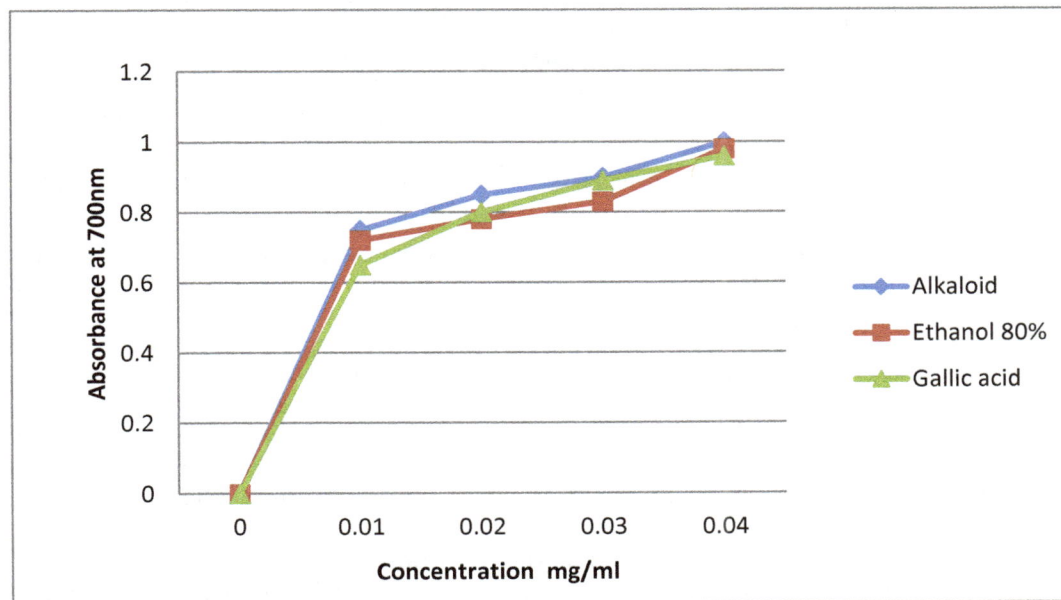

Figure 3. Reducing capacity of different amounts of ethanolic aqueous of stem bark and alkaloid extract from the root bark of R. caffra compared with gallic acid (a standard antioxidant compound) using spectrophotometric detection of Fe^{3+}----$\rightarrow Fe^{2+}$ transformation. Each value is expressed as mean $\pm$ S.E (n=3)

4. Discussion

The study showed that the aqueous and methanolic extract of the stem bark, root and leaf of R. caffra exhibit antibacterial activities against tested bacterial species. However, the methanolic extracts of the root and stem barks showed more antibacterial activity against gram negative bacteria (E. coli) than the gram positive bacteria. The ethanolic aqueous and alkaloid extracts of the roots of R. caffra, exhibited moderate to high antimicrobial activity against the tested bacterial strains. The alkaloid extracts of the roots of R. caffra were more active than all test samples with MIC values ranging from 0.625 mg /ml to 1.25 mg/ml and 5.00 mg /ml against S. aureus, E. coli, and E. faecalis, respectively. The presence of the chemical constituents, tannin, alkaloid, glycoside, saponin, flavonoids, reducing sugars in the stem and root barks of Rauvolfia caffra has added to the claim that plants possess chemical substances in their various parts. These classes of chemical compounds in the plant extracts are known to show curative effects against several pathogens. The mechanism of inhibitory action of these phytochemicals on microorganisms may be due to the impairments of variety of enzymes systems, including those involved in energy production, interference with the integrity of cell membrane and structural component synthesis (Okwu & Morah, 2007; Ali & Dixit, 2012). From previous studies about thirty two alkaloids have been reported by many workers to exhibit antimicrobial activities (Nasser & Court, 1984; Elisabetsky & Costa-Campos, 2006). This is a good indication that these extract have a mechanism of overcoming the barriers of the gram-negative cell wall. Escherichia coli was found to be more susceptible to the extract than the other organisms, this might suggest that the plant may contain some antidiarrhoeal properties. E. faecalis showed little susceptibility and only to the aqueous extract of the stem bark and the ethanolic extract of the leaf. This organism has been reported to be resistant to antimicrobial substances (Livermore, Winstanley, & Shannon, 2001).

Stem bark and the root extracts demonstrated no significant differences in antibacterial activities with root and leaf extracts at P > 0.05 but significant difference was observed in the antibacterial activities of the stem bark and leaf extract against *E. coli* at P < 0.05, Generally, the root bark extract was observed to be more potential than the stem bark and the leaf extracts. The stem bark of *R. caffra,* from previous reports, has more medicinal uses as compared with roots and leaf (Tshikalange et al., 2005; Schmelzer & Fakin, 2008). The ethanolic extract showed more antibacterial activities than the aqueous extracts. The plant may contain active ingredients that are more soluble in polar solvent and these were responsible for the activity. The higher ZOI values for methanol and ethanol extracts may be explained in terms of solvent polarity. It is likely that *R. caffra* may contain active ingredients that are more soluble in polar solvent and has been the reason for higher activity. Cowan (1999), reported that most of the antibiotic compounds already identified in plants are reportedly aromatic or saturated organic molecules which can easily be solubilized in organic solvents. This can justify the traditional use of methanol in extracting plants components in the control of pathogens (Pandit & Langfield, 2004). The activity of ethanolic extract of stem bark of this plant can compete favorably with the activity of gentamicin in this study. The antibacterial activities of the aqueous and ethanolic extracts of the stem bark, root of *R. caffra* shown in this study may justify the traditional use of the plant in the treatment of bacterial induced ailments. However, there has been no report of use synergistically the extract of the stem bark, root and leaf of *Rauvolfia caffra* in the treatment of diarrhea or related diseases and this can be an important aspect of the plant. Gentamicin has been reported to be active against many strains of Gram positive and Gram negative strains.

The alkaloid extracts demonstrated higher free radical scavenging activity than gallic acid and ethanolic aqueous extracts of the stem of *R. caffra*. Furthermore, they also exhibited good electron donating ability, which implies that alkaloids may be inhibiting some redox pathways in the bacterial cell thereby slowing their growth or causing death of microbes (Hamilton, Finlay, Stewart, & Bonner, 2009). This pharmacological property adds value to the potential antimicrobial efficacy of the alkaloids from the roots of *R. caffra*. The electron donating ability of the alkaloid is very important in the inhibition of the bacterial cell growth as the bacterial cell utilizes NADPH dependent reductase enzymes to maintain an intracellular reduced environment in the cells (Hamilton et al., 2009). The DPPH free radical scavenging activity displayed by *R. caffra* methanolic extract compare closely to the currently known scavenging bioorganic molecules and could indicate a potential source of natural anti-oxidants that can be formulated into commercial products. Naturally antioxidants have a fundamental physiological role in the human body by reducing tissue damaging free radical (Tapiero, Tew, Nguyen, & Mathe, 2002; Demo et al, 1998). The provision of these possibly free radical quenching agents from the medicinal plants can be envisaged to contribute towards the antioxidant preventive measures; when consumed, they can improve the digestive system, can function on reduction of coronary heart diseases, and some types of cancer and inflammations (Jayasri et al., 2009).

5. Conclusion

The present investigation indicates that *R. caffra* contains potential antimicrobial bioactive compounds that may be of great use for the development by pharmaceutical industries as a therapy against bacterial related diseases. The ethanol, methanol and aqueous extracts of *R. caffra* possess significant inhibitory effects against tested pathogens namely *Escherichia coli*, *Staphylococcus aureus* and *Enterococcus faecalis*. The results of study supports the Indigenous knowledge on the use of the plant as medicine along with the development of new antimicrobial drug from both roots and stems parts of this plant.

Acknowledgement

Special thanks to the Nelson Mandela African Institution of Science and Technology (NM-AIST) and Commission for Science and Technology (COSTECH) of Tanzania that supported this study. The Canadian Commonwealth Scholarship program through the Canadian Bureau for International Education (CBIE) is acknowledged for their financial support that enabled me to travel and carryout part of my research work at University of Saskatchewan,-Canada.

References

Ali, H., & Dixit, S. (2012). In vitro antimicrobial activity of flavonoids of *Ocimum sanctum* with synergistic effect of their combined form. *Asian Pacific Journal of Tropical Diseases,* 396-398. http://dx.doi.org/10.1016/S2222-1808(12)60189-3

Amole, O. O., Onabanjo, A. C., & Agbaje, E. C. (1998). Effect of bark extract of *Rauvolfia vomitoria* (Afzel.) in malaria. *Parasitology International, 47*, 283-289. http://dx.doi.org/10.1016/S1383-5769(98)81166-5

Bryant, A. T. (1996). Zulu medicine and medicine-men. *A Cape Town struck* (2nd ed.). Cape Town: SA

Burits, M., & Bucar, F. (2000). Antioxidant activity of *Nigella sativa* essential oil. *Phytotherapy Research, 14*, 323-328. http://dx.doi.org/10.1002/1099-1573(200008)14:5<323::AID-PTR621>3.0.CO;2-Q

Campbell, J. I. A., Mortensen, A., & Mølgaard, P. (2006). Tissue lipid lowering effect of traditional Nigerian antidiabetic infusion of *Rauvolfia vomitoria* foliage and *Citrus aurantium* fruit. *Journal of Ethnopharmacology, 104*, 379-386. http://dx.doi.org/10.1016/j.jep.2005.12.029 ; PMid:16455217

Cowan, M. M. (1999). Plant products as antimicrobial agents. *Clinical Microbiology Reviews, 12*, 564–582.

Cuendet, M., Hostettmann, K., & Potterat, O. (1997). Iridoid glucosides with free radical scavenging properties from *Fagraea blumei*. *Helvetica Chimica Acta, 80*, 1144-1152. http://dx.doi.org/10.1002/hlca.19970800411

Demo, A., Kefalas, P., & Boskou, D. (1998). *Nutrient antioxidant in some herbs and Mediterranean plant leaves, Food Research International, 31*, 351-354. http://dx.doi.org/10.1016/S0963-9969(98)00086-6

Dewick, P. M. (2002). Medicinal natural products: *A biosynthetic approach* (2nd ed.). John Wiley & Sons Ltd.

Doherty, V. F., Olaniran, O. O., & Kanife, U. C. (2010). Antimicrobial activity of *Aframomum melegueta* (Allegator pepper). *International Journal of Biology, 2*(2). http://dx.doi.org/10.5539/ijb.v2n2p126

Doughari, J. H., Ndakidemi, P. A., Human, I. S., & Benade, S. (2012). Antioxidant, antimicrobial and antiverotoxic potentials of extracts of *Curtisia dentata*. *Journal of Ethnopharmacology, 141*, 1040-1050 http://dx.doi.org/10.1016/j.jep.2012.03.051 ; PMid:22504170

Elisabetsky, E., & Costa-Campos, L. (2006). *The alkaloid alstanine: A review of its pharmacological properties.* eCAM; 3, 39-48.

FAO. (1986).Food and Agricultural Organization. Some medicinal forest plants of Africa and Latin America, *FAO Forestry paper, 67*, Rome, Italy.

Hamilton, C. J., Finlay, R. M. J., Stewart, M. J. G., & Bonner, A. (2009). Mycothiol disulfide reductase: A continuous assay for slow time dependent inhibitors. *Anals of Biochemistry, 388*, 91-96. http://dx.doi.org/10.1016/j.ab.2009.02.015 ; PMid:19233116

Humber, J. M. (2002). The role of complementary and alternative medicine: accommodating pluralism. *Journal of the American Medicinal Association, 288*, 1655-1656.

Jayasri, M. A., Mathew, L., & Radha, A. (2009). A report on the anti-oxidant activities of leaves and rhizomes of *Costus pictus* D. Don. *International Journal of Integrative Biology, 5*, 20-26.

Katic, M., Kusan, E., Prosek, M., & Bano, M. (1980). Quantitative densitometric determination of respine and ajmaline in *Rauvolfia vomitoria* by HPLC. *Journal of High Resolution Chromatography, 3*, 149-150. http://dx.doi.org/10.1002/jhrc.1240030311

Lai, P. K., & Roy, J. (2004). Antimicrobial and chemopreventive properties of herbs and spices. *Journal of Current Medicinal Chemistry, 11*(11), 1451-60. http://dx.doi.org/10.2174/0929867043365107

Livermore, D. M., Winstanley, T. G., & Shannon, K. P. (2001). Interpretative reading: recognizing the unusual and inferring resistance mechanisms from resistance phenotypes. *Journal of Antimicrobial Chemotherapy, 48*(Suppl S1), 87-102. http://dx.doi.org/10.1093/jac/48.suppl_1.87

Liyana-Pathirana, C. M., & Shahidi, F. (2005). Antioxidant activity of commercial soft and hard wheat (*Triticum aestivum* L.) as affected by gastric pH conditions. *Journal of Agricultural and Food Chemistry, 53*, 2433-2440. http://dx.doi.org/10.1021/jf049320i

Malik, A., & Siddiqui, S. (1979). The subsidiary alkaloids of *Rauvolfia vomitoria* Afzuelia. *Pakistan Journal of Science and Industrial Research, 22*, 121-123.

Mamta, S., & Jyoti, S. (2012). Phytochemical screening of *Acorus calamus* and *Lantana camara*. *International Research Journal of Pharmacy, 3*(5).

Mazza, G., Fukumoto, L., Delaquis, P., Girard, B., & Ewert, B. (1999). Anthocyanins, phenolics and color of Cabernet Franc, Merlot, and Pinot Noir wine from British Columbia. *Journal of Agricultural and Food Chemistry, 47*, 4009–4017. http://dx.doi.org/10.1021/jf990449f

Mbuya, L. P., Msanga, H. P., Ruffo, C. K., Birnie, A., & Tengnäs, B. O. (1994). *Useful trees and shrubs for Tanzania.* Identification, propagation and management for agriculture and pastrol communities. English Press, Nairobi, Kenya.

McMillen, H. (2004). The adapting healer: pioneering through shifting epidemiological and social cultural landscapes. *Social Science & Medicine, 59*, 889-902. http://dx.doi.org/10.1016/j.socscimed.2003.12.008

Mohammad, A. H., Khulood, A., Salim, A. R., & Zawan, A. M. (2013). Study of total phenol, flavonoids contents and phytochemical screening of various leaves crude extracts of locally grown *Thymus vulgaris*. *Asian Pacific Journal of Tropical Biomedicine, 3(9), 705-710.* http://dx.doi.org/10.1016/S2221-1691(13)60142-2

Nasser, A. M., & Court, W. E. (1984). Stem bark alkaloids of Rauvolfia caffra. *Journal of Ethnopharmacology, 11*, 99-117. http://dx.doi.org/10.1016/0378-8741(84)90099-0

National Committee for Clinical Laboratory Standard. (2002). *Performance standard for antimicrobial disc susceptibility testing.* 12th International Supplement; Approved standard M 100-S12. National committee for Clinical Laboratory standards, Wayne, Pa.

Njau, E. A. (2001). *An ethnobotanical study of medicinal plants used by the Maasai People of Manyara, Arusha-Tanzania.* M.Sc Thesis, School of Graduate Studies, Addis Ababa University.

Nkunya, M. H. (1992). Progress in the search for Antimalarials, NAPRECA. *Monograph Series No.4 NAPRECA, AAU, Addis Ababa.*

Okwu, D. E., & Morah, F. N. I. (2007). Isolation and characterization of flavanone glycoside 4,5,7 trihydroxide flavanone rhmnoglucose from Garcina kola seed. *Journal of Applied Science, 7*(2), 155-164.

Oyaizu, M. (1986). Studies on products of browning reaction prepared from glucose amine. *Japan Journal of Nutrition, 44*, 307-315. http://dx.doi.org/10.5264/eiyogakuzashi.44.307

Oyedeji, L. (2007). *Drugless Healing Secrets*: Ibadan, Panse press. Nigeria.

Pandit, K., & Langfield, R. D. (2004). Antibacterial activity of some Italian medicinal plants. *Journal of Ethnopharmacology, 82*, 135-142.

Pesewu, G. A., Cutler, R. R., & Humber, D. P. (2008). Antibacterial activity of plants used in traditional medicines of Ghana with particular reference to MRSA. *Journal of Ethnopharmacology, 116*, 102-111. http://dx.doi.org/10.1016/j.jep.2007.11.005; PMid:18096337

Rojas, J. J., Veronica, J. O., Saul, A. O., & John, F. M. (2006). Screening for antimicrobial activity of ten medicinal plants used in Colombian folkloric medicine: A possible alternative in the treatment of non-nosocomial infections. *BMC Complementary and Alternative Medicine, 6*(1), 2. http://dx.doi.org/10.1186/1472-6882-6-2

Rujjanawate, C., Kanjanapothi, D., & Pathong, A. (2003). Pharmacological effect and toxicity of alkaloids from *Gelsemium elegans* Benth. *Journal of Ethnopharmacology, 89*, 91–95. http://dx.doi.org/10.1016/S0378-8741(03)00267-8

Schmelzer, G. H., & Gumb-Fakin, A. (2008). *Medicinal plants* (pp. 480-483). Amazon Comp.UK.

Soroya, A. S. (2011). *Herbalism, phytochemistry and ethnopharmacology.* Science Publishers, Enfield, New Hampshire. Retrieved from www.scipub.net/cited 15/05/2014

Steel, R. G. D., Torrie, J. H., & Dickey, D. A. (1980). *Principles and procedures of statistics*: a biometrical approach. McGraw-Hill Inc. New York.

Tapiero, H., Tew, K. D., Nguyen, B. G., & Mathe, G. (2002). Polyphenols: do they play a role in the prevention of human pathologies? *Biomedical and Pharmacology, 56*, 200-2007. http://dx.doi.org/10.1016/0378-8741(95)01242-6

Tshikalange, T. E., Meyer, J. J. M., & Hussein, A. A. (2005). Antimicrobial activity, toxicity and the Isolation of a bioactive compound from plants used to treat sexually transmitted diseases. *Journal of Ethnopharmacology, 96*, 515-519. http://dx.doi.org/10.1016/j.jep.2004.09.057

WHO. (2004). World Health Organization, author. *The World Health Report:* The problem of antibiotic resistance. Geneva.

8

Phenotypic Identification and Phylogenetic Characterization of Uropathogenic *Escherichia coli* in Symptomatic Pregnant Women With Urinary Tract Infections in South-Western Nigeria

Aregbesola Oladipupo Abiodun[1], Oluduro Anthonia Olufunke[1], Fashina Christina Dunah[1] & Famurewa Oladiran[2]

[1] Department of Microbiology, Faculty of Science, Obafemi Awolowo University, Ile-Ife 220005, Nigeria

[2] Department of Microbiology, Ekiti State University, P.M.B. 5363, Ado-Ekiti 36001, Nigeria

Correspondence: Oluduro Anthonia Olufunke, Department of Microbiology, Faculty of Science, Obafemi Awolowo University, Ile-Ife 220005, Nigeria. E-mail: aoluduro2003@yahoo.co.uk

Abstract

The study reports the characterization of uropathogenic *E. coli* (UPEC) in urine samples of pregnant women with confirmed urinary tract infections (UTIs) in Ondo and Ekiti States, Nigeria.

Voided mid-stream urine samples were cultured on eosin methylene blue agar plates at 37°C and identified by conventional biochemical tests. Antibiotic susceptibility testing of isolates was by Kirby-Bauer's disc diffusion technique. Phylogenetic typing of the isolates was by multiplex polymerase chain reaction (PCR).

The occurrence of UPEC in pregnant women in age group 25-35 years (66.0%) was high. Two hundred and sixty four uropathogenic *E. coli* comprising 133 (50.38%) in Ondo and 131 (49.62%) in Ekiti States were recovered from 400 samlpes analyzed. In all, prevalence of UTIs with positive cultures was 66.0%. *Escherichia coli* only was 56.5%, mixed-infection (9.5%), non-*E. coli* infection (12.5%) and no growth (21.5%). Resistance to antibiotics was high with diverse multiple antibiotic resistance patterns. Greater percentage of the screened representative UPEC isolates belonged to phylogenetic group D (65.0%), group A (28.0%), group B1 (6.7%) and none to group B2.

Escherichia coli belonging to phylogenetic group D appears to be a predominant uropathogen in this study area. Presence of *chuA* gene in most of the isolates shows the significance of iron acquisition in the pathogenesis and urovirulence of UPEC.

Keywords: UPEC, Urinary tract infection, symptomatic, pregnant women, extra-intestinal pathogenic *E. coli*

1. Introduction

Urinary tract infections (UTIs) are infections caused by the presence and growth of microorganisms anywhere in the urinary tract and are among the most common bacterial infections found in humans. Uropathogenic *E. coli* are implicated in 70-90% of community acquired UTIs and 50% of nosocomial UTIs. The virulence factors and clinical picture presented by UPEC infections indicate that these pathogens are extra-intestinal pathogenic *E. coli* (*ExPEC*) strains (Johnson & Russo, 2005). Individuals with UTI will have a significant number of pathogens in the urinary system. Pathogens may be present in the bladder (cystitis), kidneys (pyelonephritis), urine (bacteriuria) or prostrate (prostatitis) (Marrs et al., 2005). Individuals with increased risk of UTIs include infants, pregnant women and the elderly. Patients with spinal cord injuries, diabetes, multiple sclerosis, urinary catheters, HIV/AIDS or underlying urologic abnormalities are also at risk (Foxman, 2002). Since the genitourinary tract is close to the rectum, faecal bacteria can ascend the urethra into the bladder. If there is a reflux of urine from the infected bladder to the ureters, the kidney may be infected. The ascending route from the faecal site is considered as the major means of transmission of UTI-inducing *ExPEC* to the urinary tract. About 20% of all UTIs cases occur in men while 50-60% of women will have at least one episode of UTI during their lifetime (Griebling, 2005). There is a tendency of recurrence of UTIs in about 25-30% of women after the initial infection due to either re-infection or recrudescence (Bower et al., 2005) *Escherichia coli* can be broadly classified into three groups: commensal *E. coli* which constitute the normal floral of the intestine; intestinal pathogenic *E. coli* which causes various infections in the intestine and *ExPEC* which elicits infections in various parts of the body excluding the intestine (Diard et al., 2010). Extra-intestinal uropathogenic *E. coli* strains are defined as *E. coli* with enhanced ability to cause infections outside the intestinal tract, such as in the bloodstream, cerebrospinal fluid or urinary tract of the host (Diard et al., 2010). Virulence factors associated with *ExPEC* include: adhesins, toxins (hemolysin and cytotoxic necrotizing factor), siderophores (aerobactin), host defense avoidance mechanisms/polysaccharides coatings (group II capsules and biofilm formation) and uropathogenic-specific protein *(usp)* (Arisoy et al., 2006; Skjøt-Rasmussen et al., 2011). Antimicrobial agents of various classes are widely used for therapeutic intervention of urinary tract infections caused by UPEC, but it is also used for prophylactic therapy. However, irrational used of antibiotics as a therapeutic agents of bacterial infections lead to the emergence of resistant bacteria (Morioka et al., 2005). Acquired resistance to antimicrobial drugs is becoming more prevalent among *E. coli* and other pathogens in this region.

There is no clear consensus in the literature on the optimal antimicrobial choice or duration of therapy for UTI during pregnancy. In light of the possible adverse effects of antimicrobials, higher quality research is needed to better understand the direct and indirect consequences of antimicrobial exposure early in life and prudent antimicrobial use is extremely important during pregnancy and early childhood (Schneeberger et al., 2014). Studies exploring cost-effective diagnostic tools at the point of care and non-antimicrobial options to prevent or treat UTIs are needed to limit unnecessary treatment of bacteriuria in pregnancy (Abbo & Hooton, 2014).

Traditional typing of *E. coli* is based on phenotypes, serotype, biotype, phage-typing or antibiotype. Molecular techniques used for the characterization of *E. coli* include; Pulsed-field gel electrophoresis (PFGE) which is considered a gold standard among molecular typing methods for a variety of clinically important bacteria, other molecular methods include: phylogenetic typing, amplified fragment length polymorphism (AFLP), random amplification of polymorphic DNA (RAPD), Variable-Number Tandem Repeat (VNTR) typing, Multi-locus sequence typing (MLST), comparative genomic hybridization, single-nucleotide polymorphisms (SNPs), optical mapping, and whole genome sequencing (Sabat et al., 2013).

Intestinal *E. coli* has been studied extensively to the molecular level. However, there is limited information on the phylogenetic lineages of uropathogenic *E. coli* in Nigeria. No attempt has been made at the molecular level to determine whether *E. coli* implicated in urinary tract infections in Nigeria are from the same or different clone, hence this study.

The study provides information on the susceptibility to various classes of antibiotics and clonal groups that exist within extraintestinal UPEC in pregnant women with UTIs in the study areas.

2. Materials and Methods

2.1 Study Area

The study areas include Ekiti and Ondo States, Southwestern Nigeria. Ekiti and Ondo States are situated entirely within the tropics. Ekiti State is located between longitudes 40°51′ and 50°451′ east of the Greenwich meridian and latitudes 70°151′ and 80°51′ north of the Equator. Ondo State lies between longitudes 4"30" and 6" East of the Greenwich Meridian, 5" 45" and 8" 15" North of the Equator. The selected hospitals in Ondo State included; State Specialist Hospital Ondo, State Specialist Hospital Akure, General Hospital, Ile-Oluji, Ondo States while those of Ekiti State included: Ekiti State University Teaching Hospital, Ado-Ekiti, Federal Medical Centre, Ido-Ekiti and Aiyegbaju community Health Centre, Aiyegbaju.

2. 2 Collection of Sample

With the permission from the Chief Medical Director and laboratory Scientists of the selected hospitals, verbal informed consent of pregnant women with confirmed urinary tract infections at the selected hospitals in Ondo and Ekiti States was obtained before sample collection. Four hundred early morning voided mid-stream urine samples comprising 200 in each state were obtained and transported to the Department of Microbiology laboratory, Obafemi Awolowo University, Ile-Ife, on ice, where the samples were analysed. Sample collection was between June, 2011 and November, 2012.

2.3 Isolation of E. coli and Antibiotic Susceptibility of Isolates

Escherichia coli isolates were presumptively identified by colonial morphology on Eosin methylene blue (EMB) agar ((Oxoid, UK)), incubated at 37°C for 24 h. Distinct greenish metallic sheen colonies on the EMB agar plates were further identified and confirmed by conventional biochemical tests (Farmer, 1999).

The antibiogram of the isolates was determined on Mueller-Hinton agar (LAB-M, UK) by the disk diffusion method (Clinical and Laboratory Standards Institute, 2012). The antibiotics tested and their concentrations (in μg) include; cefadroxil (30), ampicillin (10), nalidixic acid (30), cefepime (30), amoxicillin-clavulanate (20/10), cefuroxime (30), ceftazidime (30), cefotaxime (30) (Oxoid, UK), amoxicillin (30), gentamicin (10), ofloxacin (5), ciprofloxacin (30), tetracycline (25), augmentin (30), ceftriazole (30), nitrofurantoin (300), cotrimoxazole (30), and pefloxacin (30) (Fondos, Nigeria). The antibiotic disks were firmly placed on sterile Mueller-Hinton Agar (MHA) plates previously seeded with a 24 h old culture of the isolate (10^6 CFU/ml of 0.5 McFarland Standard). The plates were incubated at 37°C for 24 h and diameter of zones of inhibition was compared (Clinical and Laboratory Standards Institute, 2012). *Escherichia coli* ATCC 25922 was used as reference. Multiple antibiotic resistant (MAR) isolates were defined as resistance to greater than or equal to three ($\geq$ 3) classes of the antibiotics tested.

2. 4 Phylogenetic Typing

Phylogenetic grouping of the selected *E. coli* isolates was determined by triplex/multiplex PCR-based phylotyping (Clermont et al., 2000). The DNA of the selected isolates was extracted by heat lysis. The isolates were harvested from a 1.5 ml of an overnight Luria-Bertani broth culture by centrifugation in a refrigerated micro-centrifuge (Eppendorf Micro-centrifuge Model 5418, Germany) at 14000 rpm for 7 min, the supernatant was decanted and the cells were washed in 1 ml of sterile distilled water. The supernatant was decanted and the washed cells were re-suspended in 1 ml of TB buffer (Tris-Borate buffer pH 8.2), vortexed and boiled in a thermomixer incubator (Eppendorf thermomixer R mixer incubator model C108115), at 95°C for 10 min. The lysates were centrifuged in a refrigerated microcentrifuge (Eppendorf Micro-centrifuge Model 5418, Germany) at 14000 rpm for 10 min and the supernatant was transferred into a new 1.5 ml Eppendorf tube and stored at -20°C as a template DNA stock. The quality of the extracted DNA samples was checked using a nano-drop spectrophotometer (Nano-drop ND-1000 UV-Vis spectrophotometer) and the absorbance ratio of 260nm and 280 nm was estimated. The extracted DNA samples of the test isolates were amplified by multiplex polymerase chain reaction using the primers *chuA* and *yjaA* genes and the DNA fragment *TspE4.C2* with molecular weights of 279, 211 and 152 kb, respectively. A 20 µl of the PCR reaction mixture (13.15 µl of ddH$_2$O, 4 µl of master mix, 0.2 µl of each primers, 0.15 unit of *Taq* polymerase and 1.5 µl of the DNA sample) was amplified in a PCR machine (Eppendorf Mastercycler pro). The amplification conditions include: denaturation for 4 min at 94°C, 30 cycles of 5 s at 94°C and 10 s at 59°C and a final extension step of 5 min at 72°C. The PCR products were electrophoresed on a 1.5% agarose gel, stained with 1% ethidium bromide and run at 80 V for 2 h and scanned with ultraviolet trans-illuminator. The phylogenetic grouping was done using the dichotomous decision tree designed by Clermont et al. (2000), based on the presence and absence of these three markers as follows; group A (*chuA*-, *jyaA*±,TspE4.C2-), group B1 (*chuA*-, *jyaA*±, TspE4.C2+), group B2 (*chuA*+, *jyaA*+, TspE4.C2+) and group D (*chuA*+, *jyaA*-, TspE4.C2±). The primers sequences are given below:

ChuA-5'-GACGAACCAACGGTCAGGAT-3' forward

ChuA- 5'-TGCCGCCAGTACCAAAGACA-3' reverse

yjaA- 5'-TGAAGTGTCAGGAGACGCTG-3' forward

jyaA- 5'-ATGGAGAATGCGTTCCTCAAC-3' reverse

TspE4.C2- 5'-GAGTAATGTCGGGGCATTCA-3' forward

TspE4.C2- 5'-CGCGCCAACAAAGTATTACG-3' reverse

2. 5 Statistical Analysis

Significant differences and relationship between various data obtained were compared using SPSS 17 version.

3. Results

The age distribution of the pregnant women with confirmed UTIs in Ondo and Ekiti States is depicted by figure 1. The age distribution of the subjects involved in the study ranged from 19-52 years. Out of the 200 samples obtained in Ondo State, 41 (20.5%) patients were less than 25 years, 129 (24.5%) were between ages 25 – 35 while 30 (15.9%) patients were above 35 years. Similarly, in Ekiti State, 49 (24.5%) patients were below the age 25, 134 (67.0%) were between 25- 35 and 17 (8.5%) patients were above the age of 35. In all, patients below age 25 were 22.50%, age group 25-35 (66.0%) and above 35 years (11.75%). There is no significant statistical difference in the age distribution of the subjects in these study areas (P > 0.05).

The occurrence of UPEC in the samples investigated is presented in table 1. A total of 264 uropathogenic *E. coli* comprising 133 (50.38%) in Ondo and 131 (49.62%) in Ekiti States were recovered. In all, prevalence of UTIs with positive cultures was 66.0%. From the samples analyzed, *E. coli* (mono-culture) was recovered from 56.5%, mixed-infection (9.5%), non-*E. coli* infection (12.5%) and no growth (21.5%) (Table 1).

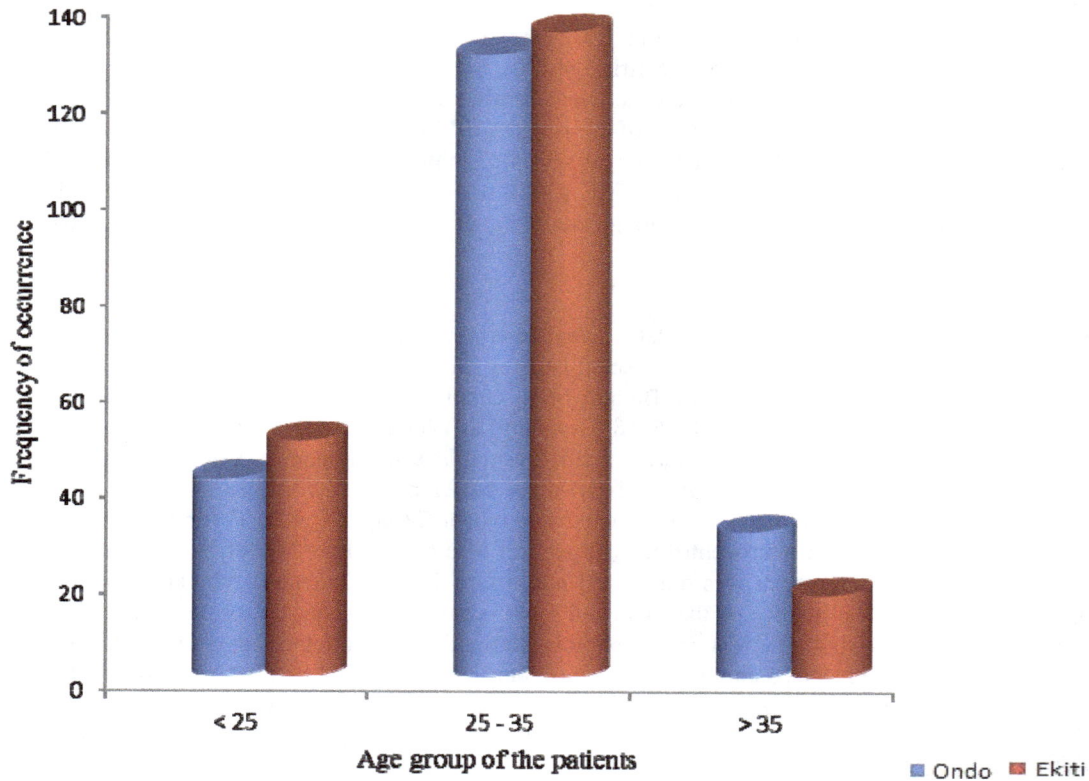

Figure 1. Age distribution of the pregnant women with confirmed urinary tract infection in Ondo and Ekiti States

Table 1. Occurrence of uropathogenic *Escherichia coli* in urine samples of pregnant women with confirmed UTIs in Ondo and Ekiti States, Nigeria

Culture	Frequency of occurrence		Total n=400	Percentage (%)
	Ondo State	Ekiti State		
Escherichia coli growth (Mono culture)	111	115	226	56.5
Escherichia coli with other bacteria (mixed culture)	22	16	38	9.5
Non-*E. coli* culture (negative culture)	29	21	50	12.5
No growth	39	47	86	21.5

n=number of samples.

The prevalence of antibiotic resistance among the UPEC isolates from both Ondo and Ekiti States is presented in table 2. In all, 76.6% of the isolates were resistant to β-lactams class of antibiotics, fluoroquinolones (62.6%), aminoglycosides (62.0%), nitrofurantoin (70.0%), tetracyclines (95.0%) and sulphonamides/trimethoprim (83.3%). There is no significant statistical difference in the prevalence of antibiotic resistance (p < 0.05) in both Ondo and Ekiti States.

Table 2. Prevalence of antibiotic resistance among UPEC isolated from urine samples of UTIs pregnant women in Ondo and Ekiti States

Classes of antibiotics	Specific antibiotics	Occurrence (n=264)		Overall (R)	% (R)	Average % (R)
		Ondo (n=133)	Ekiti (n=131)			
β-Lactams	Augmentin (30 µg)	87	91	178	67.0	
	Amoxicillin (25 µg)	90	99	189	72.0	
	Ampicillin (10 µg)	113	107	220	83.0	
	Ceftriaxone (30 µg)	101	127	228	86.4	
	Cefadroxil (30 µg)	129	130	258	98.0	
	Cefotaxime (30 µg)	85	95	180	68.0	76.6
	Cefepime (30 µg)	30	46	76	29.0	
	Ceftazidime (30 µg)	128	123	251	95.0	
	Cefuroxime (30 µg)	121	120	241	91.0	
Fluoroquinolones	Nalidixic acid (30 µg)	85	93	178	67.0	
	Ciprofloxacin (10 µg)	76	90	166	63.0	62.6
	Ofloxacin (5 µg)	78	80	158	60.0	
	Pefloxacin (10 µg)	94	101	195	60.2.	
Aminoglycosides	Gentamicin (10 µg)	62	102	164	62.0	62.0
Nitrofurantoins	Nitrofurantoin (300 µg)	89	95	184	70.0	70.0
Tetracyclines	Tetracycline (30 µg)	126	124	250	95.0	95.0
Sulphonamides/Trimethroprim	Cotrimoxazole (25 µg)	100	120	220	83.3	83.3

Key: R: resistance; UTI: urinary tract infection.

Table 3 shows the multiple antibiotic resistance (MAR) profile of UPEC recovered from pregnant women with confirmed UTI in Ondo and Ekiti States. Multiple antibiotic resistance is defined as resistance to three or more different classes of the antibiotics tested. Ninety-eight (40.2%) of the isolates were resistant to all the six classes of antibiotics tested, 87 (35.7%) to five, 44 (18.0%) and 15 (6.1%) to four and three classes of antibiotics, respectively.

The multiple antibiotic resistance patterns (MAR) exhibited by the UPEC isolates are presented in table 4. Diversities in MAR patterns were observed among the isolates. Eighteen different MAR patterns were displayed with MAR phenotype (AUG[R] NAL[R] GEN[R] TET[R] NIT[R] COT[R]) appearing the most frequent (Table 4).

Table 3. Prevalence of multiple antibiotic resistant (MAR) uropathogenic E. coli isolated from the urine samples of pregnant women with confirmed urinary tract infections in Ondo and Ekiti States

Number of classes of antibiotics	Occurrence (n=244)		Total	% of isolates with MARs
	Ondo (n=124)	Ekiti (n=126)		
6	35 (28.2)[*]	63 (50.0)	98	40.2
5	37 (29.8)	0 (39.68)	87	35.7
4	36 (29.0)	8 (6.35)	44	18.0
3	13 (10.5)	2 (1.58)	15	6.1

Key: MAR= Multiple antibiotic resistance, (%)[*] percentage of MAR on State basis.

Table 4. The multiple antibiotic resistance (MAR) phenotypes of uropathogenic *E. coli* isolated from pregnant women with confirmed UTIs in Ondo and Ekiti States

Number of classes of antibiotics	Multiple antibiotic resistance phenotypes of the isolates	Frequency	Overall (%)
	AUG NAL TET	11	
		1	
3	AUG TET COT	1	15 (6.14)
		1	
	AUG NAL COT	1	
	AUG NIT COT		
	AUG GEN TET		
	AUG NAL TET COT	16	
	AUG NAL GEN TET	4	
	AUG GEN TET COT	5	
4		2	44 (18.0)
	AUG GEN NIT TET	4	
		7	
	AUG NAL NIT COT	6	
	AUG NAL NIT TET		
	AUG NIT TET COT		
	AUG GEN NIT TET COT	5	
	AUG NAL GEN TET COT	29	
5		11	87 (35.66)
	AUG NAL GEN NIT TET	1	
		41	
	AUG NAL GEN NIT COT		
	AUG NAL NIT TET COT		
6	AUG NAL GEN NIT TET COT	98	98 (40.2)
	Total number of MAR phenotypes = 18	N=244	

Key: AUG= Augmentiin (30 µg), NaL=Nalidixic acid; GEN=Gentamicin (10 µg); NIT=Nitrofurantoin (300 µg); TET=Tetracycline (30 µg); COT=Cotrimoxazole (25 µg).

Phylogenetic distribution of 60 selected UPEC isolates in pregnant women with UTI in Ondo and Ekiti States is depicted by figure 2. Isolates were dominated by phylogenetic group D 39(65.0%), group A 17 (28.0%), group B1 4(6.7%) and none belonged to group B2. Twenty-two of the isolates in phylogenetic group D were recovered from samples in Ekiti State and 17 in Ondo State. In phylogroup A, 10 isolates were recovered in Ondo State and 7 in Ekiti State. Three out of the 4 isolates in group B1 were recovered from Ekiti States and one from Ondo State (Figure 2). Plates 1a and b show the gel electrophoresis of the amplified *ChuA*, *YjaA* and *TspE4C2* markers DNA and the molecular weight ranged from 152 to 279 bp. Fifteen (51.7%) of the isolates contained *chuA* gene, *yjaA* (13.8%) and *TspC4.c2 b* (31.0%) genes.

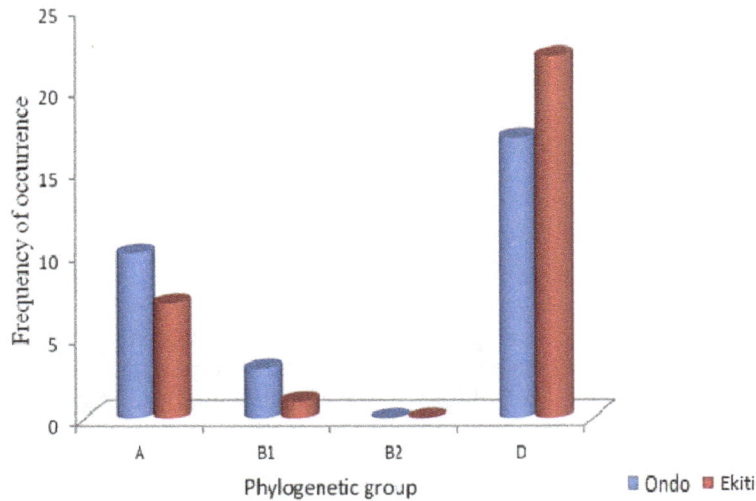

Figure 2. The distribution of the clonal types of uropathogenic *E. coli* in the study areas

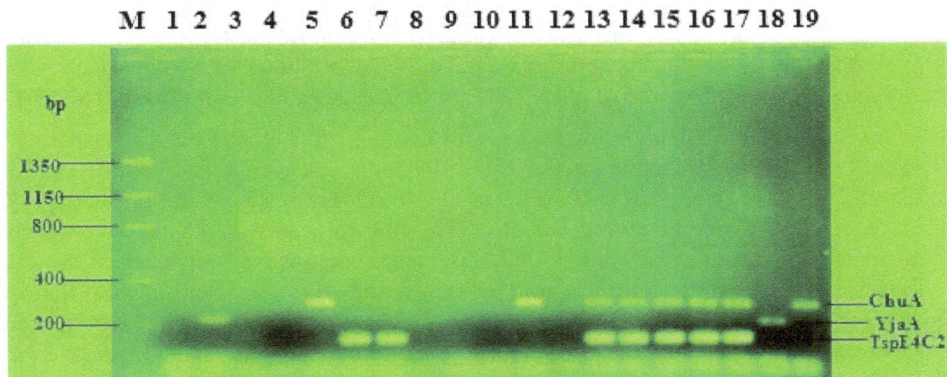

Plate 1a. Gel electrophoresis of *ChuA*, *YjaA* and *TspE4C2* markers in uropathogenic *E. coli* isolates in pregnant women with confirmed UTIs in Ondo and Ekiti States

Key: Lane M= DNA marker; Lanes 1-19= the UPEC isolates. (Isolates in lanes 1, 2, 3, 4, 8, 9, 10, 12 and 18 belong to group A), (5, 11, 13, 14, 15, 16, 17 and 19 belong to Group D) while isolates 6 and 7 belong to group B.

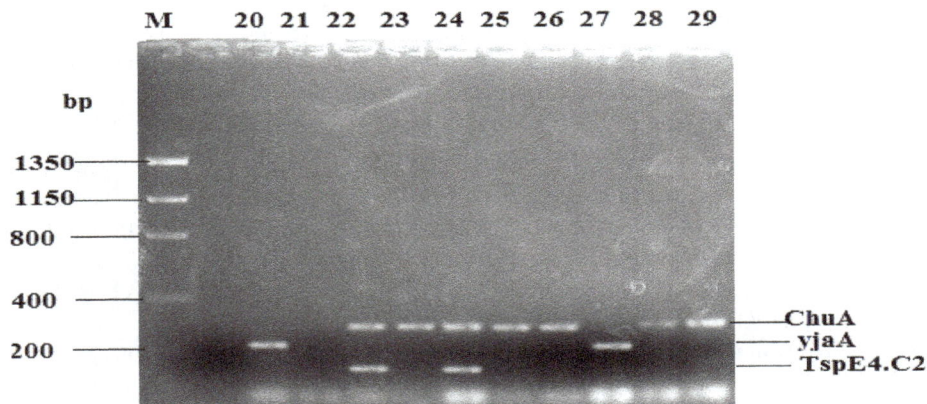

Plate 1b: Gel electrophoresis of *ChuA*, *YjaA* and *TspE4C2* markers in UPEC isolates in pregnant women with confirmed UTIs in Ondo and Ekiti States

Key: Lane M= DNA marker; Lanes 20-29 = the test isolates. Isolates on lanes 20 and 27 belong to group A while isolates on lanes 22, 23, 24, 25, 26, 28 and 29 belong to group D.

4. Discussion

Prevalence of UTI in symptomatic pregnant women in the study areas is high among the age groups considered. This is similar to the findings reported by Abid et al. (2013) among pregnant patients with UTIs in Pakistan but higher than the percentages (49.4%) reported in earlier studies by Manjula et al. (2013) in pregnant patients in India and 30% by Tamalli et al. (2013) in pregnant women who were followed up at different antenatal care clinic in Libya. The high prevalence of UTIs may be explained by sexual intercourse and pregnancy due to the normal physiologic changes induced by gestation which render pregnant women especially susceptible to these infections (National Institutes of Health, 2004; Kolawole et al., 2009).

Prevalence of UTIs in pregnancy among patients above 35 years in the study could be due to the fact that many women within these age groups are likely to have had children before the present pregnancy. Multiparty has been tagged a risk factor in acquiring bacteriuria in pregnancy (Tamalli et al., 2013). Since women in active sexual activities are believed to be prone to UTIs, sexual activity and certain contraceptive devices have been reported to increase the risk, moreover, women are mostly sexually active at the child bearing age (Sharma et al., 2009).

The recovery of UPEC (84%) in the patients sampled agrees with the findings of Ehsan et al. (2013) who reported 84% occurrence of *E. coli* in UTIs episode in Iran. Prevalence of UPEC in this study is higher than earlier report in India where *E. coli* (56.79%) predominates the pathogens recovered (Manjula et al., 2013), and 43.27% reported by Ehsan et al. (2013) as the most prevalent among pregnant women. The recovery of only species of *E. coli* from 226 of the patients suggests a mono-microbial nature of *E. coli* in UTIs.

The high resistance of the isolates to antibiotics in this study may be due to easy accessibility, prolonged use, and abuse. Studies worldwide show a noticeable increase in resistance to ciprofloxacin and other fluoroquinolones because ciprofloxacin is one of the most frequently prescribed fluoroquinolones for UTIs in adults due to its excellent activity on pathogens commonly encountered in complicated UTIs, most especially *E. coli* (Ehsan et al., 2013).

A study from Iran found 32.0% of the *E. coli* implicated in various UTI cases among pregnant patients resistant to ciprofloxacin (Kashef et al., 2010), and other studies from Singapore and Korea have reported ciprofloxacin resistance rates of about 25% (Lee et al., 2011; Bahadin et al., 2011). However, in the present study, higher incidences of resistance to these antibiotics were recorded. The low resistance incidence recorded in earlier studies may be due to better antibiotic use policies in those areas, non-abuse and inaccessibility of these antibiotics in those countries. For instance, nitrofurantoin resistance rate was significantly low in Singapore and in Italy as reported by Bahadin et al. (2011) and Caracciolo et al. (2011), respectively, which is not the case in the present study which recorded high resistance rate. Prophylactic use of antibiotics and a history of drug usage have been identified as risk factors associated with antibiotic resistance (Yuksel et al., 2006; Ehsan et al., 2013). The present study also found majority of the UPEC to be multiple antibiotic resistant particularly to fluoroquinolones and other classes of antibiotics. Fluoroquinolone resistance without concurrent resistance to other classes of antibiotics is uncommon in this study; this scenario indicates the continued declension of the activities of these antibiotics against uropathogens. High prevalence of MAR strains obtained in this study is a possible indication that very large population of *E. coli* isolates might have been exposed to several antibiotics or antibiotics with similar targets/modes of action. The implication of this finding is that most of these pathogens can be voided into the environment including water bodies and possibly enter into the food chain. People living in and/or around these premises with little or no access to safe pipe-borne water may be tempted to drink from such water bodies polluted with urine samples of affected UTI patients. This may create a time bomb of epidemic for unsuspecting and or ignorant members of the community which may ingest such MAR-UPEC into their systems and would in no time lead to devastating public health consequences.

A link between strain phylogeny and virulence has been reported. Phylogenetic analysis have shown that *E. coli* strains fall into four main phylogenic groups; A, B1, B2 and D (Herzer et al., 1990) and that virulent extra-intestinal strains of *E. coli* belong mainly to groups B2 and D while commensal *E. coli* predominantly belongs to A and B1 phylogroups (Skjøt-Rasmussen et al., 2011). Phylogenetic group D has been recognized as the cause of community acquired UTIs in adult women mainly in the United State (Smith et al., 2008) and also accounted for 51% of UTI cases at the University health centre in Michigan (Amee et al., 2001). The *chuA* gene is part of the heme transport locus, which appears to be widely distributed among pathogenic *E. coli* strains. Prevalence of *chuA* gene in most of the representative isolates in the study may be a pointer to the significance of iron acquisition in the pathogenesis and urovirulence of UPEC. Greater percentage (60%) of isolates in the study belonged to phylogroup D hence, corroborates the reports of Yanping et al. (2012) who reported that the predominant group of UPEC recovered from UTI patients in their study belonged to phylogroup D. The finding also agrees with the report of Cao et al. (2011), in a multicenter study in China, where larger percentage (54%)

the UPEC isolated from both first time and recurrent UTI cases were from phylogenetic group D. Similarly, the present study corroborates Johann et al. (2005) reports. In their findings, 35 (63%) out of the 56 UTIs *E. coli* characterized belonged to phylogroup D, 2 (4%) to group B1 and only 1 isolate belonged to group B2.

Recognizable proportion of UPEC typed in this study were in group A and few in group B1 lineage. Similar case where ExPEC strains were isolated from UTI patients in Russia were dominated by groups A and B1 (Moreno et al., 2008). Extra-intestinal uropathogenic *E. coli* belonging to groups A and B1 has been reported to preferentially infect immuno-compromised hosts (Moreno et al., 2008), and are associated with specific blood group antigens and the non-secretor phenotypes (Hooton et al., 1996). The *YjaA* gene is involved in *E. coli* cellular response to hydrogen peroxide, cadmium and acid stress as well as involved in biofilm formation (Gordon et al., 2008)

Although most of the UPEC implicated in various cases of UTIs were believed to be highly concentrated in group B2 (Moreno et al., 2009; Codruţa-Romaniţa et al., 2011; Skjøt-Rasmussen et al., 2011). Non-detection of group B2 strains in this study may be due to differences in geographical locations and temporal variation as well as specific features of the population. Moreover, it could also be attributed to the enormous diverse pool of *E. coli* species (Bailey et al., 2010), or the bacterial characteristics in different topographical arena under the influence of antibiotics usage or the host genetic factor (Duriez et al., 2001).

Resistance problem is now recognized as having a prominent clonal component attributable to the emergence and dissemination of specific antibiotic resistant clonal group of ExPEC ((Johnson et al., 2010). The report of the clonal group of UPEC in pregnant women with UTI in this study is unique and appears to be the first of its kind in the study areas.

5. Conclusion

The study shows that the prevalence of multi-antibiotic resistant uropathogenic *E. coli* mediated urinary tract infections in the study area is high and *Escherichia coli* belonging to phylogenetic group D appears to be a predominant uropathogen.

Acknowledgements

Authors thank the Staff and Head of Department of Molecular Biology and Biotechnology Division of the Nigerian Institute of Medical Research, Lagos, and the Chief Medical Directors and the laboratory scientists of the various hospitals used.

References

Abbo, L. M., & Hooton, T. M. (2014). Antimicrobial stewardship and urinary tract infections. *Antibiotics, 31*, 74-192. http://dx.doi.org/10.3390/antibiotics3020174

Abid, S., Sohail, A., & Muhammad, A. H. (2013). Prevalence and antimicrobial susceptibility of gram negative bacteria isolated from urinary tract infections. *J Infect Mol Biol, 1*(2), 35-37. http://www.nexusacademi cpublishers.com/journal/2

Amaeze, N. J., Abah, A. U., & Okoliegbe, I. N. (2013). Prevalence and antibiotic susceptibility of uropathogens among patients attending University of Abuja Teaching Hospital, Gwagwalada, Abuja. *Intern J Medicin Med Sci, 5*(10), 460-466. Retrieved from http://www.academicjournals.org/IJMMS

Arisoy, M., Aysev, D., Ozel, D., Köse, S. K., Ozsoy, E. D., & Akar, N. (2006). Detection of virulence factors of *Escherichia coli* from children by multiplex polymerase chain reaction. *International J Clin Pract, 60*, 170-173. http://dx.doi.org/ 10.1111/j.1742-1241.2005.00668.x

Bahadin, J., Teo, S., & Mathew, S. (2011). Aetiology of community-acquired urinary tract infection and antimicrobial susceptibility patterns of uropathogens isolated. *Singapore Med J, 52*(6), 415-420.

Bailey, J. K., Pinyon, J. L, Anantham, S., & Hall, R. M. (2010). Commensal *Escherichia coli* of healthy humans: a reservoir for antibiotic-resistance determinants. *J Med Microbiol, 59*, 1331-1339. http://dx/doi.org/10.1099/ jmm.0.022475-0

Bower, J. M., Eto, D. S., & Mulvey, M. A. (2005). Covert operations of uropathogenic *Escherichia coli* within the urinary tract. *Traffic, 6*, 18-31. http://dx.doi.org/ 10.1111/j.1600-0854.2004.00251.x

Cao, X., Cavaco, L. M., Lv, Y., Li, Y., Zheng, B., Wang, P., ... Aarestrup, F. M. (2011). Molecular characterization and antimicrobial susceptibility testing of *Escherichia coli* isolates from patients with urinary tract infections in 20 Chinese hospitals. *J Clin Microbiol, 49*, 2496–2501. http://dx.doi.org/ 10.1128/JCM.02503-10.

Caracciolo, A., Bettinelli, A., Bonato, C., Isimbaldi, C., Tagliabue, A., Longoni, L., & Bianchetti, M. G. (2011). Antimicrobial resistance among *Escherichia coli* that cause childhood community-acquired urinary tract infections in Northern Italy. *Italian J Pediatr, 37*, 3. http://dx.doi.org/10.1186/1824-7288-37-3

Clermont, O., Bonacorsi, S., & Bingen, E. (2000). Rapid and simple determination of the *Escherichia coli* phylogenetic group. *Appl Environ Microbiol, 66*, 4555-4558. http://dx.doi.org/10.1128/AEM.66.10.4555-4558.2000.

Clinical and Laboratory Standards Institute. (2012). *Performance standards for antimicrobial susceptibility testing*. 22nd informational supplement, M100-S22. Wayne, PA.

Codruța-Romanița, U., Grigore, L. A., Georgescu, R. M., Bâltoiu, M. C., Condei, M., & Teleman, M. D. (2011). Phylogenetic background and extraintestinal virulence genotypes of *Escherichia coli* vaginal strains isolated from adult women. *Revis Românăde Medici Laborator, 19*(1/4), 37-45.

Diard, M., Garry, L., Selva, M., Mosser, T., Denamur, E., & Matic, I. (2010). Pathogenicity-associated islands in extraintestinal pathogenic Escherichia coli are fitness elements involved in intestinal colonization. *J Bacteriol, 192*(19), 4885-4893. http://dx.doi.org/10.1128/JB.00804-10.

Duriez, P., Clermont, O., Bonacorsi, S., Bingen, E., Chaventre, A., Elion, J., ... Denamur, E. (2001). Commensal *Escherichia coli* isolates are phylogenetically distributed among geographically distinct human populations. *Microbiol, 147*, 1671–1676.

Ehsan, V., Roya, N., Ali, A., Farshid, K., Reza, N., & Rasool, H. (2013). The Last Three Years Antibiotic Susceptibility Patterns of Uropathogens in Southwest of Iran. *Jundishapur J Microbiol, 6*(4), 1-5. http://dx.doi.org/10.5812/jjm.4958

Farmer, J. J. (1999). Enterobacteriaceae: Introduction and Identification. In P. R. Murray, E. J. Baron, M. A. Pfaller, F. C. Tenover, & R. H. Yolken (Eds.). *Manual Clin Microbiol* (pp. 442-450). Washington, D.C.

Foxman, B. (2002). Epidemiology of urinary tract infections: incidence, morbidity and economic costs. *American J Medicin, 113*, 55-13. http://dx.doi.org/10.1016/S0002-9343(02)01054-9

Griebling, T. L. (2005) Urologic Diseases in American Project: Trends in Resource Use for Urinary Tract Infections in Women. *J Urology, 173*(4), 1281-1287.

Herzer, P. J., Inouye, S., Inouye, M., & Whittam, T. S. (1990), Phylogenetic distribution of branched RNA-linked multi-copy single-stranded DNA among natural isolates of *Escherichia coli. J Bacteriol, 172*, 6175-6181. http://dx.doi.org/0021-9193/90/116175-07$02.00/0.

Hooton, T. M., Scholes, D., Hughes, J. P., Winter, C., Roberts, P. L., Stapleton, A. E., ... Stamm, W. D. (1996). A prospective study of risk factors for symptomatic urinary tract Infection in young women. *New England J Medicin, 335*, 468-474. http://dx.doi.org/10.1056/NEJM199608153350703

Johann, D. D. P., Laupland, K. B., Church, D. L., Menard, M. L., & Johnson, J. R. (1996). Virulence factors of *Escherichia coli* Isolates that produce CTX-M-Type extended-spectrum β-lactamases. *Antimicrob Agents Chemothe, 49*(11), 4667–4670. http://dx.doi.org/10.1128/AAC.49.11.4667-4670.2005.

Johnson, J. R. I., & Russo, T. A. (2005). Molecular epidemiology of extraintestinal pathogenic (uropathogenic) *Escherichia coli. Int J Med Microbiol, 295*(6-7), 383-404. http://dx.doi.org/10.1128/JCM.00949-08.

Johnson, J. R., Johnston, B., Clabots, C., Kuskowski, M. A., & Castanheira, M. (2010). *Escherichia coli* sequence type ST131 as the major cause of serious multidrug-resistant *E. coli* infections in the United States. *Clin Infect Dis, 51*, 286–294. http://dx.doi.org/10.1086/653932

Kashef, N., Djavid, G. E., & Shahbazi, S. (2010). Antimicrobial susceptibility patterns of community-acquired uropathogens in Tehran, Iran. *J Infect Develop Countries, 4*, 202-206.

Kolawole, A. S., Kolawole, O. M., Kandaki-Olukemi, Y. T., Babatunde, S. K., Durowade, K. A., & Kolawole, C. F. (2009). Prevalence of urinary tract infections (UTI) among patients attending Dalhatu Araf Specialist Hospital, Lafia, Nasarawa State, Nigeria. *Intern J Medicin and Med Sci, 1*, 163-167. Retrieved from http://www.academicjournals.org/ijmms.

Lee, S. J., Lee, D. S., Choe, H. S., Shim, B. S., Kim, C. S., Kim, M. E., & Cho, Y. H. (2011). Antimicrobial resistance in community-acquired urinary tract infections: results from the Korean antimicrobial resistance monitoring system. *J Infect Chemother; 17*(3), 440-6. http://dx.doi.org/10.1007/s10156-010-0178-x.

Lilian, M., Abbo, L. M., & Hooton, T. M. (2014). Antimicrobial stewardship and urinary tract infections. *Antibiotics, 3*, 174-192. http://dx.doi.org/10.3390/antibiotics3020174

Manjula, N. G., Girish, C. M., Shripad, A. P., Subhashchandra, M. G., & Channappa, T. S. (2013). Incidence of urinary tract infections and its aetiological agents among pregnant women in Karnataka region. *Advan Microbiol, 3*, 473-478. http://dx.doi.org/http://dx.doi.org/10.4236/aim.2013.36063.

Marrs, C. F., Zhang, L., & Foxman, B. (2005). *Escherichia coli* mediated urinary tract infections: are there distinct uropathogenic *E. coli* (UPEC) pathotypes? *FEMS Microbiol Lett, 252*, 183–190. http://dx.doi.org/10.1016/j.femsle.2005.08.028

Moreno, E., Andreu, A., Pigrau, C., Kuskowski, M. A., Johnson, J. R., Prats, G. (2008). Relationship between Escherichia colistrains causing acute cystitis in women and the fecal *E. coli* population of the host. *J Clin Microbiol, 46*, 2529-2534. http://dx.doi.org/ 10.1128/JCM.00813-08

Morioka, A., Asai, T., Ishihara, K., Kojima, T. Y., & Takahashi, T. (2005). In vitro of 24 antimicrobial agents against *Staphylococcus* and *Steptococcus* isolated from diseased animals in Japan. *J Veter Med Sci., 67*, 207-210.

National Institutes of Health (NIH). (2004)). Fact Sheet: *What I need to know about urinary tract infections.* NIH Publication No. 04–4807.

Sabat, A. J., Budimir, A., Nashev, D., Sá-Leão, R., van Dijl, J. M., Laurent, F., ... Friedrich, A. W. (2013). Overview of molecular typing methods for outbreak detection and epidemiological surveillance. *Eurosurveillance, 18*(4), 20380. Retrieved from http://www.eurosurveillance.org/ViewArticle.aspx?ArticleId=20386

Schneeberger, C., Kazemier, B. M., & Geerlings, S. E. (2014). Asymptomatic bacteriuria and urinary tract infections in special patient groups: Women with diabetes mellitus and pregnant women. *Curr. Opin. Infect. Dis, 27*, 108–114. http://dx.doi.org/10.1097/QCO.0000000000000028

Sharma, M., Aparna Yadav, S., & Chaudhary, U. (2009). Biofilm production in uropathogenic *Escherichia coli*. *Indian J Pathol Microbiol, 52*, 294-294. http://dx.doi.org/http://www.ijpmonline.org/text.asp?2009/52/2/294/48960

Skjøt-Rasmussen, L., Hammerum, A. M., Jakobsen, L., Lester, C. H., Larsen, P., & Frimodt-Møller, N. (2011). Persisting clones of *Escherichia coli* isolates from recurrent urinary tract infection in men and women. *J Med Microbiol, 60*, 550-554. http://dx.doi.org/10.1099/jmm.0.026963-0.

Tamalli, M., Bioprabhu, S., & Alghazal, M. A. (2013). Urinary tract infection during pregnancy at Al-khoms, Libya. *Intern J Medicin Med Sci, 3*(5), 455-459.

Yamamoto, S. (2007). Molecular epidemiology of uropathogenic *Escherichia coli*. *J Infect Chemother; 13*, 68-73. http://dx.doi.org/10.1007/s10156-007-0506-y.

Yanping. L., Yanning, M., Qiang. Z., Leili, W., Ling, G., Liyan, Y., ... Jiyong Y. (2012). Similarity and Divergence of Phylogenies, Antimicrobial Susceptibilities, and virulence factor profiles of escherichia coli isolates causing recurrent urinary tract infections that persist or result from re-infection. *J Clin Microbiol, 50*(12), 4002–4007. http://dx.doi.org/10.1128/JCM.02086-12.

Yuksel, S., Ozturk, B., Kavaz, A., Ozcakar, Z. B., Acar, B., Guriz, H., Aysel, D., Ekim, M., Yalcinkaya, F. (2006). Antibiotic resistance of urinary tract pathogenns and evaluation of empirical treatment in Turkish children with urinary tract infections. *Intern J Antimicrob Agents, 28*(5), 413-6. http://dx.doi.org/10.1016/j.ijantimicag.2006.08.009

9

Influence of Drying Methods on Antioxidant Activities and Immunomodulatory of Aqueous Extract From Soybean Curd Residue Fermentated by *Grifola frondosa*

Dan Zhu[1], Hongyi Sun[1], Shuhong Li[1], Xuansheng Hu[1], Xi Yuan[1], Chao Han[1] & Zhenya Zhang[1]

[1] Graduate School of Life and Environmental Sciences, University of Tsukuba, Ibaraki, Japan

Correspondence: Zhenya Zhang, Graduate School of Life and Environmental Sciences, University of Tsukuba, Ibaraki 305-8577, Japan. E-mail: zhang.zhenya.fu@u.tsukuba.ac.jp

Abstract

The antioxidant activities and immunomodulatory of three aqueous extract from SCR fermented by *G. frondosa* (AE-ND, AE-OD and AE-FD) which were dried by different methods were evaluated. In this study, AE-OD exhibited higher antioxidant activities including DPPH radical scavenging activity (IC_{50} 13.03 ± 0.47 mg/mL), $ABTS^{+}$ radical scavenging activity (IC_{50} 2.15 ± 0.07 mg/mL) and reducing power (absorbance 2.39 ± 0.01, 5 mg/mL). Likewise, oven drying method was more efficient on the survival of macrophage cells and the highest cell viability was 126.09 ± 2.56% at the concentration of 80 μg/mL. However, AE-FD could effectively and chronically enhance the apoptosis of HeLa cells (52.27 ± 0.59%) even incubated after 48 h. The results indicated that aqueous extracts from SCR fermented by *G. frondosa* using different drying methods differently exhibited strong antioxidant and immunomodulatory activities. These could provide a theoretical basis for industrial production preservation of high-quality compounds in extracts from SCR.

Keywords: Soybean curd residue, *Grifola frondosa*, antioxidant activity, immunomodulatory

1. Introduction

Grifola frondosa (*G. frondosa*), a basidiomycete fungus belonging to the Polyporaceae family (Xu et al., 2010), is one kind of edible and officinal mushroom, whose fruiting body is called "Huishu hua" in Chinese and "Maitake" in Japanese. (Yang et al., 2014). "Shen nong ben cao jing" means that it has been frequently used for improving the ailment of the spleen and stomach, calming the nerve and the mind, and treating the hemorrhoids (Hsieh et al., 2006). Currently, mushrooms have become attractive on account of not only their unique and palatable edibleness as food or food-flavouring materials but also abundant and tremendous pharmacology as a source of physiologically beneficial medicine (Mau et al., 2004). Wide varieties of bioactive substances have been isolated from fruit bodies and liquid-cultured mycelium of *G. frondosa* (Fan et al., 2011; Chen et al., 2012) and exhibited remarkable biological activities such as anti-tumor (Cui et al., 2013), anti-HIV (Nanba et al., 2000), anti-hypertension (Bae et al., 2011), anti-viral (Gu et al., 2006), anti-diabetic (Kurushima, Kodama, Schar & Turner, 2000), immunomodulatory (Lee et al., 2003).

Soybean curd residue (SCR) is a by-product of bean production manufacturing and 0.8 million tons of SCR is disposed in Japan annually (Li et al., 2014). On account for characteristics of high moisture content and short shelf life which make definite restriction on the recycle of the residue, SCR is just discharged as agro-industrial waste or incinerated artificially, except a little used as feed stuff. Actually, SCR is comprising of a good source of nutrients, including protein, dietary fiber, minerals, along with monosaccharides and oligosaccharides (Van et al., 1989). Hence, submerged fermentation process, as a novel approach, has been generally carried out to cultivate strains to achieve active compounds.

Drying is a considerable procedure on dehydration of food stuffs (Doymaz, 2005), vegetable processing (Larrosa et al., 2015) and biomass treatment (Garau, Simal, Rossello, & Femenia, 2007), which guaranteed to preserve the quality and quantity of final products without enzymatic deterioration inhibition and microbial growth caused by moisture content of raw materials, to reach reduction of weight and volume, minimizing packaging, transportation and storage costs (Heras et al., 2014).

Disparate drying methods of various materials including mushrooms have been applied, such as sunlight (Kooli, Fadhel, Farhat, & Belghith, 2007), oven drying (Ali, Cone, Hendriks, & Struik, 2014), vacuum (Tang, Santamaria, Bachman, & Park, 2013), microwave (Zielinska et al., 2013) and freeze drying effect of palmitoylated alginate microencapsulation on viability of Bifidobacterium longum during freeze-drying and each method has its own characteristics (Borchani et al., 2011) since operation physical, structural, chemical, nutritional circumstances could be changed which can affect the quality attributes like texture, color, flavor and nutritional value (Di Scala & Crapiste, 2008). Nevertheless, investigations on drying methods for evaluation of G. frondosa have not been reported yet. Consequently, the purpose of this study is to assess influence of different drying methods (non-drying, oven drying and freeze drying) on the antioxidant activities as well as immunomodulatory activities of aqueous extracts from SRC fermented by G. frondosa, which could provide theoretical basis for industrial production preservation of high-quality compounds extracts from SCR.

2. Materials and Methods

2.1 Materials and Reagents

Ascorbic acid, potassium ferricyanide, trichloracetic acid, ammonium sulfae, ferrous chloride were purchased from Wako Pure Chemical Osaka, Japan. Inc. 2,2'-azinobis-(3-ethylbenzothiazoline-6-sulfonic acid) (ABTS), 2,2-diphenyl-1-picry-hydrazyl (DPPH), minimal essential medium eagle medium (DMEM), fetal bovine serum (FBS) and penicillin-streptomycin solution were purchased from Sigma Aldrich, Inc. (Saint Louis, MO, USA). MTT stock solution (5 mg/mL in D-PBS filtrated by 0.2 um filter) and dimethyl sulfoxide (DMSO) were supplied by National Institute of advanced industrial science and technology, AIST, Japan. All other chemicals and solvents were analytical grade and utilized without further purification.

2.2 Microorganism and Fermentation

Fresh SCR (75% moisture content) was obtained from Inamoto Co., Ltd. (Tsukuba, Japan). The impurities were removed from the crushed powder (1.00 g) with 80% ethanol at room temperature for 24 h. The strain of G. frondosa ACCC51616 used in this study was supported by the China Agricultural Culture Collection. The mycelium, maintained on potato dextrose agar (PDA) slants and was subcultured every three months, incubated at 25 °C for 10 days on a modified agar plate which consisted of the following: glucose 2.0%, potato extract 0.4%, agar 2.0%, mineral salt solution (KH_2PO_4 0.3%, and $MgSO_4 \cdot 7H_2O$ 0.15%). The mycelia of G. frondosa was transferred into a sterile petridish (diameter: 100 mm) containing 20 mL of PDA and incubated at 25 °C for 6 days. Afterwards, 100 mL experimental liquid inoculum whose composition was the same as that of medium above except agar was not included, was conducted in a 300 mL flask containing 100 mL medium with ten units of activated mycelia agar, which was a 5 mm × 5 mm square individually achieved by a sterilized self-designed cutter and then in order to activate the culture, it was incubated in a rotary shaker at 25 °C, 120 rpm for 7 days. Solid-state fermentation was administrated in a 200 mL flask with wet SCR as the substrate under optimal culture conditions by pipettes. All the media were autoclaved at 121 °C for 15 min prior to utilization. The whole medium including mycelium was smashed using a sterilized blender and used as the inoculums in the following experiments.

2.3. Cell Line

The macrophage cell line (RAW 264.7) and human cervical cancer cell line (HeLa) were obtained from Japanese Riken bioresource center Cell Bank and maintained in DMEM containing 10% (v/v) FBS and antibiotics (100 U/mL penicillin and 100 μg/mL streptomycin) at 37 °C in a humidified atmosphere of 5% (v/v) CO_2 in a water jacket incubator (ASTEC APC-30D CO_2 incubator, Fukuoka, Japan). Cells were cultured for 2-3 days to reach the logarithmic phase and utilized for experiments.

2.4. Aqueous Extracts from G. frondosa

The treatment of aqueous extracts was according to a literature procedure with a few modifications (Yuan et al., 2013). Aqueous extracts were obtained with hot water (1: 30 ratio of raw material to water, w/v) for 2 h and were separated from insoluble residue by centrifugation (8000 × g for 15 min, at 4 °C). The supernatant was filtered through a Whatman GF/B filter paper and concentrated in a rotary evaporator under reduced pressure at 50 °C. Ultimately, materials were acquired after lyophilization and reserved at -20 °C for further experiments.

2.5. Drying Procedures of Raw Materials

In order to evaluate the influences of drying methods on the bio-activities of substrate, it was carried out three distinctive groups: Non-drying (ND), Oven drying (OD) and Freeze drying (FD). The sample (Non-drying aqueous extract) AE-ND was carried out directly crush without any drying approach. The sample (Oven drying aqueous extract) AE-OD was implemented at 50 °C in an electro-thermostatic blast oven (EYELA WFO-700, JAPAN), which selected to preserve the quality in previous study. It was reserved in a desiccator for 2 days to maintain equilibrium of individual moisture and the final humidity about 10% was determined. The sample (Freeze drying aqueous extract) AE-FD was processed in a vacuum freeze drier (EYELA FREEZE DRYER FDU-506, JAPAN) at 35 °C heating shelf temperature, 553 Pa cavity pressure and -45 °C cold trap temperature for 48 h.

2.6 The Determinations of the Antioxidant Properties

2.6.1 DPPH Radical Scavenging Activity Assay

DPPH scavenging activities of aqueous extracts was determined according to (Nakajima et al., 2007; Yang et al., 2006) by using 1,1-diphenyl-2-picryl-hydrazyl. Concisely, Aliquots (0.5 mL) of diverse concentrations of raw materials were mixed with 2 mL (25 µg/mL) of a MeOH solution of DPPH and shaken vigorously. After 30 min of reaction in darkness, the optical density (O.D.) was determined at a wavelength of 517 nm with a spectrophotometer (SHIMADZU UV-1600, JAPAN). Decrease of the DPPH solution absorbance indicated an increase of the DPPH radical scavenging activity. Ascorbic acid was used as the positive control. The DPPH radical-scavenging activity was calculated by the following Equation (1):

$$DPPH\text{-}scavenging \ activity \ (\%) = (1 - A_{sample}/A_{control}) \times 100\% \tag{1}$$

Where $A_{control}$ is the absorbance without samples and A_{sample} the absorbance in the presence of the samples.

IC_{50} value (mg/mL) was the effectual concentration at which DPPH radicals were scavenged by 50% and was obtained by interpolation from a linear regression analysis.

2.6.2 Reducing Power Assay

The reducing power of the fractions was determined according to the method of Shi et al., (2013) with slight modifications. An aliquot of each sample (1.0 mL), containing different concentrations (0.3125-10.00 mg/mL), was mixed with 1.0 mL of phosphate buffer (0.2 M, pH 6.6) followed by 1.0 mL of 1% (w/v) potassium ferricyanide $[K_3Fe(CN)_6]$. The mixture was cooled 5 min at -20 °C after incubating for 20 min in a water bath at 50 °C. Then 1.0 mL of 1% trichloroacetic acid (TCA) was added and precipitate was centrifuged at $4000 \times g$ for 10 min. The supernatant (1.0 mL), mixed with 2.0 mL of distilled water and 0.4 mL of 0.1% ferric chloride ($FeCl_3$), was standing at ambient temperature for 15 min. Eventually, the absorbance was measured at 700 nm against a blank in the spectrophotometer, with higher absorbance values indicative of greater reducing capacity of ferric (Fe^{3+}) to ferrous (Fe^{2+}) ions. Ascorbic acid was invoked as the positive control. There is a positive correlation between absorbance value and reducing powering.

2.6.3 ABTS$^{·+}$ Radical Scavenging Activity Assay

The radical scavenging activities of the polysaccharides against radical cations (ABTS$^{·+}$) were measured using the methods with some modifications (Shi et al., 2013). When combined with an oxidant (2.45 mM potassium persulfate), ABTS (7 mM in 20 mM sodium acetate buffer) was maintained to create a stable, dark blue-green radical solution in the dark at room temperature for 12-16 h before use. The working ABTS$^{·+}$ solution was supposed to be diluted with ethanol of the stock solution to achieve an absorbance value of 0.70 ($\pm$ 0.02) at 734 nm wavelength when samples (0.15 mL) of various concentration (0.3125-10.00 mg/mL) were vigorously mixed with 2.85 mL of ABTS$^{·+}$ solution. Ascorbic acid was used as the positive control. Eventually, the absorbance was measured at 734 nm after incubation at room temperature for 10 min. The scavenging activity of the ABTS free radicals was calculated using the following Equation (2):

$$ABTS^{+} \ radical \ scavenging \ activity \ (\%) = (A_0 - A_1)/A_0 \times 100\% \tag{2}$$

where A_0 is the absorbance of control without sample and A_1 is the test sample without ABTS$^{·+}$.

IC_{50} value (mg/mL) was the effective concentration at which ABTS$^{·+}$ radical scavenging activity was scavenged by 50% and was obtained by interpolation from a linear regression analysis.

2.7 Cell Viability Assay

The viability and proliferation of murine macrophage (RAW 264.7) cell line was accomplished by MTT reduction assay (Sun et al., 2013) with minor modifications to evaluate the immunomodulatory effects of SCR fermented by *G. frondosa* using three drying methods. Briefly, RAW 264.7 cells were cultured in DMEM medium at 37 °C in a 5 % CO_2 atmosphere to logarithmic phase. Cells were harvested, and an aliquot (100 µL) of suspension (5×10^4 cells/mL) were dispensed into a 96-well plate (2×10^3 cells/well) and pre-incubated at 37 °C in a 5 % CO_2 atmosphere for 24 h. Then cells were exposed to various concentrations of extracts (0, 20, 40, 60, 80, 100 µg/mL) for 48 h. After drugs exposure, 96-well plate was removed from incubator and 10 µL MTT stock solution (0.5 mg/mL) was added to each well incubated at 37 °C, 5% CO_2 for 4 h. Afterwards, 96-well plate was removed from incubator and aspirated the solution and further added 100 µL DMSO to each well and rotated the plate for 10 min to distribute evenly. Ultimately, absorbance was measured with an ELISA reader (BIO-RAD iMark^{TM} Microplate Reader, JAPAN) at 490 nm. Cell viability rate was calculated as the percentage of MTT absorbance. The inhibition rate was calculated using Equation (3).

$$Cell\ viability\ (\%) = (A_{sample}-A_{blank1})/\ (A_{control}-A_{blank2}) \times 100\% \tag{3}$$

where A_{sample} is the absorbance of the sample; A_{blank1} is the absorbance of medium and sample; $A_{control}$ is the absorbance of control group and A_{blank2} is the absorbance of medium.

2.8 Anti-proliferation Effect of Aqueous Extracts on HeLa cells

HeLa cells were operated to detect the anti-proliferation activity of aqueous extracts from SCR. Anti-proliferation effect of different aqueous extracts was determined by the MTT assay and specific experimental procedures were basically consistent except cells and does of drugs (0, 50, 100, 150, 200, 250 μg/mL) changed. Inhibition rate was calculated as the percentage of MTT absorbance. The inhibition rate was calculated using Eq (4).

$$Inhibition\ rate\ (\%) = [1- (A_{sample}-A_{blank1})/\ (A_{control}-A_{blank2})] \times 100\% \tag{4}$$

2.9 Statistical Analysis

All treatments in the present study were performed in triplicate. Data were expressed as means ± standard deviations (S.D.) and analyzed by using a SPSS package (SPSS 19.0 for windows, SPSS Inc., Chicago, IL) one-way analysis of variance (ANOVA) test for mean differences among the samples. P-Values of < 0.05 were considered to be statistically significant.

3. Results and Discussion

3.1 DPPH Scavenging Radical Activity Assay

Table 1. Antioxidant activities (DPPH, ABTS$^{·+}$, reducing power) of aqueous extracts using three drying methods from SCR fermented by *G. frondosa*

Sample	IC$_{50}$^{d} (mg/mL)		Reducing power^e
	DPPH^e	ABTS$^{·+e}$	
AE-ND	9.86 ± 1.03^a	2.48 ± 0.02^a	1.19 ± 0.04^a
AE-OD	13.03 ± 0.47^a	2.15 ± 0.07^b	2.39 ± 0.01^b
AE-FD	108.85 ± 3.78^b	3.04 ± 0.14^c	1.14 ± 0.01^a

[a, b, c] Different superscript letters in the same column indicate significant difference (*$p < 0.05$).

[d] IC$_{50}$ value: the concentration at which the antioxidant activity was 50%.

[e] Values are expressed as means ± S.D. (standard deviation) of three parallel measurements.

AE-ND: aqueous extract without drying; AE-OD: oven drying. AE-FD: freeze drying.

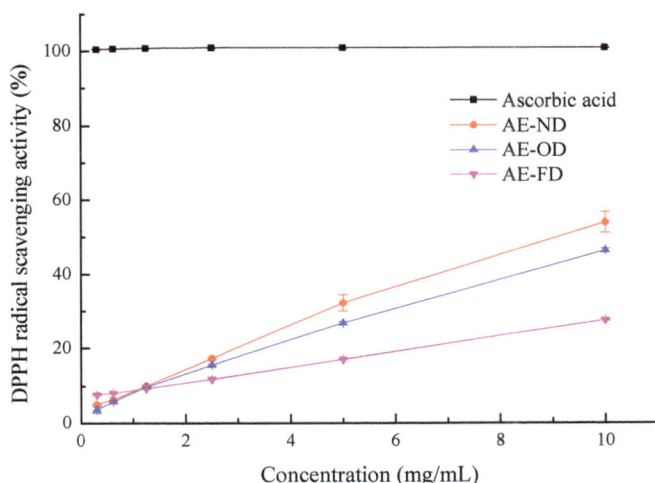

Figure 1. DPPH radical scavenging activities of aqueous extracts of SCR fermented by *G. frondosa*

To investigate efficiently the capacity to scavenge specific free radicals is a prevailing strategy to identify the antioxidant activity of definite compounds *in vitro* (Sascha, Nicklisch & Herbert, 2014). DPPH is a stable free radical used for determining the electron-donating capacity (Brand-Williams, Cuvelier & Berset, 1995). The mechanism of DPPH radical scavenging activity is that DPPH radical carrying a single electron in an alcohol solution can be exhibited a strong absorption at 517 nm of UV spectrum. However, when a free radical scavenger makes the pairing single-electron to absorb gradually disappear, the number of electrons which fade into the extent of its acceptance of the quantitative relationship can be used for rapid quantitative analysis of the spectrophotometer. In other words, the degree of decolorization is associated with free radical scavenging capacity.

Figure 1 indicates the data of the DPPH radical scavenging activities of aqueous extracts of three drying methods from *G. frondosa* with various concentrations. According to Figure 1, the scavenging rates of aqueous extracts using three drying methods constantly increased from 0.3125 to 10 mg/mL, which could be concluded that all samples were totally in a dose-dependent manner, consistent with the researches of bioactivity of *G. frondosa*. Nevertheless, compared with ascorbic acid, the positive group, the highest value obtained by AE-ND was $54.02 \pm 2.76\%$ at the concentration of 10 mg/mL, on account of various compounds mixed. Among three drying methods, AE-ND displayed more significant DPPH radical scavenging activity except under extremely low concentration which implied drying process for *G. frondosa* played a negative role by means of destroying interior construction and impairing scavenging capacity. IC_{50} values, half maximal (50%) inhibitory concentration, were calculated likewise and outlined in Table 1. IC_{50} of AE-ND, AE-OD and AE-FD were 9.79 mg/mL, 13.05 mg/mL and 108.98 mg/mL, respectively. Therefore, it was demonstrated that the effects of drying methods including non-drying and oven drying were significant on DPPH radical scavenging activity.

3.2 Reducing Power Assay

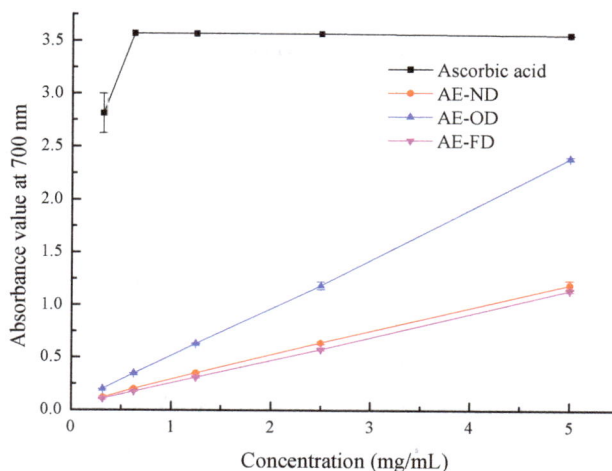

Figure 2. Reducing power of aqueous extracts of SCR fermented by *G. frondosa*

The reaction mechanism reducing power assay, in essence, is the process of measuring ferrous ion (Fe^{2+}) formation monitored spectrophotometrically at 700 nm, based on the theory of transfer ability from ferric (Fe^{3+}) to ferrous (Fe^{2+}) ion through the donation of an electron (Lue et al., 2010). An assay like reducing power could support a basic theoretical foundation for efficient electrons contribution to a concrete antioxidant, performing on the termination of free radical chain reactions.

In summary, it could be clearly observed from Figure 2 that three drying methods had fantastic reducing power and all samples were totally in a dose-dependent manner as same as the effect of DPPH radical scavenging activity they possessed. At the concentration of 5 mg/mL, absorbance values at 700 nm of AE-ND, AE-OD and AE-FD were 1.19 ± 0.04, 2.39 ± 0.012 and 1.14 ± 0.014, respectively, which indicated that aqueous extracts from SCR fermented by *G. frondosa* could be further isolated and purified for obtaining higher active and complicated ingredients. Besides, AE-OD exhibited the strongest reducing power at any given concentration as compared with other two samples, of which one line almost coincided with another. Depend on this result, various drying methods could indeed affect the antioxidant activities of raw materials, also consistent with previous research (Fan, Li, Deng & Ai, 2012; Li et al., 2014 & Suvarnakuta, Chaweerungrat & Devahastin, 2011). Hence, effects of drying methods including oven drying and freeze drying were noteworthy on hydroxyl radical scavenging activity.

3.3 ABTS$^{\cdot+}$ Radical Scavenging Activity Assay

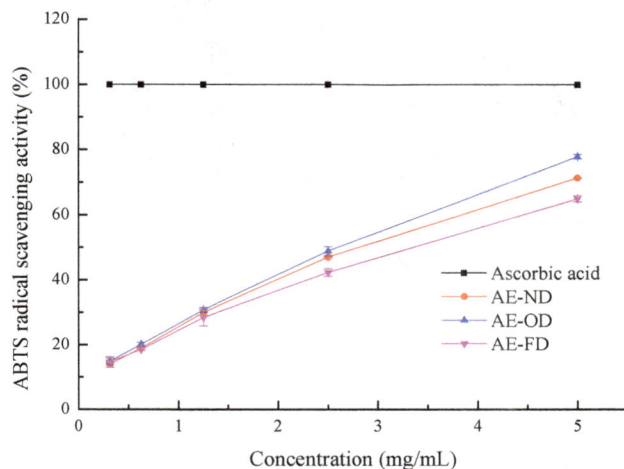

Figure 3. ABTS$^{\cdot+}$ radical scavenging activities of aqueous extracts of SCR fermented by *G. frondosa*

In specification, the ABTS assay based on the generation of a blue/green ABTS$^{\cdot+}$ that can be reduced by antioxidants (Floegel et al., 2011). In particular, ABTS is oxidized with appropriate oxidants, such as hydrogen peroxide, to form a blue-green water-soluble ABTS$^{\cdot+}$. While antioxidants inhibited the generation and enhanced restoration of ABTS$^{\cdot+}$, the color of the solution could lighten the absorbance value at 734nm absorbance band and that meant the sample cleared ABTS$^{\cdot+}$ radicals possessed the antioxidant activity.

ABTS$^{\cdot+}$ radical scavenging activities of aqueous extracts of three drying methods from *G. frondosa* were summarized in Figure 3. Generally, all samples using different drying methods were ascended steadily and were completely in a dose-dependent manner. And there was a delicate difference among three samples above 1.25 mg/mL. Specifically, the order of three samples was as followed: AE-OD, AE-ND and AE-FD and at the concentration of 5 mg/mL, the inhibition rates were 77.88 ± 0.68%, 71.37 ± 0.09%, 65.02 ± 0.94%, respectively. From Table 1, IC$_{50}$ values of individual sample were 2.15 mg/mL, 2.48 mg/mL and 3.05 mg/mL, which declared that oven drying method was more advantageous than the other two drying processes on ABTS$^{\cdot+}$ radical scavenging.

3.4 Immunological Activities of Aqueous Extracts on RAW 264.7 Cells

Inflammation is a protective response mechanism (Kim, Hwang & Park, 2014) to the release of a large amount of inflammatory mediators, such as pro-inflammatory cytokines, NO, iNOS, and COX-2 (Kwon et al., 2013) due to bacteria and viruses invaded into the body (Choi, Kim & Han, 2014) which causes chronic inflammatory diseases such as arthritis, asthma, multiple sclerosis, and atherosclerosis. (Lee, Ryu, Lee & Lee, 2012). Murine macrophage (RAW 264.7) cell with strong adhesion and the ability to engulf the antigen, belonging to the immune system, played multiple roles in the study of phagocytosis, cell-mediated immunity and molecular immunology (Oh et al., 2012).

The immunological effects of aqueous extracts from SCR fermented by *G. frondosa* on cell viabilities of macrophage cells were shown in Figure 4. It could be demonstrated that three kinds of drying methods enhanced the proliferation of RAW 264.7 cells for 24 h, especially at the concentration of 40 µg/mL treated with AE-ND, at the concentration of 80 µg/mL treated with AE-OD and AE-FD, which stated that various drying methods exhibited different degrees of proliferation of macrophage cells also meant a higher immunological effect achieved via drying methods at a constant concentration. Oven drying method was more efficient on the survival of macrophage cells on account of the highest cell viability (126.09 ± 2.56%) was obtained at the concentration of 80 µg/mL when treated with AE-OD. Furthermore, after this the stimulation effect lowered with increase of dosage due to the toxicity of samples themselves. As for incubation of 48 h, cells treated with samples were expressed varying levels of descent which proved long-term excessively exposed to SCR was not suitable for the culture of RAW 264.7 cells.

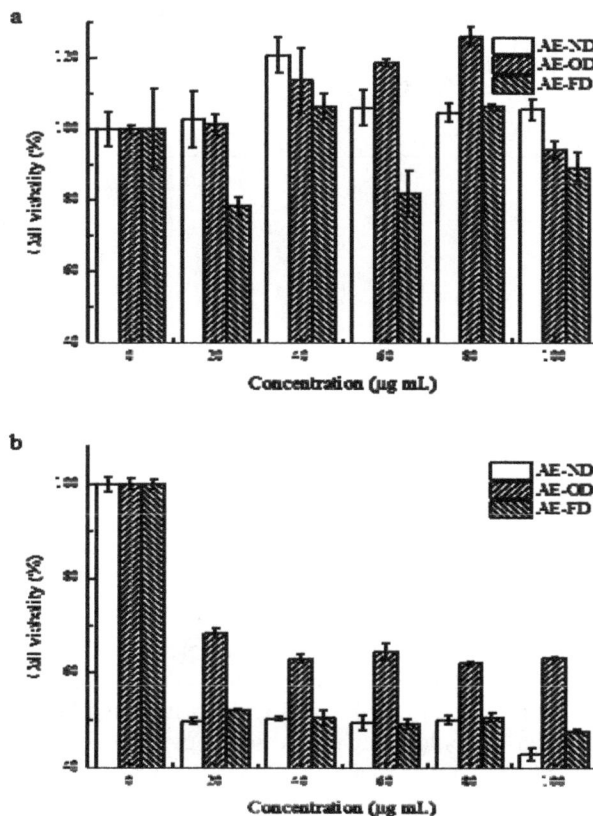

Figure 4. Immunological activities of aqueous extracts from SCR fermented by *G. frondosa* on RAW 264.7 cells. a: cells were incubated with various concentrations (0,20, 40, 60, 80, 100 μg/ml) of AE-ND, AE-OD and AE-FD for 24 h. b: cells were incubated with various concentrations (20, 40, 60, 80, 100 μg/ml) of AE-ND, AE-OD and AE-FD for 48 h. Data are expressed as means ± S.D. (n = 3) (p < 0.05 in comparison with control)

3.5 Anti-proliferation of Aqueous Extracts from SCR Fermented by G. frondosa on HeLa Cells

Current therapies including surgery, chemotherapy and radiotherapy (Feng & Chien, 2003) have been applied for the treatment of advanced stages of cancer, which is the malignant and bloodcurdling disease threatening the survival and development of humanity (Mohanty & Sahoo, 2010). Nevertheless, scientists are searching for substituted compounds extracted from natural plants or biomass since above treatments are generally associated with serious side effects. (Kwon et al., 2007).

HeLa cell line, the transformation of human papilloma virus (Human Papillomavirus 18 or HPV18), has been a very important tool and widely used in cancer research, biological experiments or cell culture to evaluate and determine anti-inflammatory (Hilmi et al., 2003), cytotoxicity (Parthiban et al., 2011) and proliferation and apoptosis (Chen et al., 2014). Figure 5 showed the apoptosis of HeLa cells incubated with drugs at the concentrations of 0, 50, 100, 150, 200, 250 μg/ml of for 24 h and 48 h. At first 24 h, three samples were expressed preeminent anti-proliferation activity in a dose-dependent manner. The inhibition rates of AE-ND, AE-OD and AE-FD were 73.41 ± 3.22%, 45.79 ± 0.50% and 67.53 ± 0.93% at the concentration of 250μg/ml, respectively. Contrast to the results of antioxidant activities of previous study, the aqueous extract using freeze drying showed higher ability which might most of deterioration and microbiological reactions are stopped which gives a final product of excellent quality due to absence of liquid water and the low temperature required for the process (Ratti, 2001). After incubation for 48 h, inhibition rates still kept at a high level which indicated it was meaningful to investigate the extracts from SCR because of enduring immunomodulatory effect.

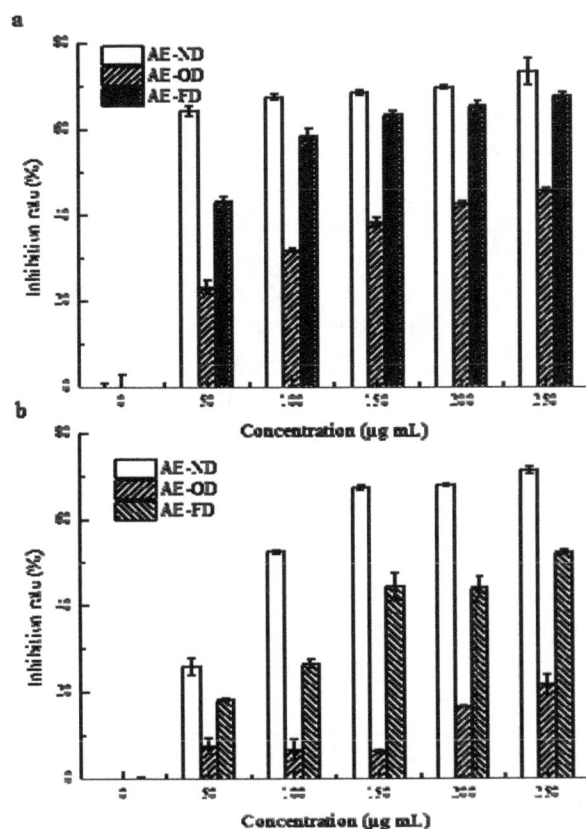

Figure 5. Anti-proliferation of aqueous extracts from SCR fermented by *G. frondosa* on HeLa cells. a: cells were incubated with various concentrations (0, 50, 100, 150, 200, 250 μg/ml) of AE-ND, AE-OD and AE-FD for 24 h. b: cells were incubated with various concentrations (0, 50, 100, 150, 200, 250 μg/ml) of AE-ND, AE-OD and AE-FD for 48 h. Data are expressed as means ± S.D. (n = 3) (p < 0.05 in comparison with control)

4. Conclusions

The results of the present work indicated that three aqueous extracts obtained from soybean curd residue fermented by *G. frondosa* showed different levels of strong antioxidant activities and also differently exhibited the immunomodulatory activities The results showed that three extracts exhibited antioxidant activities in a concentration-dependent manner. Among three extracts, AE-ON had a higher scavenging effects on DPPH free radical, $ABTS^{+}$ free radical and had potential reducing power. AE-FD had a higher capacity on the proliferation of RAW 264.7 cells and the apoptosis of HeLa cells. These results suggested proper drying methods should be supposed to adjust to various bio-activities. Hence, further research is concentrated on isolation, purification and function effects of SCR using various drying methods.

References

Ali, M., Cone, J. W., Hendriks, W. H., & Struik, P. C. (2014). Oven-drying reduces ruminal starch degradation in maize kernels. *Animal Feed Science and Technology, 193*, 44-50. http://dx.doi.org/10.1016/j.anifeedsci.2014.04.013

Bae, I. Y., Kim, H. Y., Lee, S. Y., & Lee, H. G. (2011). Effect of the degree of oxidation on the physicochemical and biological properties of *Grifola frondosa* polysaccharides. *Carbohydrate Polymers, 83*(3), 1298-1302. http://dx.doi.org/10.1016/j.carbpol.2010.09.037

Borchani, C., Besbes, S., Masmoudi, M., Blecker, C., Paquot, M., & Attia, H. (2011). Effect of drying methods on physico-chemical and antioxidant properties of date fibre concentrates. *Food Chemistry, 125*(4), 1194-1201. http://dx.doi.org/10.1016/j.foodchem.2010.10.030

Brand-Williams, W., Cuvelier, M. E., & Berset, C. (1995). Use of a free radical method to evaluate antioxidant activity. *LWT - Food Science and Technology, 28*(1), 25-30. http://dx.doi.org/10.1016/S0023-6438(95)80008-5

C Ratti. (2001). Hot air and freeze-drying of high-value foods: a review. *Journal of Food Engineering, 49*(4), 311-319. http://dx.doi.org/10.1016/S0260-8774(00)00228-4

Chen, G. T., Ma, X. M., Liu, S. T., Liao, Y. L., & Zhao, G. Q. (2012). Isolation, purification and antioxidant activities of polysaccharides from *Grifola frondosa. Carbohydrate Polymers, 89*(1), 61-66. http://dx.doi.org/10.1016/j.carbpol.2012.02.045

Chen, L., Zhang, X., Chen, J., Zhang, X. Z., Fan, H. H., Li, S. C., & Xie, P. (2014). NF-κB plays a key role in microcystin-RR-induced HeLa cell proliferation and apoptosis. *Toxicon, 87*(1), 120-130. http://dx.doi.org/10.1016/j.toxicon.2014.06.002

Choi, E. Y., Kim, H. J., & Han, J. S. (2014). Anti-inflammatory effects of calcium citrate in RAW 264.7 cells via suppression of NF-κB activation. *Environmental Toxicology and Pharmacology*, In Press, Accepted Manuscript. http://dx.doi.org/10.1016/j.etap.2014.11.002

Cui, F. J., Zan, X. Y., Li, Y. H., Yang, Y., Sun, W. J., Zhou, Q., Yu, S. L., & Dong, Y. (2013). Purification and partial characterization of a novel anti-tumor glycoprotein from cultured mycelia of *Grifola frondosa. International Journal of Biological Macromolecules, 62*, 684-690. http://dx.doi.org/10.1016/j.ijbiomac.2013.10.025

Debnath, T., Park, P. J., Nath, N. C. D., Samad, N. B., Park, H. W., & Lim, B. O. (2011). Antioxidant activity of *Gardenia jasminoides* Ellis fruit extracts. *Food Chemistry, 128*(3), 697-703. http://dx.doi.org/10.1016/j.foodchem.2011.03.090

Doymaz, İ. (2005). Drying behaviour of green beans. *Journal of Food Engineering, 69*(2), 161-165. http://dx.doi.org/10.1016/j.jfoodeng.2004.08.009

Fan, L. P, Li, J. W., Deng, K.Q., & Ai, L. Z. (2012). Effects of drying methods on the antioxidant activities of polysaccharides extracted from Ganoderma lucidum. *Carbohydrate Polymers, 87*(2), 1849-1854. http://dx.doi.org/10.1016/j.carbpol.2011.10.018

Fan, Y. N., Wu, X. Y., Zhang, M., Zhao, T., Zhou, Y., Han. L., & Yang, L. Q. (2011). Physical characteristics and antioxidant effect of polysaccharides extracted by boiling water and enzymolysis from *Grifola frondosa. International Journal of Biological Macromolecules, 48*(5), 798-803. http://dx.doi.org/10.1016/j.ijbiomac.2011.03.013

Feng, S. S., & Chien, S. (2003). Chemotherapeutic engineering: application and further development of chemical engineering principles for chemotherapy of cancer and other diseases. *Chemical Engineering Science, 58*(8), 4087-4114. http://dx.doi.org/10.1016/S0009-2509(03)00234-3

Floegel, A., Kim, D. O., Chung, S. J., Koo, S. I., & Chun, O. K. (2011). Comparison of ABTS/DPPH assays to measure antioxidant capacity in popular antioxidant-rich US foods. *Journal of Food Composition and Analysis, 24*(7), 1043-1048. http://dx.doi.org/10.1016/j.jfca.2011.01.008

Garau, M. C., Simal, S., Rosselló, C., & Femenia, A. (2007). Effect of air-drying temperature on physico-chemical properties of dietary fibre and antioxidant capacity of orange (*Citrus aurantium* v. Canoneta) by-product. *Food Chemistry, 104*(3), 1014-1024. http://dx.doi.org/10.1016/j.foodchem.2007.01.009

Gu, C. Q., Li, J. W., & Chao, F. H. (2006). Inhibition of hepatitis B virus by D-fraction from *Grifola frondosa*: Synergistic effect of combination with interferon-α in HepG2 2.2.15. *Antiviral Research, 72*(2), 162-165. http://dx.doi.org/10.1016/j.antiviral.2006.05.011

Heras, R. M. L., Heredia, A., Castelló, M. L., & Andrés, A. (2014). Influence of drying method and extraction variables on the antioxidant properties of persimmon leaves. *Food Bioscience, 6*, 1-8. http://dx.doi.org/10.1016/j.fbio.2014.01.002

Hilmi, F., Gertsch, J., Bremner, P., Valovic, S., Heinrich, M., Sticher, O., & Heilmann, J. (2003). Cytotoxic versus anti-inflammatory effects in HeLa, jurkat t and human peripheral blood cells caused by guaianolide-Type sesquiterpene lactones. *Bioorganic & Medicinal Chemistry, 11*(17), 3659-3663. http://dx.doi.org/10.1016/S0968-0896(03)00346-8

Hsieh, C., Liu, C. J., Tseng, M. H., Lo, C. T., & Yang, Y. C. (2006). Effect of olive oil on the production of mycelial biomass and polysaccharides of *Grifola frondosaunder* high oxygen concentration aeration. *Enzyme and Microbial Technology, 39*(3), 434-439. http://dx.doi.org/10.1016/j.enzmictec.2005.11.033

Kim, H. Y., Hwang, K. W., & Park, S. Y. (2014). Extracts of Actinidia arguta stems inhibited LPS-induced inflammatory responses through nuclear factor–κB pathway in Raw 264.7 cells. *Nutrition Research, 34*(11), 1008-1016. http://dx.doi.org/10.1016/j.nutres.2014.08.019

Kooli, S., Fadhel, A., Farhat, A., & Belghith, A. (2007). Drying of red pepper in open sun and greenhouse conditions.: Mathematical modeling and experimental validation. *Journal of Food Engineering, 79*(3), 1094-1103. http://dx.doi.org/10.1016/j.jfoodeng.2006.03.025

Kurushima, H., Kodama, N., & Nanba, H. (2000). Activities of polysaccharides obtained from *Grifola frondosa* on insulin-dependent diabetes mellitus induced by streptozotocin in mice. *Mycoscience, 41*(5), 473-480. http://dx.doi.org/10.1007/BF02461667

Kwon, D. J., Ju, S. M., Youn G. S., Choi, S. Y., & Park, J. (2013). Suppression of iNOS and COX-2 expression by flavokawain A via blockade of NF-κB and AP-1 activation in RAW 264.7 macrophages. *Food and Chemical Toxicology, 58*, 479-486. http://dx.doi.org/10.1016/j.fct.2013. 05.031

Kwon, H. J., Bae, S. Y., Kim, K. H., Han, C. H., Cho, S. H., Nam. S. W., Choi, Y. H., & Kim, B. W. (2007). Induction of apoptosis in HeLa cells by ethanolic extract of *Corallina pilulifera. Food Chemistry, 104*(1), 196-201. http://dx.doi.org/10.1016/j.foodchem.2006.11.031

Larrosa, A. P. Q., Jr., T. R. S. C., & Pinto, L. A. A. (2015). Influence of drying methods on the characteristics of a vegetable paste formulated by linear programming maximizing antioxidant activity. *LWT - Food Science and Technology, 60*(1), 178-185. http://dx.doi.org/10.1016/j.lwt.2014.08.003

Lee, B. C., Bae, J. T., Pyo, H. B., Choe, T. B., Kim, S. W., Hwang, H. J., & Yun, J. W. (2003). Biological activities of the polysaccharides produced from submerged culture of the edible Basidiomycete *Grifola frondosa. Enzyme and Microbial Technology, 32*(5), 574-581. http://dx.doi.org/10.1016/S0141-0229(03)00026-7

Lee, H. S., Ryu, D. S., Lee, G. S., & Lee, D. S. (2012). Anti-inflammatory effects of dichloromethane fraction from Orostachys japonicus in RAW 264.7 cells: Suppression of NF-κB activation and MAPK signaling. *Journal of Ethnopharmacology, 140*(2), 271-276. http://dx.doi.org/10.1016/j.je2012.01.016

Li, S. H., Sang Y. S., Zhu, D., Yang, Y. N., Lei, Z. F., & Zhang, Z. Y. (2013). Optimization of fermentation conditions for crude polysaccharides by *Morchella esculenta* using soybean curd residue. *Industrial Crops and Products, 50*, 666-672. http://dx.doi.org/10.1016/j.indcrop.2013.07.034

Li, S. H., Wang, L. B., Song, C. F., Hu, X. S., Sun, H. Y., Yang, Y. N., Lei, Z. F., & Zhang, Z. Y. (2014). Utilization of soybean curd residue for polysaccharides by *Wolfiporia extensa (Peck) Ginns* and the antioxidant activities in vitro. *Journal of the Taiwan Institute of Chemical Engineers, 45*(1), 6-11. http://dx.doi.org/10.1016/j.jtice.2013.05.019

Li, X. B., Feng, T., Zhou, F., Zhou, S., Liu, Y. F., Li, W., Ye, R., & Yang, Y. (2014). Effects of drying methods on the tasty compounds of Pleurotus eryngii. *Food Chemistry, 166,* 358-364. http://dx.doi.org/10.1016/j.foodchem.2014.06.049

Lue, B. m., Nielsen, N. S., Jacobsen, C., Hellgren, L., Guo, Z., & Xu, X. B. (2010). Antioxidant properties of modified rutin esters by DPPH, reducing power, iron chelation and human low density lipoprotein assays. *Food Chemistry, 123*(2), 221-230. http://dx.doi.org/10.1016/j.foodchem.2010.04.009

Mau, J. L., Chang, C. N., Huang, S. J., & Chen, C. C. (2004). Antioxidant properties of methanolic extracts from *Grifola frondosa, Morchella esculenta* and *Termitomyces albuminosus* mycelia. *Food Chemistry, 87*(1), 111-118. http://dx.doi.org/10.1016/j.foodchem.2003.10.026

Mohanty, C., Sahoo, K. S. (2010). The *in vitro* stability and *in vivo* pharmacokinetics of curcumin prepared as an aqueous nanoparticulate formulation. *Biomaterials, 31*(25), 6597-6611. http://dx.doi.org/10.1016/j.biomaterials. 2010.04.062

Nakajima, Y., Sato, Y., & Konishi, T. (2007). Antioxidant small phenolic ingredients in *Inonotus obliquus* (persoon) Pilat (Chaga). *Chemical and Pharmaceutical Bulletin, 55*(8), 1222-1226. http://dx.doi.org/10.1248/cpb.55.1222

Nanba, H., Kodama, N., Schar, D., & Turner, D. (2000). Effects of Maitake (*Grifola frondosa*) glucan in HIV-infected patients. *Mycoscience, 41*(4), 293-295. http://dx.doi.org/10.1007/BF024 63941

Oh, Y. C., Cho, W. K., Im G. Y., Jeong, Y. H., Hwang, Y. H., Liang, C., & Ma, J. Y. (2012). Anti-inflammatory effect of Lycium Fruit water extract in lipopolysaccharide-stimulated RAW 264.7 macrophage cells. *International Immunopharmacology, 13*(2), 181-189. http://dx.doi.org/10.1016/j.intimp.2012.03.020

Parthiban, P., Pallela, R., Kim, S. K., Park, D. H., & Jeong, Y. T. (2011). Synthesis of polyfunctionalized piperidone oxime ethers and their cytotoxicity on HeLa cells. *Bioorganic & Medicinal Chemistry Letters, 21*(22), 6678-6686. http://dx.doi.org/10.1016/j.bmcl.2011.09.063

Sascha, C. T., Nicklisch, J., & Herbert W. (2014). Optimized DPPH assay in a detergent-based buffer system for measuring antioxidant activity of proteins. *MethodsX, 1*, 233-238. http://dx.doi.org/10.1016/j.mex.2014.10.004

Scala, K. D., & Crapiste, G. (2008). Drying kinetics and quality changes during drying of red pepper. *LWT - Food Science and Technology, 41*(5), 789-795. http://dx.doi.org/10.1016/j.lwt.2007.06.007

Shi, M., Zhang, Z. Y., & Yang, Y. N. (2013). Antioxidant and immunoregulatory activity of *Ganoderma lucidum* polysaccharide (GLP). *Carbohydrate Polymers, 95*(1), 200-206. http://dx.doi.org/10.1016/j.carbpol.2013.02.081

Sun, H. Y., Wang, S. F., Li, S. H., Yuan, X., Ma, J., & Zhang Z. Y. (2013). Antioxidant Activity and Immunomodulatory of Extracts From Roots of *Actinidia kolomikta. International Journal of Biology, 5*(3), 1-12. http://dx.doi.org/10.5539/ijb.v5n3p1

Suvarnakuta, P., Chaweerungrat, C., & Devahastin S. (2011). Effects of drying methods on assay and antioxidant activity of xanthones in mangosteen rind. *Food Chemistry, 125*(1), 240-247. http://dx.doi.org/10.1016/j.foodchem.2010.09.015

Tang, H. Y., Santamaria, A. D., Bachman, J., & Park, J. W. (2013). Vacuum-assisted drying of polymer electrolyte membrane fuel cell. *Applied Energy, 107*, 264-270. http://dx.doi.org/10.1016/j.apenergy.2013.01.053

Van der Riet, W. B., Wight, A. W., Cilliers, J. J. L., & Datel, J. M. (1989). Food chemical investigation of tofu and its byproduct okara. *Food Chemistry, 34*(3), 193–202. http://dx.doi.org/10.1016/03088146(89)90140-4

Xu, H., Liu, J. H., Shen, Z. Y., Fei, Y., & Chen, X. D. (2010). Analysis of chemical composition, structure of *Grifola frondosa* polysaccharides and its effect on skin TNF-α levels, lgG content, T lymphocytes rate and caspase-3 mRNA. *Carbohydrate Polymers, 82*(3), 687-691. http://dx.doi.org/10.1016/j.carbpol. 2010.05.035

Yang, B., Wang, J., Zhao, M., Liu, Y., Wang, W., & Jiang, Y. (2006). Identification of polysaccharides from pericarp tissues of litchi (*Litchi chinensis* Sonn.) fruit in relation to their antioxidant activities. *Carbohydrate Research, 341*(5), 634-638. http://dx.doi.org/10.1016/j.carres.2006.01.004

Yang, H. L., Zhang, L., Xiao, G. N., Feng, J. B., Zhou, H. B., & Huang, F. R. (2014). Changes in some nutritional components of soymilk during fermentation by the culinary and medicinal mushroom *Grifola frondosa. LWT - Food Science and Technology, In Press, Corrected Proof.* http://dx.doi.org/10.1016/j.lwt.2014.05.027

Yuan, X., Hu, X. S., Liu, Y., Sun, H. Y., Zhang Z. Y., & Cheng, D. L. (2014). *In vitro* and *In vivo* Anti-Diabetic Activity of Extracts From *Actinidia kolomikta. International Journal of Biology*, 6(3), 1-10. http://dx.doi.org/10.5539/ijb.v6n3p1

Zielinska, M., Zapotoczny, P., Alves-Filho, O., Eikevik, T. M., & Blaszczak, W. (2013). A multi-stage combined heat pump and microwave vacuum drying of green peas. *Journal of Food Engineering, 115*(3), 347-356. http://dx.doi.org/10.1016/j.jfoodeng.2012.10.047

10

Modern Problems of Energy Exchange in Humans and Mammals

K. P. Ivanov[1]

[1] Institute of Physiology, Russian Acad. Sci., Sankt-Petersburg, Russia

Correspondence: K. P. Ivanov, Institute of Physiology, Russian Acad. Sci., Nab. Makarova, 6, Sankt-Petersburg 199034, Russia. E-mail: kpivanov@nc2490.spb.edu

Abstract

In a living organism 72% of energy exchange occurs in the inner organs, which comprise only 5-6% of the total body mass. Other energy expenditures occur at the expense of skin, bones, connective tissue, muscles at rest. The level of energy consumption determines the general physiological state of a human organism, serves for the diagnostics of various diseases, especially of the endocrine system diseases, of disruptions of thermoregulation, disruptions of peptide exchange, of carbohydrate and lipometabolism, etc. We emphasize that in modern textbooks of physiology and biology the problem of energy exchange in humans and animals is covered inadequately. Usually it takes only 2-2.5% of the volume. Whereas new problems of energy exchange appeared recently, which were not advanced before. These are, for example, the reasons and mechanisms of a high energy consumption under conditions of metabolism, the physiological significance of the efficiency of a human organism, special processes of the heat exchange between an organism and environment, physiological and social components of human energy exchange. There is also a problem of theoretical possibility of life without energy.

Keywords: energy exchange in humans, heat exchange in humans, efficiency of an organism, life without energy

Energy exchange is the central and the most general problem of life and, consequently, of modern physiology, biology, and medicine. It plays the main role in all the stages of generation and development of life. All the functions of an organism from the brain activity to the growth of nails and hair depend on the availability and consumption of energy. Unfortunately, nowadays the interest to theoretical and practical problems of energy exchange decreased. In the old and modern textbooks on physiology and pathophysiology only 2-2.5% of the volume is usually assigned to the energy exchange. In the table of contents of some textbooks on physiology the problems of energy exchange are totally absent. Such is, for example, the textbook "Fundamental and clinical physiology" (2004), 1050 pages in volume published by "Academia" publishing house. Whereas the process of energy exchange must occupy an important place in medical investigations, since the level of energy exchange determines the physiological state of an organism and is an important criterion of health. A decrease or an increase in the energy exchange are important for the diagnostics of various diseases, for determining the changes in the organism metabolism, for observation over the development of physiological adaptation to various factors of the environment (low or high temperature, the deficit of nutrition, a reinforced muscle load, etc.). Modern development of the studies of energy exchange is associated with the fact that nowadays new problems appeared in this region of science, which were never advanced before. This is due to the natural development of science. It is interesting that a number of important problems, which seemed apparent, were simply ignored by the specialists for a long time.

For a modern physician the widening of the knowledge of energy exchange is very important since energetics is rapidly developing now. Our small paper is devoted to the problems of energetics of a living organism forgotten to some extent or pushed out of the way along which the most important stages of studying life go.

The First Problem. Under the conditions of metabolism a man spends about 1800 kcal per day. The question arises, for which work this energy is spent, if a man is in complete physical and mental rest, on an empty stomach, at a comfort environmental temperature. Energy expenditures for respiration, heart and kidney activity were assumed to account for these energy expenditures. However, it appeared to be completely wrong. Under metabolism only 22-23% of the energy budget of a human organism is spent on respiration, heart and kidney activity. This is about 405 from 1800 kcal (Schmidt-Nielsen, 1972). The question is for which work the remaining 1400 kcal are spent. It is interesting that this question almost never arose during the whole époque of physiological investigations. The examination of this phenomenon must be started from the fact that all the tissues of a living organism are continuously disrupting and regenerating according to the laws of thermodynamics. Both the disruption of molecules, having served their time, and the creation of new molecules require energy. The peptide synthesis is well studied by modern science (Lehninger, 1972). The disruption of each peptide bond releases about 5 kcal per mole of an albumine. However the synthesis of such bond requires about 30 kcal/mol. Therefore, the efficiency of this work is only 16%. It is believed that every day about 400 g of protein is disintegrated and regenerated in an organism of adult man (Lehninger, 1972). However the energy expenditure is difficult to calculate since a fraction of molecules is repeatedly renovated and a fraction is newly synthesized. Moreover, a large number of compounds results from the synthesis of a protein, their synthesis costs many kilocalories of energy. These are a minimum of 20 activating enzymes, 70 ribosome proteins, 4 ribosome RNA, no less than 20 transport RNA, no less than 10 auxiliary enzymes. All this additional mass of reagents is synthesized only for several episodes of synthesis. Then they are synthesized again. It is best to judge about the continuous synthesis of proteins and other products of metabolism by the actual data. For example, we know that the restoration of the human liver by half occurs during 5-6 days. If the liver weighs 1.5 kg, this means that every day 300 g of the substance of liver is restored (in a mouse the liver is restored by half during 1-1.5 days). By the data of H.W.Hochachka and G.N.Somero (Hochachka & Somero, 2002) 52% of the energy budget of the liver is spent for the synthesis and restoration of proteins. About 17% of the energy budget of liver is spent for the synthesis of carbohydrates, and the most important of the functions of liver – the urea synthesis consumes only 3% of energy.

In the human brain the restoration of proteins occurs also very intensively (Palladin, Belik, & Polyakova, 1972) and is highly competitive with liver. The brain consumes 16-18% of the total energy budget of the human organism. This energy is spent for maintaining the electrical potentials in the brain cells, but predominantly for the restoration and replacement of the protein molecules. Energy consumption is quite the same at a complete mental rest, during sleep, and during intensive intellectual work. The brain in such a case can be compared with a computer, in which all the energy is spent for maintaining the anodic voltage (in the brain – for the exchange of proteins). The very calculation by a computer almost does not use energy, as in the brain the intellectual activity is almost not accompanied by the energy consumption. The brain tissue in itself is very sensitive to the energy deficit. A decrease in the consumption of oxygen by 15-20% results in heavy disruptions in the brain functions (Ivanov, 1993).

It is interesting that the heart consumes a large amount of energy – about 10.7% of the total energy budget of an organism. The contractions of myocardium are believed to be the main reason for a high energy consumption. However, the arrested, not contracting heart is known to continue consuming from 20 to 30% of the energy of beating heart (Ivanov, 2007-2008). It is believed that this energy expense comprises the "heart metabolism" and is consumed for restoration of protein structures. We must note that in other organs there are indications to a rapid restoration of proteins. They are estimated by the duration of the cell life. Thus, for example, the lifetime of the cells of the mucous membrane of the stomach is 1.8 days; of the small intestine – 1.3 days; of the liver – 10 days; of the large intestine – 10 days; of the rectum – 5 days; skin epithelium – 5-6 days. The erythrocytes live for 100-120 days, leukocytes – 1-2 days etc.

Therefore, a living organism is a substance with dynamically restored composition according to the laws of thermodynamics. The basis of this restoration is the protein composition of an organism. From birth to death an organism seems to be restored very many times. The dynamic restoration of the body composition is energetically very expensive. That is why even at a complete rest and visual inactivity an organism spends a lot of energy. If we calculate the energy spent by mankind for maintaining life during a year, it appears to be equal or exceed the amount of electric energy produced by all the power plants of modern world per year (Ivanov, 2007-2008). This is not an abstraction. This is a modern view on life and physiology of a living organism. We think that every physician and every student must know this most important law of the living world.

The Second Problem. This problem deals with a special role of adenosine triphosphoric acid (ATP) in an organism of any living organism and humans. The matter is that almost all the flow of energy in a living organism passes ATP, since only the energy released upon ATP hydrolysis is the free energy capable of producing physiological work and of taking part in the synthesis of other macroergs. On this basis we may say that almost all the energy of human metabolism (1800 kcal/day) is spent for the ATP synthesis with the help of ATP-synthetase enzyme. It is used for the work of the synthesis of various compounds in an organism necessary for its vital activity, for the ion transport against the concentration gradient in a cell and intercellular medium, it is necessary for muscle contractions, for restoration of cellular populations etc.

The fact that all the flow of energy proceeds via the only carrier is very important. This allows the energy to be accumulated in a cell at the expense of only one carrier and its expenditure to be regulated in the most efficient manner. However the synthesis of ATP has a number of problems, which have been discovered and studied by the science only recently. First, the work of ATP synthesis has the efficiency of only 40% (Lehninger, 1972). This means that from 1800 kcal only 40%, i.e. only 720 kcal, place at an organism disposal the free energy capable of producing physiological work. The remaining 60% is the bound energy, which cannot produce any physiological work. Second, the most important special feature of ATP is its very small energy capacity. Upon hydrolysis of a mole of ATP to give ADP and P only 10 kcal/mol is released. Consequently, with the aim of providing a human organism with energy for a day 72 energy full-value moles of ATP must be synthesized. But each mole has a mass of 506 g. In such a case every day 506·72 = 36 432 g of ATP must be synthesized. It is interesting that nobody ever calculated the reproduction of ATP mass. We were the first to make such a calculation in 1972 and showed that the reproduction of ATP mass comprises 36 kg per day for a man, i.e. more than a half of his body mass. If we apply to comparative physiology, we can obtain quite fantastic numbers. In a rabbit, for example, the mass of ATP synthesized during a day will be almost equal to the body mass of an animal. In a rat it will be twice as much as the body mass. In a dwarfed mouse it will exceed the body mass by a factor of 10-12. All this points to the fact that we still scarcely know the price of life and vital activity. From medical point of view such knowledge seems to be very important for physicians. They allow the quantitative estimations to be made of the energy processes necessary for life and health.

We emphasize that 780 kcal are not used by the tissues completely. The work of using the energy of ATP also has an efficiency, which is about 50% (Alyukhin & Ivanov, 1984). In such a case only 360 kcal or one fifth of the total energy consumed by a man really is spent for restoration of destroying molecules, for maintaining the normal composition of the cells, and for the work of heart, liver, and lungs, their activity being necessary for fulfilling this work at the level of metabolism. This is a very important fact in the total energetics of a living organism. Unfortunately, it is not mentioned in the textbooks on physiology, it is unknown for students and often for their teachers. This fact suggests that the efficiency of an organism, taking into account the energy expenditures of an organism per day being 1800 kcal, is comparatively low - about 15-18%.

The Third Problem. Individual Energy Expenditures of a Man. How much energy is consumed by a man during his physiological and social activity? An intelligent person spends about 2500 kcal during a day of his scientific or administrative activity. Approximately 700 kcal is spent for motive activity. Other 1800 kcal are the expenditures for metabolism, i.e. for respiration, heart and liver activity, energy consumption by the brain and other various organs, for restoration of continuously disintegrating living structures. However it must be taken into account the fact that each calorie, which is received by a man with food costs by a factor of 10 more in the agriculture (Odum, 1983). Consequently, the life of a man will cost by a factor of 10 more and will comprise 25000 kcal per day. Though this is, of course, the external energy, but it is included into the energy value of food, is consumed by a man, and, consequently, increases his energy balance. But the matter is that a man for his normal life must have a roof over his head, closes, heating and lighting of his dwelling, hot water supply, transport, and in a number of other everyday things and services. An American economist Odum (Odum, 1983) believes that the expenditures of this external energy necessary for the normal life of a man increase his energy balance by a factor of 10 more. In such a case a man will cost to nature and industry 250 000 kcal or 290 kWt-hour per day. This is a very large amount of energy, but we often neglect this. In Moscow underground, for example, during rush hours about 2500 passengers are sitting, staying, or moving to the doors of 8 wagons of a train. Each of them has the power of 160 Wt (Fanger, 1970). This means that all the passengers release the total power of 400 kWt. This energy is close to the intake of electric motors setting the train in motion. Without these simple calculations it would be difficult to conceive the amount of energy consumed by the passengers of an underground train.

We believe that any scientific worker must know the limits of energy consumed by a man. For a physician this gives the notion about energy price of life of each man.

The Fourth Problem. Acquisition and Emission of Energy. As is noted in physiological textbooks, the quantity of consumed energy in calories per unit of time must be equal to the quantity of energy in calories, which an organism releases to the medium during the same time. Only in such a case, as is written in the textbooks, the body temperature may be maintained at a constant level. However, it appears incomprehensible and to some extent absurd that a man releases to the medium as much energy as he obtains. The question arises: if a man releases all the obtained energy, at what expense he maintains his life. It is interesting that this problem had been neglected for about two centuries – from the moment of determining the energy expenditures of a living organism made by Lavoisier and up to now. At any rate, there is no adequate explanations of this enigma either in the old or modern textbooks on physiology or in the special works on the problems of energy exchange in a living organism. It seems that A.Lehninger (Lehninger, 1972) was the first biologist to speak up about this problem very cautiously. He wrote that an organism consume free energy for physiological work, but releases it into the medium in the form less applicable for use. Now the main features of this problem are known. As has been noted above the whole energy used by an organism goes through the synthesis of adenosine triphosphoric acid. The efficiency of physiological work of this synthesis is about 40% (Lehninger, 1972). Then, if the energy of ATP hydrolysis is 10 kcal/mol (($\Delta G°{\sim}{-}10$kcal/mol), the energy consumption for the synthesis of 1 mole of ATP will be 25 kcal/mol. On this basis the calculations were made of the quality of energy obtained by a living organism and of the energy released into the medium. If we adhere to classical thermodynamics, the total balance of energy for a system or a body is given as:

$$\Delta H = \Delta G + T\Delta S$$

where ΔH is the total quantity of energy, ΔG – the free energy, i.e. energy, which can be used to produce work, ΔS – the change in the entropy, which gives $T\Delta S$ – the so called bound energy, the heat, which cannot produce work.

An organism receives energy from food, where, according to efficiency there is 40% of free energy and 60% of bound energy. Than the total balance of energy may be written as:

$$\Delta H = \Delta G + T\Delta S$$

40%   60%

After energy transformations and use of the free energy ΔG for various kinds of physiological work in an organism the quantity of free energy decreases and the quantity of bound energy increases. It is quite clear that after using the free energy a somewhat depreciated energy is released into the medium, energy deprived of a fraction of energy potential. Therefore, an organism lives at the expense of free energy, which it spends for various kinds of physiological work (Odum, 1983). The free energy fulfills a certain work and transforms into heat, whereas the calories of food preserving their heat equivalent lose a fraction of their energy potential. The free energy transforms into bound energy. This phenomenon is called the change in "the quality" of energy (Odum, 1983). These relationships are well explained in the textbook of V.O.Samoilov (Samoilov, 2004). Therefore we can say that an organism receives a full-value energy with a large energy potential with food and releases energy to some extent devoid of energy potential. In this case the heat identity between various calories is preserved, but there are differences in the energy potential. In such a way a puzzling equality between the energy received by an organism with food and the energy released into the medium disappears. The concept of "the quality of energy" is an energy excess from the point of view of biology and physiology, this phenomenon almost never was considered in a living organism. However, in the industrial energetics it is excessively popular. It was calculated that 1 calorie of the sun emission is equal to 0.0005 calories obtained from combustion of coal by its energy potential. 100 calories of phytodetritus (green mass) are energetically equal to 0.05 calories of coal combustion (Odum, 1983).

These numbers are quite unknown to biologists, physiologists, and physicians. It is necessary to elucidate such special features of energetics in the reference books. These numbers must be illustrated in the textbooks and lectures. Then students and their teachers will have a notion about the relationships between free and bound energy according to modern ideas.

The Fifth Problem. Life Without Energy. This is one more special feature of energy exchange. It has no direct practical relation to human physiology. But it has a great significance for the whole animate world from microorganisms to lowest vertebrates and even for some mammals.

A living tissue is known to be destroyed and lost either from necrosis or from apoptosis when deprived of energy. This happens during myocardium and brain ischemia, limb gangrene. As was noted above, the renewal of the tissue composition is the most important function of a living organism. But is life possible without this renewal and influx of energy? Cold abruptly decreases the metabolism and almost excludes the disruption and restoration of the cell composition of the tissues. At the temperature of minus 100-130°C the viability of an organism can be preserved for an uncertain time, if, of course, the method will be found of its rewarming without damaging the tissues. However the matter is that at the temperature higher than zero some organisms are capable of switching off the normal energy exchange partially or completely and preserve the viability of an organism in doing so. Thus, for example, microorganisms inhabiting the crust at a depth of 2-3 km from the surface can preserve their viability switching off their metabolism almost completely at comparatively high temperature given almost complete absence of oxygen and nutrients. They are divided only once per 100 years. During this time before division they accumulate the energy potential. This potential is not spent before the division (Ehrich, 1996) (Fredrickson & Onstott, 1996). Many insects, fly, for example, switch off their metabolism at the temperature higher than zero and exist almost without metabolism for several months.

Even some vertebrates demonstrate such a property. African Dipnoi fish buries itself into the slime during dry period and completely stops its energy exchange for a period from one year to 5 and even, by some data, to 10 years (Hochachka & Somero, 2002). It preserves the store of carbohydrates and viability in doing so. Even the nitrogen exchange is not destroyed in these fish during such a sleep. These are, of course, amazing animals. Some mammals are capable of an abrupt decrease in the energy influx for a certain period and of preserving life. The so called Widdell seal can spend 1 h and 20 min under water without respiration. It appears to be able to decrease its energy exchange by factors of 2-3 and stay alive (Hochachka & Somero, 2002). Such cases are unknown for humans. However the mechanisms of such safe for life detachments of the energy exchange are extremely interesting from scientific and practical points of view. These phenomena are studied continuously. However the mechanism of switching off the metabolism with preservation of certain stores of nutrients and maintaining the viability remain an enigma.

Each science strives for development. There are few references in this paper. This is the result of the fact that there are very few works in those directions of biology and physiology of animals and humans, which we discussed and tried to reveal their essence. We considered, of course, not all the strategically important features of energy exchange. However I believe that the small material given here nevertheless will allow us a lift by one-two steps for better understanding of life and energetics of a living organism.

References

Alyukhin, Y. S., & Ivanov, K. P. (1984). Relationships between work and energy expenditures in the isolated muscle of a mammal. *Dokl. AN SSSR, 282*, 983-987.

Ehrich, N. L. (1996). *Geomicrobiology*. Marseill: Dekker.

Fanger, P. O. (1970). *Thermal Comfort*. Copenhagen: Danish Technical Press.

Fredrickson, J. K., & Onstott, T. S. (1996). Microbes deep inside the earth. *Nature, 275*(4), 42-47.

Hochachka, H. W., & Somero, G. N. (2002). *Biochemical adaptation*. New York, Oxford: Oxford Univ. Press.

Ivanov, K. P. (1993). *Principles of energetics of an organism. V.2. Biological oxidation*. S. Petersburg: Nauka.

Ivanov, K. P. (2007-2008). *Principles of energetics of an organism. V.5. Energetics of animate world*. S.Petersburg: Nauka.

Komkov, A. V. (Ed.). (2004). *Fundamental clinical physiology*. Textbook. Moscow, Akademia.

Lehninger, A. (1972). *Principles of biochemistry*. New York: Worth Publishers INC.

Odum, E. P. (1983). *Basic Ecology* (Vol. 1). Philadelphia: Saunders college Publishing.

Palladin, A. V., Belik, Y. I., & Polyakova, N. M. (1972). *Proteins of brain and their exchange*. Kiev: Naukova Dumka.

Samoilov, V. O. (2004). *Medical biophysics*. S. Petersburg: Spets. Lit.

Schmidt-Nielsen, K. (1972). *Animal physiology*. Cambridge: Cambridge Univ. Press.

11

Some Physiological and Ethological Effects of Nicotine; Studies on the Ant *Myrmica sabuleti* as a Biological Model

Marie-Claire Cammaerts[1], Zoheir Rachidi[1] & Geoffrey Gosset[1]

[1] Zoheir Rachidi, Geoffrey Gosset, Faculté des Sciences, Université Libre de Bruxelles, Bruxelles, Belgium

Correspondence: Marie-Claire Cammaerts, Zoheir Rachidi, Geoffrey Gosset, Faculté des Sciences, Université Libre de Bruxelles, Bruxelles, Belgium. E-mail: mtricot@ulb.ac.be

Abstract

Nicotine is one of the most consumed alkaloids. Its effects still lead to mixed conclusions. Having recently observed that ants could be used as biological models, we examined several physiological and ethological effects of nicotine using the ant *Myrmica sabuleti* as a biological model. We pointed out that nicotine increases the individuals' locomotion, decreases the precision of their reaction, decreases their response to pheromones, does not impact their audacity, and reduces their food consumption. Nicotine largely decreases the individuals' learning abilities when the reward does not contain this alkaloid but increases the rapidness of acquiring visual as well as olfactory conditioning when the reward contains nicotine. In the latter case, the final conditioning score is only very slightly higher than the control score, and there is no memorization. Under nicotine consumption, ants' cognitive abilities seem of better quality, the ants in fact acting more rapidly. There is no habituation to nicotine consumption. No dependence on this alkaloid exists, as long as the individuals feel well, but dependence appears as soon as the individuals need something, are deprived of something. After nicotine consumption ends, a first slow, than a short rather quick, and finally again a slow decrease of its effects occurs. As for other alkaloids, no dependence is associated to no habituation and to rather slow vanishing of the effects. However, since the effects of nicotine quickly vanish during a short time, dependence might develop if the individual then encounters any problem or need. All the effects, here revealed using ants as biological models, are in agreement with those actually known for nicotine. Some effects could even be better defined, i.e. the impact on cognition, the occurrence of dependence. The present work brings so some new information about this important alkaloid and shows, once more, that ants could be used as biological models, for a first, not expensive step of a more general, biological or medical study.

1. Introduction

Nicotine is one of the most consumed, and the most easily available alkaloid in the word. It can be consumed by inhalation, ingestion or contact. Several effects due to nicotine consumption are known (Waldum et al., 1996; Frenk & Dar, 2005; Mayer, 2013). For instance, it is known that this alkaloid has an effect on the brain, acts rapidly and imitates acetylcholine, what induces feeling some satisfaction, some well-being. It is commonly admitted that nicotine has an anti-depressive action, helps reducing anxiousness, stimulates the brain functioning and may act as an appetite suppressant. It might lead to dependence, this being yet controversy, and the underlying mechanisms being not yet entirely elucidated. Substances are often added to products containing nicotine in order to increase this potential dependence. All the effects of nicotine are probably not yet known, and all of them are probably not related to people. Scientific experiments on several living organisms are lacking, though a lot of invertebrates and vertebrates can be used, as biological models, for studying biological problems (Kolb & Wishaw, 2002; Wehner & Gehring, 1999; Deutsch, 1994; Devineni & Heberlein, 2013; Russell & Burch, 2014). Among the invertebrates, insects, and especially social hymenoptera, are advantageously used as biological models (Andre, Wirtz, & Das, 1989; Doring & Chittka, 2011). Among others, bees are often used (for instance, Abramson, Wells, & Janko, 2007). We have recently shown that ants could be used as biological models for examining some physiological and ethological effects of elements, treatments, and factors (Cammaerts, Rachidi, Bellens, & De Doncker, 2013, Cammaerts & Gosset, 2014, Cammaerts, Rachidi, & Gosset, 2014). Indeed, several nests, with hundreds of ants, can easily be maintained in laboratories, at low cost and very conveniently, throughout the entire year. These insects are highly evolved as for their morphology, their physiology, their social organization and their behavior. They have a unique resting position of their labium, mandibles and maxilla (Keller, 2011a, b), and numerous glands emitting efficient compounds (Billen & Morgan, 1998). Their societies are highly organized with a strong division of labor, an age-based polyethism and social regulation (Hölldobler & Wilson, 1990). Their behavior is incomparable: they care for their brood, build sophisticated nests, chemically mark the inside of their nest, and, differently, their nest entrances, their nest surroundings and their foraging area (Passera & Aron, 2005). They generally use an alarm signal, a trail pheromone, and a recruitment signal (Passera & Aron, 2005); they are able to navigate using memorized visual and olfactory cues (cf. Cammaerts, 2012, Cammaerts & Rachidi, 2009; Cammaerts, Rachidi, Beke, & Essaadi, 2012a); they efficiently recruit nestmates where, when and as long as it is necessary (Passera, 2006), and, finally, they provide their area with cemeteries (Keller & Gordon, 2006). We have already largely studied the species *Myrmica sabuleti* Meinert 1861. Among others, we actually know its ecology, eye morphology, visual perception, visual and olfactory conditioning, navigation system, recruitment strategy, responses to pheromones, acquisition by callow ants of adults' cognitive abilities (Cammaerts & Cammaerts, 2014; Cammaerts et al., 2013; Cammaerts, 2013 a, b, 2014, in press). Consequently, we aimed to study, on *M. sabuleti*, as many effects of nicotine as we could, in order to known some more about this alkaloid, and to check again the potential use of ants as biological models. More precisely, we intended to examine if nicotine impacts the individuals' activity, behavior, audacity, food consumption, learning capability, memory, and cognitive ability, and in which circumstances these physiological and ethological functions may be affected. We also wanted to precise if nicotine leads to habituation, and/or to dependence (addiction) and, if so, in which circumstances. A last objective would be defining the decrease of the effects of nicotine after its consumption ends. According to the experience we have about *M. sabuleti*, we thought being able to achieve the here above enumerated objectives. We planned several observations, on a few nests, but must then make checking and supplementary experiments consequently to first results. Here, we present the carried out experiments, the obtained results, and the deducting conclusions and thought.

2. Functions examined and carried out experimental planning.

2.1 Functions

The effect of nicotine on the following physiological and ethological functions was examined:

- the individuals' general activity, via their locomotion, and more precisely their linear and their angular speed.

- the precision of their reaction via their orientation towards a source of their alarm signal

- their response to their pheromone via their trail following behavior

- their audacity via the numbers of them coming onto a risky apparatus

- their consumption of food via the number of them eating meat food

- their learning and memory abilities via their acquisition of visual and olfactory conditioning, and their retaining of the acquired conditioning

- their habituation to nicotine consumption via their locomotion after 7 days of nicotine consumption

- their dependence on nicotine via the numbers of them preferring food containing this alkaloid

- a cognitive ability requiring no memory via their behavior in an adequate experimental apparatus

- the decrease, with time, of the effects of nicotine, by assessing the ants' orientation to an alarm signal, in the course of time, after the alkaloid had been removed from the food.

2.2 Experimental Planning

The following experiments were performed, in such an order for obtaining adequate controls, testing not yet experimented individuals, checking first results and looking for complementary ones. Details are given below, in the 'Material and Methods' section, but let us already inform that six ant nests (labeled I to VI) were used, nests I and II for control and test experiments, nests III and IV for test experiments, and nests V and VI for control experiments.

- control experiments were made on nests I and II fed with lactose solution: the ants' locomotion, orientation, trail following behavior, audacity, and food consumption were assessed; visual and olfactory conditioning, and memory, were assessed, the reward being the meat food free of nicotine; habituation to and dependence on lactose were examined.

- test experiments were made on nests I and II fed with lactose + nicotine solution: the ants' locomotion, orientation, trail following behavior, audacity, and food consumption were assessed; the ants' habituation to and dependence on nicotine were examined.

- test experiments were made on nests III and IV fed with lactose + nicotine solution: the ants' visual and olfactory conditioning, and memory, were assessed, the reward being the meat food free of nicotine; the ants' dependence on nicotine was examined.

- an ants' ability requiring no memory was assessed on nests I and II fed with nicotine, and on nests V and VI having never received this alkaloid.

- test experiments were made on nests III, IV, I and II fed with lactose + nicotine solution: the ants' visual (on nests III and IV) and olfactory (on nests I and II) conditioning, and memory, were assessed, the reward being the sugar food containing nicotine.

- the decrease of the effects of nicotine was examined on nests I and II by assessing, in the course of time, the ants' orientation to an alarm signal, once the ants stopped eating nicotine.

3. Material and Methods

3.1 Collection and Maintenance of Ants

The study was made on six colonies of *M. sabuleti* collected, in summer 2013, in an old quarry located at Treigne (Ardenne, Belgium). The ants were nesting under stones, on a field covered with small, often odorous plants. The six collected colonies were demographically identical, containing about 600 workers, one or two queens and brood (at larval stage). They were maintained in the laboratory in artificial nests made of one to three glass tubes half-filled with water, with a cotton-plug separating the ants from the water. The glass tubes were deposited in trays (34 cm x 23 cm x 4 cm), the sides of which were covered with talc to prevent the ants from escaping. The trays served as foraging areas, food being delivered into them (Figure 1A). The ants were fed with sugar-water provided *ad libitum* in a small glass tube plugged with cotton, and with cut *Tenebrio molitor* larvae (Linnaeus, 1758) provided twice a week on a glass slide. Temperature was maintained between 18°C and 22°C, humidity at about 80%, these conditions remaining constant over the course of the study. Lighting had a constant intensity of 330 lux while caring for the ants, training and testing them. During other time periods, the lighting was dimmed to 110 lux. The electromagnetic field had an intensity of 2-3 $\mu W/m^2$.

3.2 Acquisition of the Alkaloids; Realization of Aqueous Solutions for Ants

One gram of lactose and 15 gram of lactose (8) + nicotine (2) were provided, via the pharmacist Mr. Cardon J. (1050, Brussels), by the manufacturer UNDA dolisos (1140, Brussels). The two products were provided as a white bright powder, at the highest level of purity possible. Using a precise balance, 60 mg of lactose, as well as 75 mg of lactose (8) + nicotine (2), were prevailed and dissolved in 15 ml of a saturated solution of brown sugar, the ants' common liquid food. The concentration in nicotine of the solution was thus 15 mg in 15 ml, so 1/1,000. These solutions were given to the ants, like their usual liquid food, in a small glass tube plugged with cotton, this cotton being refreshed each two days.

3.3 Assessment of the Ants' Linear Speed, Angular Speed and Orientation

Ants' linear and angular speed was assessed for detecting excitation in the animals. This assessment was made on ants freely moving on their foraging area. Ants' orientation towards an isolated congener's head (the source of the ants' alarm pheromone) was assessed for examining the ants' precision of reaction. An isolated worker's head, with widely open mandibles, is a source of alarm pheromone identical to that of an alarmed worker, in terms of the dimensions of the emitting source (the mandibular glands opening) and the quantity of pheromone emitted (Cammaerts-Tricot, 1973).

Each time, such assessment was made on ants of two colonies having never consumed nicotine but being fed with the solution of lactose, then on ants of these two same colonies having consumed the solution of lactose + nicotine during two days. For each assessment, the movement of 10 ants of each two colonies (N = 20 ants) was analyzed.

Trajectories were recorded manually, using a water-proof marker pen, on a glass slide placed on the top of the experimental tray, set horizontally 3 cm above the area where the tested individuals were moving. A metronome set at 1 second was used as a timer for assessing the total time of each trajectory. Each trajectory was recorded during 5 to 10 seconds or until the ant reached the stimulus. All the trajectories were then traced (copied) with a water-proof marker pen onto transparent polyvinyl sheets (Figure 1 B, C). These sheets could then be affixed to a PC monitor screen. The trajectories were then analyzed using specifically designed software (Cammaerts, Morel, Martino & Warzée, 2012b). Briefly, each trajectory was defined in the software by clicking as many points as needed with the mouse. Then, the total time of the trajectory (assessed using the metronome) was entered, and feature of the trajectory could be measured (linear speed, angular speed, orientation).

The three variables used to characterize the trajectories were defined as follows:

The linear speed (V) of an animal is the length of its trajectory divided by the time spent moving along this trajectory. It was measured in mm/s.

The angular speed (S) (i.e. the sinuosity) of an animal's trajectory is the sum of the angles, measured at each successive point of the trajectory, made by the segment 'point i → point i – 1' and the segment 'point i → point i + 1', divided by the length of the trajectory. This variable was measured in angular degrees/cm.

The orientation (O) of an animal towards a given point (here an empty piece of paper or an ant's head) is the sum of the angles, measured at each successive point of the registered trajectory, made by the segment 'point i of the trajectory → given point' and the segment 'point i →point i + 1' divided by the number of measured angles. This variable was measured in angular degrees. When such a variable (O) equals 0°, the observed animal perfectly orients itself towards the point; when O equals 180°, the animal fully avoids the point; when O is lower than 90°, the animal has a tendency to orient itself towards the point; when O is larger than 90°, the animal has a tendency to avoid the point.

Each distribution of 20 variables was characterized by its median and its quartiles since it was not Gaussian (Table 1, lines 1, 2), and the distribution of values obtained for ants having consumed alkaloid was statistically compared to that previously obtained for the ants having never consumed alkaloid using the non-parametric χ^2 test (Siegel & Castellan, 1989, p 111-116). The significance threshold was set to $\alpha = 0.05$.

Table 1. Effect of nicotine on five physiological and ethological functions

Functions	Variable assessed	No nicotine	+ nicotine	Statistics
Activity n = 20	Linear speed mm/sec	17.4(14.5-20.2)	20.9(20.1-23.5)	P < 0.001
(locomotion)	Sinuosity ang.deg./cm	107(86-119)	129(109-152)	P=0.05
Precision of a reaction n=20	Orientation to an alarm signal; ang.deg.	32.0(21.6-56.59)	53.7(43.3-63.8)	P ≈ 0.001
Response to pheromones n = 40	Trail following behavior	C: 1 (1 - 1.3)	C: 1 (1 - 2)	NS
	(n° of arcs walked)	T:12.0 (8.8-17.3)	T:6.5 (3.0-9.3)	P < 0.001
Audacity n = 10	n° of ants on a tower	1.3 (1 - 2)	1.6 (0 - 3)	P = 0.35
Food consumption n = 10	n° of ants eating meat (mean, extremes)	3.7 (1 - 8)	1.4 (0 - 3)	P = 0.05

Details are given in the text. Briefly, nicotine increased ants' linear and angular speed, unchanged their audacity, and decreased their precision of reaction, their response to pheromones as well as their food consumption.

3.4 Assessment of the Ants' Trail Following Behavior

This behavior was assessed for examining the ants' response to their pheromones. The trail pheromone of *Myrmica* ants is produced by the workers' poison gland. So, ten of these glands were isolated in 0.5 ml (500µl) hexane and stored for 15 min at -25 °C. To perform one experiment, 0.05 ml (50µl) of the solution was deposited, using a normograph pen (a pen used for drawing, hexane extract being poured inside the pen instead of ink), on a circle (R = 5 cm) pencil drawn on a piece of white paper and divided into 10 ang. deg. arcs. One minute after being prepared, the piece of paper with the artificial trail was placed in the ants' foraging area. When an ant came into contact with the trail, its movement was observed. Its response was assessed by the number of 10 ang. deg. arcs it walked without departing from the trail, even if it turned back on the trail. If an ant turned back when being in front of the trail, its response was assessed as "zero arcs walked"; when an ant crossed the trail without following it, its response equaled "one walked arc". Before testing the ants on a trail, they were observed on a circumference imbibed with 50µl of pure hexane and the control numbers of walked arcs were so obtained. For each control and test experiment, 20 individuals of each two used colonies were observed; 40 numbers of walked arcs were so each time recorded. Each distribution of values was characterized by its median and its quartiles, since it was not Gaussian (Table 1, line 3). The experiments were performed on two colonies fed with the solution of lactose, then on the two same colonies fed since three days with the solution of lactose + nicotine. The distributions of values obtained for ants having consumed nicotine were statistically compared to the corresponding ones obtained for ants having never consumed that substance, this by using the non parametric χ^2 test (Siegel & Castellan, 1989). On such experimental trails, *Myrmica* workers do not deposit their trail pheromone because they do so only after having found food or a new nest site.

3.5 Assessment of the Ants' Audacity

A tower built in strong white paper (Steinbach ®) (h= 4cm; diam= 1.5cm) was set on the foraging area of two colonies fed with the solution of lactose, and then of the same two colonies fed during three days with the solution of lactose + nicotine (Figure 1 D). The ants present on the tower were counted 10 times. The mean and the extreme values of the obtained values were established each time and the two series of values were compared using the non parametric Mann-Whitney U test (Siegel & Castellan, 1989, p 128-137; Table 1, line 4).

3.6 Assessment of the Ants' Consumption of Food

The observations were made on two colonies fed with the solution of lactose, and then on the two same colonies fed for five days with the solution of lactose + nicotine. Each time, the workers present on the *T. molitor* larva were counted 10 times. The numbers obtained under the two kinds of food consumption were statistically compared using the Mann-Whitney U test (same reference as above) and the mean as well as the extreme values of the recorded numbers were established (Table 1, line 5).

3.7 Assessment of the Ants' Ability in Acquiring Operant Either Visual or Olfactory Conditioning, and of the Ants' Either Visual or Olfactory Memory

Briefly, at a given time, either a green hollow cube or pieces of thyme were set above the *T. molitor* larva, this time tied to the supporting piece of glass, or, in a checking experiment, either a yellow hollow cube or pieces of onion were set above or aside the sugar food containing nicotine. The ants so underwent, each time, either visual or olfactory operant conditioning. Each time, tests were performed, in the course of time, while the ants were expected acquiring conditioning, then, after having removed the green (or the yellow) cube or the pieces of thyme (or of onion), while the ants were expected partly losing their conditioning.

In detail, ants were collectively visually trained to a hollow green (or yellow) cube constructed of strong paper (Canson ®) according to the instructions given in Cammaerts and Nemeghaire (2012) and set over the meat food (or the sugar food) which served as a reward. The color has been analyzed to determine its wavelengths reflection (Cammaerts, 2007). The ceiling of each cube was filled unlike the four vertical faces, this allowing the ants entering the cubes. Choosing the green (or yellow) cube was considered as giving the 'correct' choice when ants were tested as explained below. The ants were olfactory conditioning by setting four pieces of thyme (or of onion) aside the tied dead *T. molitor* larvae (or the source of sugar food, respectively). Choosing the pieces of thyme (or of onion) was considered as giving the 'correct' choice when ants were tested as explained below

Ants were individually tested in a Y-shaped apparatus constructed of strong white paper according to the instructions given in Cammaerts, Rachidi and Cammaerts (2011), and set in a small tray (30 cm x 15 cm x 4 cm), apart from the experimental colony's tray (Figure 1A, E, F). Each colony had its own testing design. The apparatus had its own bottom and the sides were covered with talc to prevent the ants from escaping. In the Y-apparatus, the ants deposited no trail since they were not rewarded. However, they may utilize other chemical secretions as traces. As a precaution, the floor of each Y-apparatus was changed between tests. The Y-apparatus was provided with either a green (or yellow) cube, or four pieces of thyme (or of onion), in one or the other branch (Figure 1 E, F). Half of the tests were conducted with the cube, or the odorous plant, in the left branch and the other half with the cube, or the odorous plant, in the right branch of the Y maze, and this was randomly chosen. Control experiments had previously been made on never conditioned ants. In the present work, conditioning experiments were made on ants of two colonies having never received nicotine but fed with the solution of lactose (see above), and on ants of two other, identical, collected colonies having received the solution of lactose + nicotine during seven days (Table 2). These experiments were made using green cubes and thyme. Then, when supplementary experiments had to be made, the same colonies were used but the cues were replaced by yellow cubes and onion. Indeed, once an animal is conditioned to a given stimulus, it becomes no longer naïve for such an experiment. It was thus impossible to perform, on the same ants, conditioning without then with nicotine; the only solution was to use four similar colonies, two for conditioning without nicotine, and two for conditioning under nicotine consumption. And after that, it was impossible to make a checking experiment on the same colonies using the same cues; the only solution was to perform identical experiments but with different (other color, other odor), though similar (cube, pieces), cues (Table 4).

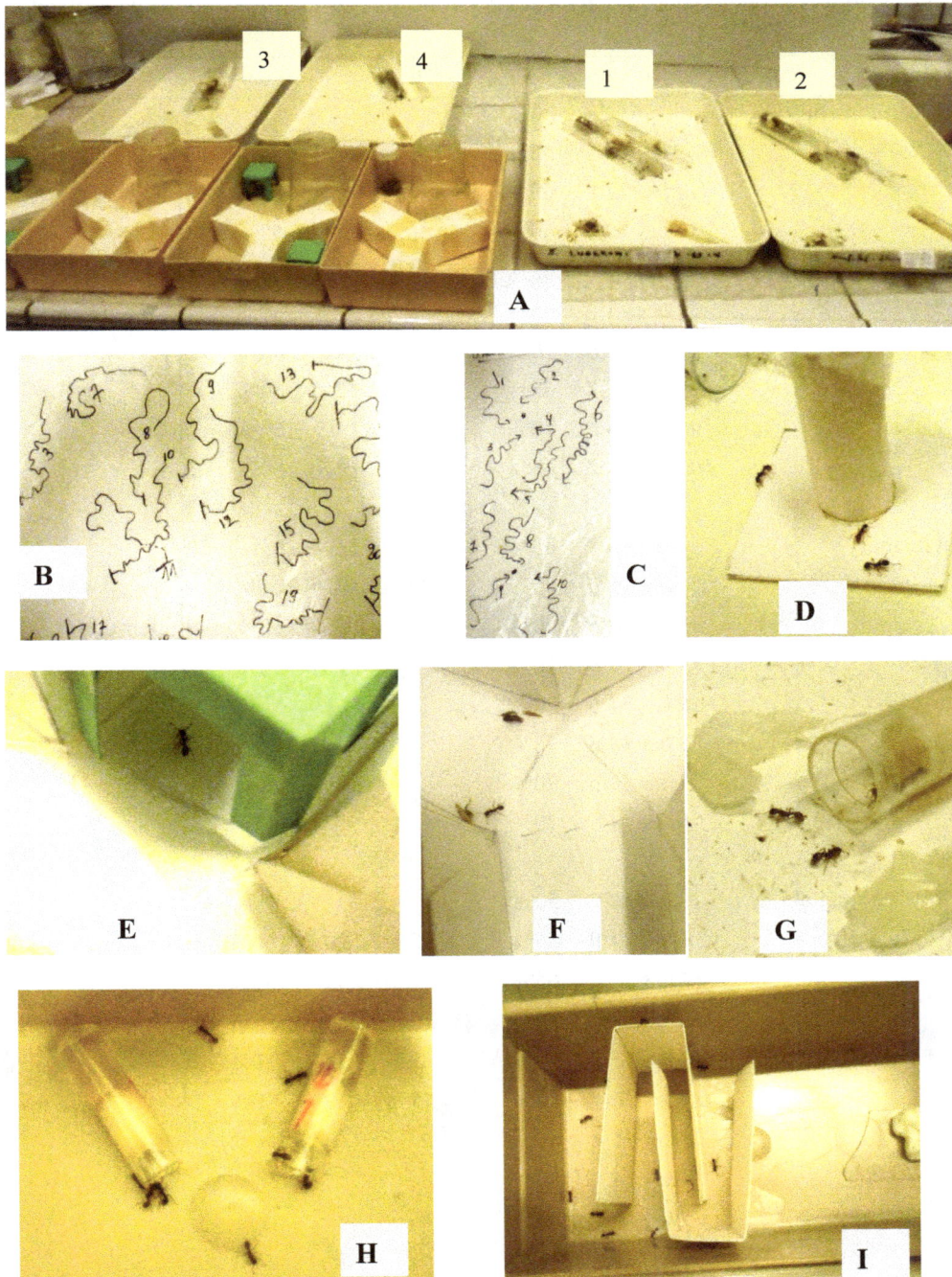

Figure 1. Some views of the experiments. **A**: four nests of *Myrmica sabuleti* used, and the Y mazes prepared for experimenting. **B**: ants' trajectories under nicotine consumption; the ants' linear and angular speed were large. **C**: ants trajectories, under nicotine consumption, in the vicinity of an alarm signal (the black points); not all the ants correctly moved towards the signal. **D**: ants, under nicotine consumption, confronted with a risky apparatus; as usual, the ants were not inclined to climb on the tower. **E**: an ant correctly responding to a green hollow cube, during a test in a Y maze. **F**: an ant correctly responding to thyme, during a test in a Y maze. **G**: ants, under nicotine consumption, trained to pieces of onion, and rewarded with sugar water containing nicotine. **H**: ants, under nicotine consumption, not hungry and feeling well, confronted with an aqueous solution of sugar + lactose and one of sugar + lactose + nicotine (on the right); the ants did not prefer the food containing nicotine; they were not dependent on that alkaloid consumption. **I**: ants, under nicotine consumption, tested in an apparatus made of a narrow loggia (on the left), a maze, and a large area (on the right); the ants, initially set in the narrow loggia, soon found their way through the maze, towards the large area

Table 2. Effect of nicotine on conditioning acquisition and on memory, using meat food (free of nicotine) as a reward

Studied function	no nicotine		+ nicotine		statistics
Visual learning ability	* C: 61/59	50%			
n= 20 + 20 = 40	7 hrs 11/9	55%	7 hrs 11/9	55%	
	24 hrs 12/8	60%	24 hrs 12/8	60%	
	30 hrs 12/8	60%	30 hrs 12/8	60%	
	48 hrs 13/7	65%	48 hrs 12/8	60%	
	55 hrs 13/7	65%	55 hrs 12/8	60%	P = 0.016
	72 hrs 14/6	70%	72 hrs 12/8	60%	
	79 hrs 13/7	65%	79 hrs 12/8	60%	
	96 hrs 14/6	70%	96 hrs 12/8	60%	
	103 hrs 13/7	65%	103 hrs 12/8	60%	
Visual memory	7 hrs 13/7	65%	7 hrs 11/9	55%	
n= 20 + 20 = 40	24 hrs 15/5	75%	24 hrs 10/10	50%	
	30 hrs 14/6	70%	30 hrs 10/10	50%	P = 0.063
	48 hrs 14/6	70%	48 hrs 10/10	50%	
	55 hrs 14/6	70%			
	72 hrs 14/6	70%			
	79 hrs 14/6	70%			
	96 hrs 14/6	70%			
Olfactory learning ability	* C: 61/59	50%			
n = 20 + 20 = 40	7 hrs 11/9	55%	7 hrs 10/10	50%	
	24 hrs 13/7	65%	24 hrs 9/11	45%	
	30 hrs 14/6	70%	30 hrs 11/9	55%	P = 0.031
	48 hrs 15/5	75%	48 hrs 10/10	50%	
	55 hrs 16/4	80%	55 hrs 10/10	50%	
	72 hrs 16/4	80%			
	79 hrs 16/4	80%			
	96 hrs 16/4	80%			
Olfactory memory	7 hrs 13/7	65%			
N= 20 + 20 = 40	24 hrs 12/8	60%			
	30 hrs 12/8	60%			
	48 hrs 11/9	55%			
	55 hrs 11/9	55%			

Ants were trained to a green hollow cube or to pieces of thyme set above or aside the meat food. Then, during tests, the numbers of ants giving the correct and the wrong response were counted, and so, the percentage of correct response for the ant population was determined. The percentages obtained for ants consuming nicotine were compared to those obtained for ants having never consumed this alkaloid using the non parametric Wilcoxon test; N, T and P are given, according to the nomenclature of Siegel and Castellan, 1989. These results are graphically presented in Figure 2, upper graphs. * = results obtained by Cammaerts et al., 2011.

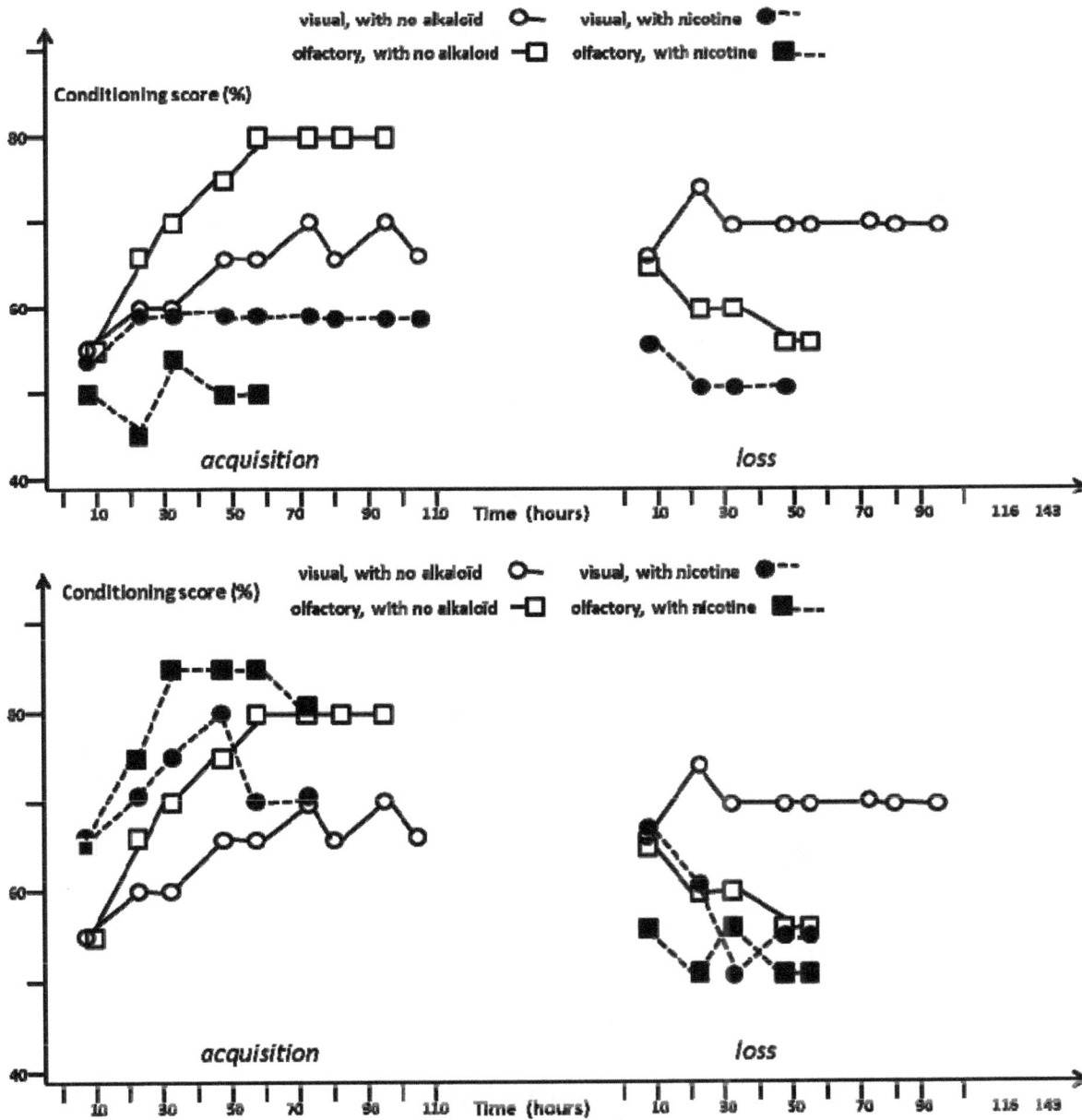

Figure 2. Response of ants conditioned to a visual (circles) or an olfactory (squares) cue, consuming (black forms) or not (empty forms) nicotine, and rewarded with food free of nicotine (upper graphs) or food containing nicotine (lower graphs). Rewarded without nicotine, the ants could scarcely acquire conditioning; rewarded with nicotine, they soon acquired conditioning but retained nearly nothing of it

To conduct a test on a colony, 10 workers of that colony - randomly chosen from the workers of that colony - were transferred one by one to the area at the entrance of the Y-apparatus. Each transferred ant was observed until it turned either to the left or to the right in the Y-tube, and its choice was recorded. Only the first choice of the ant was recorded and this only when the ant was entirely under the cube, i.e. beyond a pencil drawn thin line indicating the entrance of a branch (Figure 1 E, F). Afterwards, the ant was removed and transferred to a polyacetate cup, in which the border was covered with talc, until 10 ants were so tested, this avoiding testing the same ant twice. All the tested ants were then placed back on their foraging area. For each experiment, the numbers of ants, among 10 + 10 = 20, which turned towards the "correct" green (or yellow) cube or thyme (or onion), or went to the "wrong" empty branch of the Y were recorded. The percentage of correct responses for the tested ant population was so established (Tables 2, 4). The results obtained for ants that have consumed nicotine were compared to those obtained for ants that have never consumed such a substance using the non parametric test of Wilcoxon (Siegel & Castellan, 1989, 87-95). For such a test, the value of N, T, and P, according to the nomenclature of the here above cited authors, are given in the results section.

3.8 Assessment of an Ant'S Ability Requiring No Memory

This assessment required a) two nests consuming nicotine (we used nests I and II) and two other ones having never consumed this alkaloid (we used nests V and VI, very similar to nests I to IV, collected the same day, on the same field), b) a novel experimental apparatus. This last one, schematically presented in Figure 3 and visible in Figure 1 I, consisted in a small tray (15 cm x 7 cm x 4.5 cm) inside of which two pieces of white extra strong paper (Steinbach ®, 12 cm x 4.5 cm), duly twice folded, were inserted in order to create a maze between a narrow initial space (initial loggia) and a larger area (free area). Two such experimental apparatus were built and used at the same time. Nest I vs nest V, and then nest II vs nest VI were experimented. Each time, for each nest, 15 ants were set in the initial loggia of the apparatus, and those located in this loggia as well as in the free area were counted after 5, 10, 15 and 20 min. The numbers obtained for ants consuming nicotine (nests I and II) were statistically compared to those obtained for ants having never received such an alkaloid (nest V and VI) using the non parametric test of Wilcoxon (Siegel & Castellan, 1989).

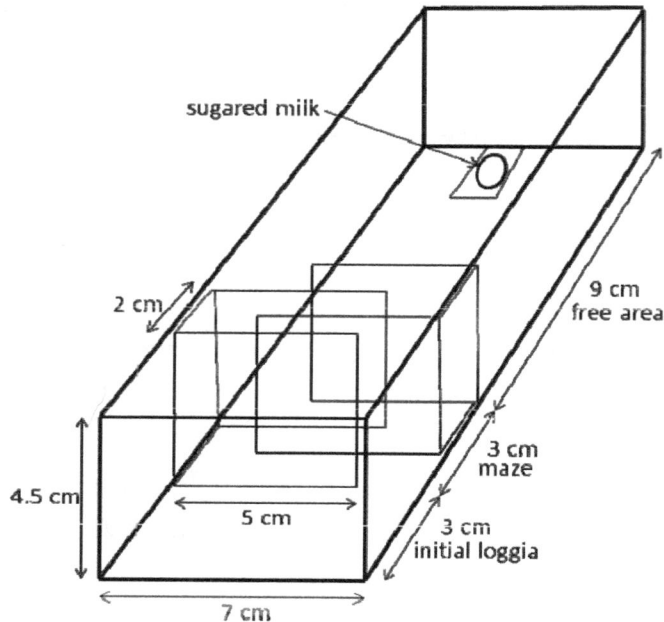

Figure 3. Schema of an experimental apparatus which allows assessing a cognitive ability. Fifteen individuals (e.g. ants) are set in the initial loggia; they tried to escape; the numbers of individuals still in the initial loggia, and those having reached the free area, are counted in the course of time

3.9 Ants' Habituation to the Consumed Alkaloid

Four to seven days after ants (of two previously tested colonies) had continuously consumed either the solution of lactose or that of lactose + nicotine, the ants' linear and angular speed were again assessed. The results were compared to those obtained at the beginning of the experiment, after one day of lactose or lactose + nicotine consumption, using the non parametric χ^2 test (Siegel & Castellan, 1989; Table 5, line 1).

3.10 Ants' Dependence on the Consumed Alkaloid

After the ants (of colonies I, II, III, IV. because the experiment had to be repeated) had continuously consumed the solution of lactose or that of lactose + nicotine during five to nine days, an experiment was performed to examine if the ants had acquired some dependence on the consumed alkaloid. Each time, fifteen ants were transferred into a small tray (15 cm x 7 cm x 5 cm), the borders of which had been covered with talc and in which laid two tubes (h = 2.5cm, diam. = 0.5cm), one containing sugar water, the other sugar water + either lactose or lactose + nicotine (nicotine being at the concentration 1/1,000), the tubes being plugged with cotton (Figure 1 H). In one of the trays, the tube containing the alkaloid was located on the right; in the other tray, it was located on the left. The ants drinking each liquid food were counted 12 times, and the proportion of ants drinking the solution of lactose, or of lactose + nicotine, was established each time (Table 5, line 2). The obtained values were statistically compared to those expected if ants randomly went drinking each kind of food, using the non parametric goodness of fit χ^2 test (Siegel & Castellan, 1989, 45-51).

3.11 Duration of the Effect of the Alkaloid

Fourteen days after the ants had continuously consumed the solution of nicotine, the liquid food containing this alkaloid was removed from the ants' tray and replaced by a solution of sugar water + lactose, free of alkaloid. This change was made at a given recorded time. After that, the ants' orientation towards an isolated worker's head was assessed after successive given time periods. The results revealed the decrease of the effects, on ants, of the consumed alkaloid (Table 5, line 3; Figure 4). Their statistical significance could be estimated via the non parametric χ^2 test (Siegel & Castellan, 1989, 111-124). This experiment was not performed on ants fed with the solution of lactose since this solution appeared having no effects on the ants.

4. Results

4.1 Effect on Ants' Locomotion

Comparatively with ants consuming a solution of lactose, those consuming a solution of lactose + nicotine walked more quickly (df = 2, χ^2 = 13.98, P < 0.001), and more sinuously (df = 2, χ^2 = 5.9, P = 0.05) (Table 1, line 1). This was obvious while testing the ants (Figure 1 B).

4.2. Effect on Ants' Precision when Reacting

While ants consuming a solution of lactose perfectly oriented themselves towards a source of alarm pheromone, those consuming a solution of lactose + nicotine not well oriented themselves towards such a source (Figure 1 C). This result was statistically significant ((Table 1, line 2; df = 2, χ^2 = 13.61, P ≈ 0.001).

4.3. Effect on Ants' Response to Their Pheromones

Ants consuming a solution of lactose, as well as those consuming a solution of lactose + nicotine, did not move along a line on which no trail pheromone had been deposited (Table 1, line 3, C (control): df = 1, χ^2 = 2.73, 0.05<P<0.10, NS).

Ants consuming a solution of lactose well followed a trail, meanly moving along about 12 angular arcs. Ants consuming a solution of lactose + nicotine not so well followed such a trail, meanly moving along only about 6 angular arcs (Table 1, line 3, T (test)). This results was statistically significant (df = 3, χ^2 = 17.92, P < 0.001).

4.4 Effect on Ants' Audacity

The numbers of ants coming onto a risky apparatus were only slightly, and statistically not significantly, higher after the ants had consumed nicotine for a few days than before they consumed this alkaloid (Table 1, line 4, U = 165, Z = -0.933, P = 0.35). The ants' behavior in front of the apparatus was very similar whatever the presence of nicotine in their sugar food: they hesitated, seldom came onto the apparatus, and soon went away from it.

4.5 Effect on Ants' Consumption of Food

Soon after ants consumed nicotine, the numbers of individuals coming onto the meat food, and eating that food, somewhat decreased. The ants did not stop eating; they went on coming on the provided food; but meticulous observations and counts revealed that their food consumption was in fact reduced (Table 1, line 5; U = 126.5, Z = 1.97, P = 0.05).

4.6 Effect on Ants' Visual and Olfactory Conditioning Capability, and on Their Memory, the Reward Being Free of Nicotine

The ants receiving a solution of lactose, and trained to a green cube set above the meat food, acquired a conditioning score of 70% in 72hrs, and kept that score for more than four days after the cue removal (Table 2, Figue 2). When fed with a solution of lactose + nicotine, the ants, identically trained, reached a score of only 60% and kept nothing of their conditioning (Table 2, Figure 2). The observed differences were statistically significant (acquisition: N = 9-3=6, T = -21, P =0.016; loss: N = 4, T = -10, P = 0.063).

Ants fed with a solution of lactose, and trained to pieces of onion set aside the meat food, reached a conditioning score of 80% in about 55hrs (about two days) and, after the cue removal, kept 5% of their olfactory conditioning (Table 2, Figure 2). After having consumed nicotine, and being submitted to identical conditioning, the ants could never acquire olfactory conditioning (Table 2, Figure 2). This unexpected result was statistically significant (N = 5,

T = -15, P 0.031). Given this failure in acquiring conditioning, the ants' olfactory memory under nicotine consumption was not examined.

The results obtained for ants consuming lactose were very similar to those previously obtained on the same species, in similar circumstances, the ants then receiving only sugar water (Cammaerts et al., 2011). The results obtained on ants consuming lactose + nicotine were unexpected, and might be due to the fact that the reward was the meat food, free of nicotine, the green cube and the pieces of thyme being set above or aside the *T. molitor* larva. It was thus check if ants consuming nicotine could present some cognitive ability requiring no memory, in the course of an experiment using no reward, and then, given the result of that checking experiment, conditioning was again tempted on ants consuming nicotine but setting, this time, the visual and the olfactory cues above or aside the sugar food containing nicotine (see the two following paragraphs).

4.7 Effect on an Ants' Cognitive Ability Requiring no Memory

When set in the small initial loggia of an adequate apparatus (Figure 1 I, Figure 3), ants having never consumed nicotine stayed there a long time and slowly found their way, through the maze, to the large free area. Among the 15 + 15 = 30 ants set in the initial loggia, 22 were still there after 15 min and 17 after 20 min, while only 2 and 6 ants were in the free area after the same time periods (Table 3). Ants having consumed nicotine, tested in an identical experimental apparatus, behaved otherwise: they soon found their way through the maze, towards the free large area. Among the 15 + 15 = 30 ants set in the small initial loggia, only 16 were still there after 15 min and 11 after 20 min, while 4 and 10 ants were in the free area after the same time periods (Table 3). Such a result was statistically significant (initial loggia: N = 8, T = 36, P = 0.004; free area: N = 7, T = 28, P = 0.008). Given these results, conditioning experiments were reproduced, with other cues of course, placing this time the cue near the food containing nicotine. To save time during which ants consumed nicotine, visual conditioning (to a yellow cube) was tempted on nests III and IV while (at the same time as) olfactory conditioning (to pieces of onion) was conducted on nests I and II (see the following paragraph).

Table 3. Effect of nicotine on a cognitive ability requiring no memory.

Experiment	Time (min)	Nest 4 or 5		Nest 1 or 2	
		initial loggia	free area	initial loggia	free area
nest 4 (control) nest 1 (nicotine)	5	11	0	6	1
	10	12	0	9	1
	15	10	2	8	2
	20	8	4	7	3
nest 5 (control) nest 2 (nicotine)	5	13	0	7	5
	10	12	0	7	4
	15	12	0	8	2
	20	9	2	4	7

Ants having never consumed nicotine and ants having done so were tested in the apparatus shown in Figure 1 D, and schematically presented in Figure 3. Under nicotine consumption, ants were more able to escape, through a maze, from the small loggia. Details and statistics are given in the text.

4.8 Effect on Ants' Visual and Olfactory Conditioning Capability, and on Their Memory, the Reward Containing Nicotine

Under nicotine consumption, ants of nests III and IV, trained to a yellow cube, responded correctly to that visual cue with a score of 65%, 70%, 75% and 80% (this last one having been assessed being blind) after 7 hrs, 24 hrs, 30 hrs and 48 hrs respectively, while ants having never consumed this alkaloid presented the scores of 55%, 60%, 60% and 65% after the same time periods (Table 4). So, under nicotine consumption, ants more rapidly acquired visual conditioning (N = 6-1=5, T = +15, P = 0.031; Figure 2, lower graphs, acquisition, circles). However, they reached a nearly similar conditioning score (70%-75% *vs* 65%-70%). After removal of the visual cue, the ants consuming nicotine more rapidly forgot the learned cue, and retained about 5% instead of 20% of their learning (Table 4; N = 5-1=4, T = -10, P = 0.063; Figure 2, lower graphs, loss, circles).

Under nicotine consumption, ants of nests I and II, trained to pieces of onion (Figure 1 G), responded correctly to that olfactory cue with a score of 65%, 75%, 85% and 85% (this last one having been assessed being blind) after 7 hrs, 24 hrs, 30 hrs and 48 hrs respectively, while ants having never consumed this alkaloid presented the scores of 55%, 65%, 70% and 75% after the same time periods (Table 4). So, under nicotine consumption, ants more quickly acquired olfactory conditioning (N = 6-1=5, T = +15, P = 0.031; Figure 2, lower graphs, acquisition, squares). But their final conditioning score was only slightly higher (80%-85% vs 80%). After the olfactory cue removal, the ants under nicotine consumption more rapidly forgot the learned cue and retained only about 2.5% (nearly nothing) of their learning (Table 4; N = 5, T = -15, P = 0.031; Figure 2, lower graphs, loss, squares).

Table 4. Effect of nicotine on conditioning acquisition and on memory, using sugar food (containing nicotine) as a reward

Studied function	no nicotine		+ nicotine		statistics
Visual learning ability	* C: 61/59	50%			
n= 20 + 20 = 40	7 hrs 11/9	55%	7 hrs 13/7	65%	
	24 hrs 12/8	60%	24 hrs 14/6	70%	
	30 hrs 12/8	60%	30 hrs 15/5	75%	P = 0.031
	48 hrs 13/7	65%	48 hrs 16/4	80%	
	55 hrs 13/7	65%	55 hrs 14/6	70%	
	72 hrs 14/6	70%	72 hrs 14/6	70%	
Visual memory	7 hrs 13/7	65%	7 hrs 13/7	65%	
n= 20 + 20 = 40	24 hrs 15/5	75%	24 hrs 12/8	60%	
	30 hrs 14/6	70%	30 hrs 10/10	50%	P = 0.063
	48 hrs 14/6	70%	48 hrs 11/9	55%	
	55 hrs 14/6	70%	55 hrs 11/9	55%	
	72 hrs 14/6	70%			
	79 hrs 14/6	70%			
	96 hrs 14/6	70%			
Olfactory learning ability	* C: 61/59	50%			
n = 20 + 20 = 40	7 hrs 11/9	55%	7 hrs 13/7	65%	
	24 hrs 13/7	65%	24 hrs 15/5	75%	
	30 hrs 14/6	70%	30 hrs 17/3	85%	P = 0.031
	48 hrs 15/5	75%	48 hrs 17/3	85%	
	55 hrs 16/4	80%	55 hrs 17/3	85%	
	72 hrs 16/4	80%	72 hrs 16/4	80%	
Olfactory memory	7 hrs 13/7	65%	7 hrs 11/9	55%	
N= 20 + 20 = 40	24 hrs 12/8	60%	24 hrs 10/10	50%	
	30 hrs 12/8	60%	30 hrs 11/9	55%	P = 0.031
	48 hrs 11/9	55%	48 hrs 10/10	50%	
	55 hrs 11/9	55%	48 hrs 10/10	50%	

Control experiments were those previously made, on ants having never consumed nicotine (column 2). Ants having consumed nicotine were then trained to a yellow hollow cube (visual conditioning) or to pieces of onion (olfactory conditioning), these cues being set, this time, above or aside the sugar water containing nicotine. During tests, made in the course of time, the numbers of ants giving the correct response were assessed, and were compared to the control corresponding ones using the non parametric test of Wilcoxon (Siegel & Castellan, 1989; N, T, P, according to their nomenclature, being given). These results are also shown in Figure 2, lower graphs.

4.9 Ants' Habituation to Nicotine

No habituation was observed, on the ants, to lactose consumption: the ants' locomotion did not change after five days of that sugar consumption (Table 5, line 1; linear speed: df = 2, χ^2 = 3.93, 0.10<P<0.20; angular speed: df = 3, χ^2 = 4.76, 0.10<P<0.20). Even more, no habituation could be detected, on the ants, to nicotine consumption. The ants' locomotion assessment was made being blind, and very similar values were obtained after five days and after one day of nicotine consumption (Table 5, line 1; linear speed: df = 2, χ^2 = 1.18, 0.50<P<0.70; angular speed: df = 2, χ^2 = 1.37, P $\approx$ 0.50). In other words, nicotine soon increased the ants' linear and angular speed, and went on having exactly the same effect after five days of the alkaloid consumption.

Table 5. Habituation to, and dependence on, lactose, and lactose + nicotine; duration of the effects of lactose + nicotine

Effect studied	Variable assessed	Numerical results
Habituation (after 7 days)	Lactose: linear speed mm/sec	19.0 (16.8 – 20.2) *vs* 1 day: NS
	sinuosity ang.deg./cm	91 (84 – 105) *vs* 1 day: NS
	Lactose + nicotine: linear speed mm/sec	20.3 (19.2 – 22.8) *vs* 1 day: NS
	sinuosity ang.deg./cm	125 (106 – 133) *vs* 1 day: NS
Dependence (12 counts)	Lactose:choices of sugar water +lactose *vs* sugar water	nest 1: 26 ants *vs* 32 ants = 26/32 = 44.8%
		nest 2: 19 ants *vs* 21 ants = 19/41 = 46.3%
	Lactose+nicotine: choice of sugar water+lactose+nicotine *vs* sugar water+lactose nests I and II: not hungry	nest 1: 30 ants *vs* 30 ants = 30/60 = 50.0%
		nest 2: 11 ants *vs* 11 ants = 11/22 = 50.0%
	choice of sugar water+lactose+nicotine *vs* sugar water: nest III, hungry	nest 3: 39 ants *vs* 14 ants = 39/53 = 73.6%
	nest IV, fed up	nest 4: 3 ants *vs* 6 ants = 3/9 = 33.3%
	nest III again, no longer hungry	nest 3: 24 ants *vs* 23 ants = 24/47 = 51.1%
	nest II, starved	nest 2: 18 ants *vs* 0 ants = 18/18 = 100%
Duration of effect	Lactose+nicotine: orientation towards an alarm signal (angular degrees) in the course of time	1 hrs: 69.0 5(53.4 – 78.7)
		3 hrs: 60.2 (51.8 – 91.9)
		5 hrs: 59.5 (50.5 – 63.7)
		8 hrs: 61.3 (39.9 – 72.2)
	control: 32.0 (21.6 - 56.59)	11 hrs: 49.5 (34.7 – 76.2)
		15 hrs: 43.4 (31.8 – 61.7)
		25 hrs: 35.6 (29.8 – 46.3)
		29 hrs: 35.8 (26.3 – 38.5)

Details are given in the text. Briefly, ants did not become habituated to nor dependent on lactose consumption; they did not become habituated to nicotine, and did not become dependent on this alkaloid as long as they felt well but developed dependence as soon as they were hungry or starved. The effects of nicotine sigmoidally decreased in the course of time, ending in about 30 hours. The latter observation is graphically shown in Figure 4, and statistical analysis is reported in the 'Results' section.

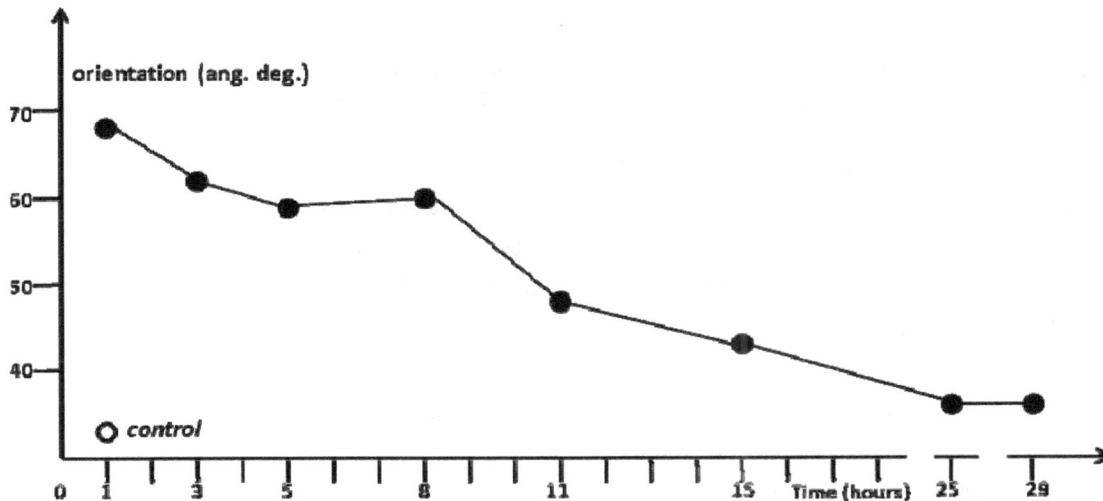

Figure 4. Decrease of the nicotine effects, after its consumption ended. The assessed parameter was the ants' orientation towards an alarm signal, an ethological response affected by the alkaloid consumption. The effects of nicotine first slowly decreased, then, from 8 hrs to 11 hrs after its consumption ended, quickly decreased, and thereafter went on slowly decreasing until a total of about 30 hours. Since the effects of nicotine slowly decreases with time, there is normally no dependence on this alkaloid consumption. But since there is a quick decrease, during a short time, some dependence might develop if, during that time period, there exist(s) any other perturbing factor(s) (f.i. not enough food)

4.10 Ants' Dependence on Nicotine

No dependence could be detected on lactose consumption: meanly about 45% of ants chose this sugar food *vs* food containing no lactose (Table 5, line 2; df = 1, χ^2 = 6.64, P = 0.01 in favor of sugar water without lactose).

No dependence on nicotine consumption could be detected for ants of nest I and II normally fed: meanly 50% of the tested ants chose the food containing nicotine, and 50% that free of the alkaloid (Table 5, line 2). But ants of nest III, tested just before they eat, presented a strong preference for the food containing nicotine: about 73% of ants chose that food (Table 5, line 2; df = 1, χ^2 = 11.79, P < 0.001). On the contrary, ants of nest IV, tested just after they had eaten, did not present such a preference: only about 33% of ants chose the food containing nicotine (df = 1, χ^2 = 1, 0.30<P<0.50). Two checking experiments were then performed. Ants of nest III were fed: being no longer hungry, the ants no longer showed any dependence on nicotine consumption: only 51% of ants chose the food containing the alkaloid (Table 5, line 2; df = 1, χ^2 = 0.02, 0.80<P<0.90). The ants of nest II were starved for 3½ days: under starvation, these ants presented a very strong dependence on nicotine: 100% of the tested ants chose the food containing nicotine (Table 5, line 2; df = 1, χ^2 = 18, P < 0.001). These results lead to wonder about human's nicotine consumption and dependence.

4.11 Time Period During Which Consumed Nicotine Affected the Ants

After ants ended eating nicotine, their precision of reaction – affected by the alkaloid – firstly slowly increased (from 69 to 61 angular degrees) in the course of the first 8 hrs. Let us recall that a high value for an orientation means that the orientation is of poor quality. Then, during the following 7 hrs, the effects of nicotine rather quickly decreased (from 61 to 43 angular degrees). After that, in the course of the 10 following hours, the alkaloid effects went on slowly decreasing, from 43 to 35 angular degrees (Table 5, line 3). The ants response became statistically similar to that exhibited before consuming nicotine between 8 hrs and 11 hrs after they ended consuming the alkaloid (8 hrs: df = 1, χ^2 = 6.66, P = 0.01; 11 hrs and 15 hrs: .df = 1, χ^2 = 2.51, 0.05<P<0.10). After a total of 29 hrs, the ants' response, e.g. their precision of reaction, was very similar to the control one (df = 1, χ^2 = 0.11, 0.70<P<0.80). During this experiment, it was also observed that the ants' behavior (locomotion, food consumption, trail following) became again usual, identical to that of ants having never consumed nicotine, according to the kinetic revealed by the ants' precision of reaction assessment. So, the effects of nicotine appeared to sigmoidally vanish in the course of time, with a first slow decrease, then a short, rather rapid one, and finally again a slow decrease (Figure 4).

5. Discussion

5.1 Synopsis

Nicotine, present in Solanaceae plants, is one of the most consumed alkaloid in the word. All its effects are not entirely researched, understood and/or revealed; even its lethal dose is yet under discussion (Mayer, 2013). Presuming that ants could be used as biological models, we looked for the effects of nicotine using the ant *Myrmica sabuleti*, a very well known species, as a biological model. The experimental work showed that nicotine:

— increases the ants' linear and angular speed, so in fact, the animals' general activity

— reduces the ants' precision of response (i.e. their orientation towards an alarm signal)

— decreases their response to their pheromones (i.e. their trail following behavior)

— does not affect their audacity

— reduces their food consumption

— largely reduces their learning ability if they are rewarded with food free of nicotine

— increases their cognitive ability in navigating through a maze

— increases their rapidness in acquiring visual or olfactory conditioning if they are rewarded with food containing nicotine

— does not valuably increase their visual and olfactory conditioning score

— decreases the time during which they retain a learned cue

— impacts their visual and their olfactory memory

— does not lead to habituation

— does not lead to dependence if the insects feel well, need nothing

— leads to dependence as soon as the individuals feel no longer well or are deprived of something (for instance of food)

— presents a sigmoid decrease of its effects, in about 25 hrs, with a slow, then a short quick, then again a slow decrease of its effects.

5.2 Discussion

Remarks must now be made about some experimental protocols, and then several results will be discussed.

5.2.1. Remarks about Some Experimental Protocols

– a total of 30 experiments were conducted. To do so, we used five large colonies, and performed each experiment on at least 10 to 20 individuals. First, the five used colonies were identical: they were collected the same day, on the same field, and were identically maintained in the same laboratory; they were demographically very similar. Secondly, nicotine somewhat impacted the ants' survival: after a few weeks, the ants appeared tired; they often slept, and were agitated in the mean time. We so minimized the amount of colonies and ants used. Our samples were not very large, though being not too small comparatively with common ethological studies. Of course, according to our sample size and the not Gaussian character of the obtained distributions of values, we used non parametric statistics for analyzing our results.

– pure nicotine could be obtained only mixed to lactose. We so made controls with lactose what allowed confirming previous results (for instance, the kinetics of the ants' acquisition and loss of visual and olfactory conditioning, their positive orthokinesis towards a source of alarm pheromone ...) and showing that lactose has no detectable effects on the animals.

– attention have been taken about the fact that, once animals have been conditioned to a given cue, they are no longer naïve for that cue. We so adequately used different but similar ant colonies or cues.

– we are perfectly conscious that ants are no mammals, and that experiments on ants could only be a first, cheap, easy, quick step of any further pharmacological or medicinal researches.

5.2.2. Discussion of Several Results

– we obtained, on ants, no habituation at all to nicotine consumption. This results was obtained being blind, on 20 ants, and was highly significant. Note that if habituation exists for some drugs, it can easily be reveal by experimenting on ants (i.e. habituation exists for cocaine and was revealed by tests on ants: Cammaerts et al., 2014). This means that, normally, no larger amount but identical amount of nicotine should be consumed for obtaining the same physiological and ethological effects. If larger amounts are consumed, progressively, in the course of time, the cause is not habituation to nicotine; but another one.

– an important result is that ants consuming nicotine never acquired valuable conditioning as long as the provided reward was free of nicotine. When the reward contained nicotine, the ants could acquire conditioning, and they acquired conditioning more quickly than usually. This means that, under nicotine consumption, what is commonly considered as being a reward (f.i. food) is no longer so considered. The only valuable reward is become the nicotine, itself. This result obtained on ants is in agreement with what occurs for human beings: nicotine replaces acetylcholine in the brain, and so induces feeling well-being (see the introduction). Consequently, ants are really good biological models, and individuals consuming nicotine may be deprived from many normally rewarding elements.

– during our experimentation, nicotine appeared to increase individuals' cognitive abilities. The experiment lasted only 20 min. So, in fact, the individuals consuming nicotine solved more quickly the problem than those having never consumed this alkaloid. This also occurred during the conditioning experiment: under nicotine consumption, the individuals did not really reach better score, they only reached the score more rapidly, and, having been quick in making the task, they retained less of what they had learned. On the basis of our experiments, we presume that the usually admitted enhancement of cognition due to nicotine may be punctual, momentarily, and not persistent. In other words, nicotine might momentarily enhance some cognitive abilities, but might, *in fine*, reduce the memorization of acquired knowledge.

– ants, in a normal physiological and ethological state, did not present any dependence on nicotine. But as soon as they no longer were in good state, they immediately developed dependence. Ants seemed thus suitable biological models for pointing out potential dependence. Nicotine may be 'in the nick' of inducing dependence; some abnormality, some need, some privation, and perhaps some compounds, could soon confer to nicotine the faculty of inducing dependence. This result may explain the mixed conclusions of researchers on the subject (Benowitz, 1996).

– nicotine leads to no habituation, and to no dependence, and its effects slowly decease in the course of time after its consumption ends. This confirms a previous observation (Cammaerts et al., 2014): no dependence occurs when there is no habituation to the 'drug', and when its effects slowly decrease with time.

– though decreasing slowly, in about 25 hrs, the effects of nicotine however present a rather quick decrease between 8 hrs and 15 hrs after this alkaloid consumption ended. This rather quick decrease, which lasts about 7 hrs, might be perceived by nicotine consumers, and might induce dependence in individuals exposed, during this short time period, to any other perturbing situations.

– under nicotine consumption, ants' food consumption largely decreased. This is in agreement with the fact that food free of nicotine is no longer considered as a reward, as a beneficial or pleasant element. Such a result is objective since obtained on ants, naïve about the conducted experimentation. People consuming nicotine often eat less. This is in favor of the use of ants as biological models.

5.3 Conclusion

A final conclusion should now be tempted to be advanced. Small amounts of nicotine do not imperil the survival of the individuals, but yet have adverse effects. The alkaloid increases, without usefulness, the agitation, the excitation; it decreases the precision of response, the food consumption, the memory, while apparently (since only momentarily) increasing the cognitive abilities. No habituation occurs. No dependence normally appears except if the individuals do not feel fully well. Nicotine acts in place of rewarding elements, and so deprives the individuals of usual pleasant elements. All these cumulated effects, essentially if large amount of the alkaloid is consumed, finally impact the individuals' health. These physiological and ethological effects, revealed on ants, are in agreement with those actually known for nicotine. Findings on ants are more objective, more precise, brings explanation and even suggest solution (i.e. results about habituation, decrease of the effects, impact on cognition, occurrence of dependence, dependence could be suppressed by feeding the ants). On the basis of our experimentation, we might cautiously advance some advices. Eventually, addition of nicotine in the consumed food may stop the decrease of food consumption; several very rewarding elements may be taken, progressively, instead of nicotine; any need, lack, problem should be solved. Doing so, very probably, nicotine consumption will decrease in the course of time, and dependence on nicotine will vanish. In any way, it can be deduced, from the present work, that ants – i.e. very well known ant species – could be used, in a first step, for studying physiological and ethological effects of compounds, treatments, or factors.

Acknowledgements

We sincerely thank Dr R. Cammaerts who applied the Mann-Whitney U tests for us, and helped us during the most time consuming experiments.

References

Abramson, C. I., Wells, H., & Janko, B. (2007). A social insect model for the study of ethanol induced behavior: the honey bee. In R. Yoshida (Ed.), *Trends in Alcohol Abuse and Alcoholism Research* (pp. 197-218). Nova Sciences Publishers, Inc.

Andre, R. G., Wirtz, R. A., & Das, Y. T. (1989). Insect Models for Biomedical Research. In A. D. Woodhead (Ed.), *Nonmammalian Animal Models for Biomedical Research*. Boca Raton, FL: CRC Press.

Benowitz, N. L. (1996). Pharmacology of Nicotine: Addiction and Therapeutics. *Annual Review of Pharmacology and Toxico*logy, *36*, 597-613. http://dx.doi.org/10.1146/annurev.pa.36.040196.003121

Billen, J., & Morgan, E. D. (1998). Pheromone communication in social insects - sources and secretions. In R. K. Vander Meer, M. D. Breed, K. E., Espelie, & M. L. Winston (Eds.), *Pheromone Communication in Social Insects : Ants, Wasps, Bees, and Termites* (pp. 3-33). Boulder, Oxford: Westview Press.

Cammaerts, M. C. (2007). Colour vision in the ant *Myrmica sabuleti* MEINERT, 1861 (Hymenoptera: Formicidae). *Myrmecological News, 10,* 41-50.

Cammaerts, M. C. (2012). Navigation system of the ant *Myrmica rubra* (Hymenoptera, Formicidae). *Myrmecological News, 16,* 111-121.

Cammaerts, M. C. (2013a). Ants' learning of nest entrance characteristics (Hymenoptera, Formicidae). *Bulletin of Entomological Research, 6.* http://dx.doi.org/10.1017/S0007485313000436

Cammaerts, M. C. (2013b). Learning of trail following behaviour by young *Myrmica rubra* workers (Hymenoptera, Formicidae). *ISRN Entomology.*

Cammaerts, M. C. (2014). Performance of the species-typical alarm response in young workers of the ant *Myrmica sabuleti* is induced by interactions with mature workers. *Journal of Insect Sciences* (in press).

Cammaerts, M. C. (ND). Learning of foraging area specific marking odor by ants (Hymenoptera, Formicidae). *Journal of Entomological Research* (in press).

Cammaerts, M. C., & Cammaerts, D. (2014). Comparative outlook over physiological and ecological characteristics of three closely-related *Myrmica* species. *Biologia, 69* (in press).

Cammaerts, M. C., & Gosset, G. (2014). Impact of age, activity and diet on the conditioning performance in the ant *Myrmica ruginodis* used as a biological model. *International Journal of Biology, 6*(2), 10-20.

Cammaerts, M. C., & Nemeghaire, S. (2012). Why do workers of *Myrmica ruginodis* (Hymenoptera, Formicidae) navigate by relying mainly on their vision? *Bulletin de la Société Royale Belge d'Entomologie, 148,* 42-52.

Cammaerts, M. C., & Rachidi, Z. (2009). Olfactive conditioning and use of visual and odorous elements for movement in the ant *Myrmica sabuleti* (Hymenoptera, Formicidae). *Myrmecological news, 12,* 117-127.

Cammaerts, M. C., Morel, F., Martino, F., & Warzée, N. (2012b). An easy and cheap software-based method to assess two-dimensional trajectories parameters. *Belgian Journal of Zoology, 142,* 145-151.

Cammaerts, M. C., Rachidi, Z., & Cammaerts, D. (2011). Collective operant conditioning and circadian rhythms in the ant *Myrmica sabuleti* (Hymenoptera, Formicidae). *Bulletin de la Société Royale Belge d'Entomologie, 147,* 142-154.

Cammaerts, M. C., Rachidi, Z., & Gosset, G. (2014). Physiological and ethological effects of caffeine, theophylline, cocaine and atropine; study using the ant *Myrmica sabuleti* (Hymenoptera, Formicidae) as a biological model. *International Journal of Biology, 3,* 64-84.

Cammaerts, M. C., Rachidi, Z., Beke, S., & Essaadi, Y. (2012a). Use of olfactory and visual cues for traveling by the ant *Myrmica ruginodis* (Hymenoptera, Formicidae). *Myrmecological News, 16,* 45-55.

Cammaerts, M. C., Rachidi, Z., Bellens, F., & De Doncker, P. (2013). Food collection and response to pheromones in an ant species exposed to electromagnetic radiation. *Electromagnetic biology and medicine, 32*(3), 315-332. http://dx.doi.org/10.3109/15368378.2012.712877

Cammaerts-Tricot, M. C. (1973). Phéromone agrégeant les ouvrières de *Myrmica rubra. Journal of Insect Physiology, 19*, 1299-1315.

Deutsch, J. (1994). *La drosophile: des chromosomes aux molécules.* John Libbey Eurotext.

Devineni, A. V., & Heberlein, U. (2013). The evolution of *Drosophila melanogaster* as a model for alcohol addiction. *Annual Review of Neurosciences, 36*, 121-138. http://dx.doi.org/10.1146/annurev-neuro-062012-1 70256

Døring, T. D., & Chittka, L. (2011). How human are insects and does in matter? *Formosan Entomologist, 31*, 85-99.

Frenk, H., & Dar, R. (2005). *Dépendance à la nicotine: critique d'une théorie.* préface du Pr Robert Molimard, Paris, Belles Lettres, traduction de *A critique of nicotine addiction*, Kluwer, 2000.

Hölldobler, B., & Wilson, E.O.. (1990). *The ants.* Harvard University Press, Springer-Verlag Berlin.

Keller, L., & Gordon, E. (2006). *La vie des fourmis.* Odile Jacob, Paris.

Keller, R. A. (2011a). A phylogenetic analysis of ant morphology (Hymenoptera: Formicidae) with special reference to the Poneromorph subfamilies. *Bulletin of the American Museum of Natural History, 355*, 99.

Keller, R. A. (2011b). Ants protect their mouthparts by locking them in place. *Colloque organized at Banyuls/mer.*

Kolb, B., & Whishaw, I. Q. (2002). *Neuroscience & cognition: cerveau et comportement.* New York, Basing Stoke: Worth Publishers.

Mayer, B. (2013). How much nicotine kills a human? Tracing back the generally accepted lethal dose to dubious self-experiments in the nineteenth century [archive], *Archives of Toxicology.* http://dx.doi.org/10.1007/s0020 4-013-1127-0

Passera, L. (2006). *La véritable histoire des fourmis.* Librairie Fayard.

Passera, L., & Aron, S. (2005). *Les fourmis: comportement, organisation sociale et évolution.* Les Presses Scientifiques du CNRC, Ottawa, Canada.

Russell, W. M. S., & Burch, R. L. (2014). *The Principles of Humane Experimental Technique.* Johns Hopkins University

Siegel, S., & Castellan, N. J. (1989). *Nonparametric statistics for the behavioural sciences.* McGraw-Hill Book Company, Singapore.

Waldum, H. L., Nilsen, O. G., Nilsen, T., Rørvik, H., Syversen, V., Sanvik, A. K., ... Brenna, E. (1996). Long-term effects of inhaled nicotine. *Life Science, 58,* 1339-1346. http://dx.doi.org/10.1016/0024-3205 (96)00100-2.

Wehner, R., & Gehring, W. (1999). *Biologie et physiologie animales.* De Boek Université, Thieme Berlag, Paris, Bruxelles.

12

Brain Reorganization Allowed for the Development of Human Language: Lunate Sulcus

Kwang Hyun Ko[1]

[1] Hanyang University, Korea
Correspondence: Kwang Hyun Ko, Hanyang University, Korea.

Abstract

This article presents the hypothesis of a connection between eidetic memory and the lunate sulcus, a feature that was repositioned during evolution. Humans have evolved from ape-like ancestors for 7 million years. Along with a prominent increase in brain size, the reorganization of the brain marked by the sulcus generated the evolutionary momentum toward the development of human language. This article reviews the reorganization of the human brain using an interdisciplinary approach of examining animal behavior and anthropological and biological studies. This brain reorganization must have occurred during early maturity and is thought to be responsible for eidetic imagery in some adolescents and superior short-term memory in chimpanzees. During early development, the neural connections in prefrontal cortex and posterior parietal lobe rapidly expand, while visual memory capacity of human brain would become limited. Biological studies have demonstrated that the lunate sulcus is subject to white matter growth, and dental fossil and tomography studies have shown that the brain organization of Africanus is pongid-like.

Keywords: eidetic memory, lunate sulcus, Einstein, chimps, gray matter

1. Introduction

The principles that govern the evolution of brain structure are controversial. The timeline of hominin evolution spans approximately 7 million years, and the re-arrangement of the brain during human evolution is thought to have been more organizational than volumetric (Balter, 2007). Brain volume remained relatively stable, whereas particular milestone changes in the position of surface anatomical features, such as the simian sulcus, suggest that the brain underwent internal reorganization (Bruner, 2014). It is argued that the evolutionary expansion of the frontal areas of the lunate sulcus would have caused a shift in the location of the fissure (Rincon, 2004).

2. Animal Behavior Studies

In set of short-term memory tasks, such as memorizing the sequential order of numerals and recalling them, the performance of young chimps was shown to exceed that of human adults (Matsuzawa et al., 2007). Chimpanzees are apparently superior to humans at achieving instantaneous memorization. Matsuzawa indicates an evolutionary tradeoff between the eidetic/photographic memory of these chimps and the higher cognitive abilities of humans, such as the advanced capability for complex language (Choi, 2007). In this regard, the simian sulcus, also known as the lunate sulcus, is an anatomical fissure between the temporal and occipital lobes that is found in primates (Allen et al., 2006). Intriguingly, the sulcus lunatus lies in the back of human brains and has a more frontal location in chimpanzees.

Moreover, chimpanzees at the California Institute of Technology played a hide-and-seek computer game and bested African villagers and Japanese undergraduate students. The computer program was based on game theory and examined the abilities of individuals to predict their opponent's move (Martin et al., 2014). The chimps were more successful than humans at this game because they have strong short-term memory and talents for pattern analysis.

The human brain has greater capacity than the chimp brain. Humans have the largest and most complex brains of any living primate. The brain capacity of the modern human is three times that of apes, with the human brain having a larger structure and more neurons (Lieberman, 2011); however, increased brain size in humans cannot solely explain the superior cognitive ability of chimpanzees described above. The stronger short-term memory demonstrated in chimps compared to humans must be due to structural differences in the brains of these two species. Holloway stated the importance of the reorganization of the human brain during evolution, referring to the posterior location of the lunate sulcus (Balter, 2007).

3. Child Behavior Studies

It has been long been thought that adults surpass their younger childhood selves in every intellectual, neurological, and biological measure. Surprisingly, researchers have observed eidetic imagery in children 6 to 12 years old, while photographic memory abilities have generally not been documented in adults (Searleman, 2007). This particular skill fades as children mature. Interestingly, no associations were found between the ability to generate photographic memories and any cognitive or emotional level.

In a similar context, several studies revealed that six-year-olds were better than adults at distinguishing discrete information (Nardini et al., 2010). In fact, adults performed worse at answering whether the angle of a particular disc was the same as that of other discs. That study demonstrated that adults combine multiple pieces of visual data into single pieces of data, while kids perceive visual data separately. This integrative capability obtained in adulthood has a cost. Vladimir Sloutsky of Ohio State University states that people categorize information as they become mature and intelligent, resulting in lower accuracy of memories of individual characteristics (Wahl, 2013).

In another psychology experiment, test groups consisting of adults and children as young as 5 years old were shown pictures of animals. After scientists provided a dummy task of looking for "beta cells" in the animal pictures, the subjects were again shown 28 more images and were asked if they had seen the image before. To the researcher's surprise, the young participants performed better than the adults. These results were attributed to visual confusion caused by integrative processing in adults and also provide and explanation for the remarkable eidetic memories of some children (Sloutsky & Fisher, 2004).

The size of the brain size and the number of neurons it contains increase as children enter adulthood (Blakemore, 2012). Nevertheless, the ability to identify separate information and eidetic memory are more prominent in children than adults. As previously mentioned, chimps have also been shown to have superior performance over adult humans in short-term memory tests. Eidetic memory and individual identification abilities are lost as children enter adulthood and the brain grows.

I hypothesize that evolutionary pressures resulted in the human brain undergoing internal reorganization to develop the capability of human language. Furthermore, this reorganization must have been implemented during early maturity and is likely responsible for eidetic imagery in some adolescents. Specific details of rearrangement of different brain structures, such as white matter, the lunate sulcus, and the prefrontal cortex, will be mentioned in the following descriptions of biological studies.

4. Biological Studies

Einstein died on 17 April 1955 and made an elegantly famous statement at the time of his death, "I have done my share; it is time to go" (Cohen & Graver, 1990). This great 20th century physicist donated his brain for medical use. For decades, the brain of this distinguished scientist has attracted avid researchers and curious audiences, as it was the first historical brain to be dissected, and its dissection generated an astounding result. The brain autopsy suggested that regions involving speech and language are small compared to areas involved in numerical and spatial processing. Other studies have supported Einstein's claim that he thought visually rather than verbally. The lateral sulcus (Sylvian fissure) was absent in Einstein's brain, and importantly, the lunate sulcus was not prominent in Einstein's brain (Falk et al., 2013).

4.1 Explaining the Sulcus

Sulci are depressions or grooves in the cerebral cortex; they surround a gyrus and create the characteristic wrinkled appearance of the brain in humans and other mammals (Carlson, 2009). They are relatively shallow grooves that surround a gyrus compared to bigger grooves called fissures that divide the brain into lobes. The gyri, fissures, and sulci generate a larger surface area of the human brain and other mammalian brains. Furthermore, the folded structure of the brains of humans and some intelligent mammals generates a larger surface area, which allows for greater cognitive power.

It has been stated that the development of the sulcus varies greatly between individuals; furthermore, the structure of the sulcus varies greatly with age. However, the factors that determine the iconic shape of the gyri and sulci in the human brain are not entirely clear.

4.2 What may be the Reason

Researchers that refuted traditional models of sulci development in the human brain, described different growth rates of gray and white matter as key to understanding the development of sulci. Comparatively faster growth of gray matter (top layer) would pin down the white matter to mold the shape of the cortex (Tallinen et al., 2014). Surprisingly, human infants, unlike baby chimps, experience a developmental stage during which white matter that establishes neural connections in the brain shows dramatic expansion, and this causes reorganization of sulci, in this case, the lunate sulcus.

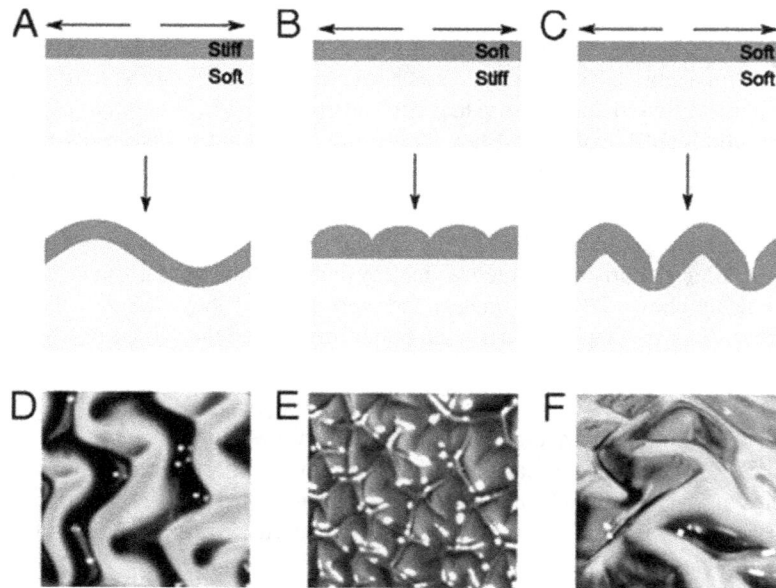

Figure 1. Sulcification in a layered material based on differential growth of white matter and gray matter. The top layer stands for gray matter, and the bottom represents white matter.

Source: Tallinen, Chung, Biggins, and Mahadevan (2014).

4.3 Gray/White Matter

The gray and white matters of the brain have important functions. The gray matter is in charge of cognition and transports nutrients and energy to the brain. Conversely, the white matter orchestrates action potentials and coordinates, or communicates, between brain regions (Fields, 2008). The sulcus lunatus is an anatomical fissure between the temporal and occipital lobes that is also observed in chimpanzee brains, but in a more anterior location (Falk, 1983). In chimps, the fissure is marked by a smaller amount of white matter growth, allowing for freer connections between the two lobes and allowing for the storage of visual memories. The dramatic expansion of the prefrontal lobe and posterior lobe that developed during the evolution of humans and occurs during a child's growth, resulted in a more posterior location of the sulcus.

Although brain volume increases during child growth, gray matter wanes through a pruning process. Fast growing white matter replaces the waning total volume of gray matter, allowing interactions between different parts of the brain. MRI comparison studies have shown that brain regions in children have short, localized connections that become longer in adults (Petersen et al., 2009). The process of replacement of gray matter in adolescence is called synapse-pruning and is described in the vernacular simply as, 'use it or lose it.' In humans, the temporal lobe forms longer connections to the posterior parietal lobe (Wernicke's area) and prefrontal cortex (Broca's area), areas essential to language-learning and overall cognition, rather than to the occipital lobe, which lies adjacent to the temporal lobe. More surprisingly, memory studies in hunter-gatherer societies indicated that pre-literate tribal groups excel in visual memory (Madden et al., 2006). An aboriginal forage society without a writing system would rely more on eidetic memories for information processing and, thus, survival. In contrast, in contemporary societies, eidetic memory becomes obsolete and children's brains are devoted the learning of language, numbers, and social cognition. As a direct result of synapse-pruning in humans living in modern societies, the specific location of the sulcus in humans, which is set by explosive growth of white matter, results in limited connections between the temporal and occipital lobes.

Chimpanzee   Egyptian   Egyptian

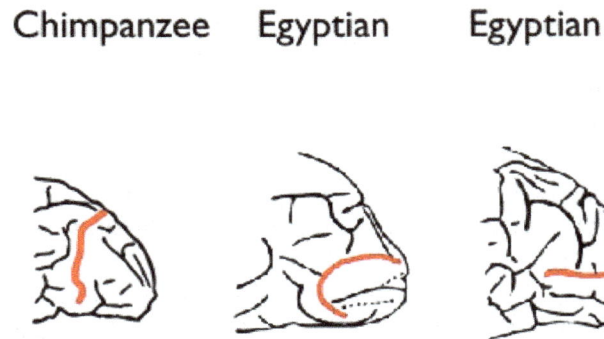

Figure 2. Specifically, fast growing white matter in adolescence was the key change in the evolution of intelligence. This molds the brain making it capable of language while comparatively allowing for fewer connections between the prefrontal cortex and occipital lobe and, thus, resulting in a less active occipital lobe. The growth of white matter also accounts for sulcus variation with age

Source: Falk (2014).

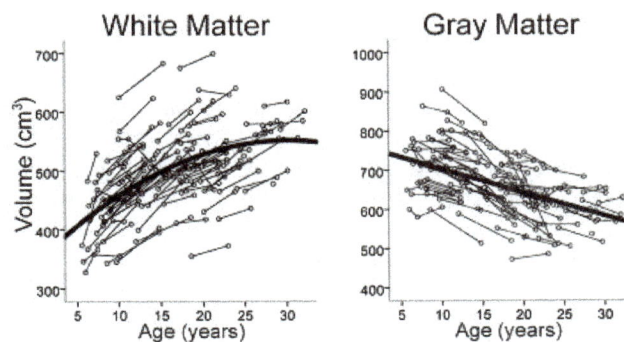

Figure 3. Gray matter decrease offsets white matter increase in brain

Source: Lebel and Beaulieu (2011).

5. Conclusion

Superior memory abilities have been demonstrated in children and chimps whose brain capacities and number of neurons are far less than those of the adult human brain. Hominin brain size has dramatically increased over millions of years, and studies have confirmed that evolutionary changes in the brain occur during child development. Active development of particular brain regions in the posterior parietal cortex (sense of self), prefrontal cortex (social cognition), and temporal lobe (language interpretation) occur in the early years of life (Gogtay, 2004). Fast growth of white matter in specific areas replaces localized regional linkages of brain sections and is accompanied by a shift of the sulcus and the loss of the superb short-term memory observed in chimps and human children.

Many animals besides chimps depend on eidetic memories for survival. For example, eastern gray squirrels excel at remembering landmarks and retrieving food from caches. However, our ancestors eventually did not require photographic memories. Evolutionary changes, intricate hunting tools, and the use of fire, shelter and recordkeeping gradually enabled satisfactory survival independent of short-term visual memories. Humans have built rich, protected shelters and do not need to pinpoint the exact location of predators or food.

5.1 Answering the Lunate Sulcus

The debate about the interpretation of a depression in australopithecine between physical anthropologists Holloway and Falk is ongoing. Holloway argues that sulcal patterns in australopithecine indicate cerebral organization more like that in modern hominins, while Falk insists that the lunate sulcus is in a position that indicates an ape-like pattern (Falk, 1987). The argument extends from the AL 162-28 endocast to all australopithecine fossils. Holloway states the presence of present-day hominid sulcal features, while Falk maintains that the features are pongid in nature.

In a recent tomography study of the Taung Australopithecus africanus fossil, Carlson and colleagues failed to find any signs of human-infant skull features (Holloway et al., 2014). Previous studies suggested that features of the Taung specimen would allow the child's brain to grow rapidly, similar to brain growth in modern home sapiens. Nonetheless, new brain CT scans of the Taung Australopithecus africanus fossil have shown that it lacks these features.

The timing of the evolution of dramatic changes in the hominin brain has been indicated by dental fossil studies. Researchers, including Christopher Dean, analyzed growth patterns in the enamel of fossil teeth recovered from a wide array of early human ancestors (Dean et al., 2001).

The scientists did expect that Homo erectus— indicative by its name "upright man," the first human ancestor to show modern human-like characteristics in terms of body proportions, weight and jaws --would show evidence of a modern human-like postnatal growth period.

However, the results proved otherwise and suggest that the prolonged interval of childhood may have occurred in the recent development of the large human brain—a period of growth that is considered a key event in human evolution by increasing the time available for learning.

Overall, studies have indicated that the hominin brain growth observed in modern humans may not have evolved until recently in the course of evolution, and this confirms Falk's hypothesis that australopithecine fossils demonstrate a pongid pattern rather than a modern hominin pattern.

5.2 Further Biological Implications of Human Language Apart from Lunate Sulcus

With regards to further biological implications of human language, the posterior parietal cortex (the region of the parietal neocortex posterior to the primary somatosensory region) contains cortical fields that determine 'the sense of self' and plan movements by coding the locations of objects both within and outside of the body frame (Krubitzer & Disbrow, 2010). Increasing the 'sense of will' in the posterior parietal lobe naturally embedded meaning and intention to language while creating structures to aid in the representation of language. The principal activity of the prefrontal cortex is thought to be the management of social cognition, which is the performance of executive functions in accordance with goals (Miller et al., 2002). Social cognition, therefore, has given humans talent for analyzing social context and situations. The temporal lobe works to interpret language, emotions, and memory (Smith, 2007) and serves to store sounds and meanings of language for possible interpretation.

Certain diseases have also helped clarify the functions of different brain regions. Broca's aphasia (prefrontal cortex damage) is a condition under which individuals know what they want to say but cannot speak the words (Purves, 2008). These patients can generally comprehend words and simple sentences but are incapable of producing fluent speech. Other problems associated with Broca's aphasia include articulating, finding, and repeating words and difficulty understanding sentences with complex grammatical structures (Friedmann et al., 2006). Specifically, patients have trouble expressing their 'free will' due to a lack of social cognitive abilities.

However, damage to posterior parietal cortex (Wernicke's area) results in fluent yet incoherent speech. In other words, the particular individual with aphasia will be able to fluently articulate words, but the spoken sentences will be meaningless (Manasco, 2014). That is, the language is spoken with social cognition but without intention.

References

Allen, J. S., Bruss, J., & Damasio, H. (2006). Looking for the Lunate Sulcus: A Magnetic Resonance Imaging Study in Modern Humans. *The Anatomical Record Part A: Discoveries in Molecular, Cellular, and Evolutionary Biology, 288A*, 867-876. http://dx.doi.org/10.1002/ar.a.20362

Balter, M. (2007). In study of human brain evolution, zeal and bitter debate. Profile of scientist at work, Ralph Holloway. *New York Times, 27*, F2. Retrieved from http://www.nytimes.com/2007/11/27/science/27prof.html?_r=2&

Blakemore, S. (2012). Imaging brain development: The adolescent brain. *NeuroImage, 61*(2), 397-406. http://dx.doi.org/10.1016/j.neuroimage.2011.11.080

Bruner, E. (2014). *Human paleoneurology*. Springer.

Campbell, I. G., & Feinberg, I. (2009). Longitudinal trajectories of non-rapid eye movement delta and theta EEG as indicators of adolescent brain maturation. *Proceedings of the National Academy of Sciences, 106*(13), 5177-5180. http://dx.doi.org/10.1073/pnas.0812947106

Carlson, N. R. (2009). Physiology of behavior. Upper Saddle River, NJ: Pearson Education.

Choi, C. Q. (2007). Chimps Do Numbers Better than Humans. Retrieved December 17, 2014. http://www.livescience.com/7444-chimps-numbers-humans.html

Cohen, J. R., & Graver, L. M. (1990). The ruptured abdominal aortic aneurysm of Albert Einstein. *Surg Gynecol Obstet, 170*, 455.

Dean, C., Leakey, M. G., Reid, D., Schrenk, F., Schwartz, G. T., Stringer, C., & Walker, A. (2001). Growth processes in teeth distinguish modern humans from Homo erectus and earlier hominins. *Nature, 414*(6864), 628-631. http://dx.doi.org/10.1038/414628a

Fair, D. A., Cohen, A. L., Power, J. D., Dosenbach, N. U., Church, J. A., Miezin, F. M., ... Petersen, S. E. (2009). Functional Brain Networks Develop from a "Local to Distributed" Organization (O. Sporns, Ed.). *PLoS Computational Biology, 5*(5), E1000381. http://dx.doi.org/10.1371/journal.pcbi.1000381

Falk, D. (1983). The Taung endocast: A reply to Holloway. *American Journal of Physical Anthropology, 60*(4), 479-489. http://dx.doi.org/10.1002/ajpa.1330600410

Falk, D. (1987). Hominid Paleoneurology. *Annual Review of Anthropology, 16*(1), 13-30. http://dx.doi.org/10.1146/annurev.anthro.16.1.13

Falk, D. (2014). Interpreting sulci on hominin endocasts: Old hypotheses and new findings. *Frontiers in Human Neuroscience, 8*. http://dx.doi.org/10.3389/fnhum.2014.00134

Falk, D., Lepore, F. E., & Noe, A. (2013). The cerebral cortex of Albert Einstein: A description and preliminary analysis of unpublished photographs. *Brain, 136*(4), 1304-1327. http://dx.doi.org/10.1093/brain/aws295

Fields, R. D. (2008). White Matter Matters. *Scientific American, 298*(3), 54-61. http://dx.doi.org/10.1038/scientificamerican0308-54

Fields, R. D. (2009). The other brain: From dementia to schizophrenia, how new discoveries about the brain are revolutionizing medicine and science. New York: Simon & Schuster.

Friedmann, N., Gvion, A., & Novogrodsky, R. (2006). Syntactic movement in agrammatism and S-SLI: Two different impairments. *Language acquisition and development*, 205-218.

Gogtay, N. (2004). From the Cover: Dynamic mapping of human cortical development during childhood through early adulthood. *Proceedings of the National Academy of Sciences, 101*(21), 8174-8179. http://dx.doi.org/10.1073/pnass0402680101

Holloway, R. L., D. C. Broadfield, and K. J. Carlson. (2014) New high-resolution computed tomography data of the Taung partial cranium and endocast and their bearing on metopism and hominin brain evolution. *Proceedings of the National Academy of Sciences, 111*(36), 13022-13027.

Inoue, S., & Matsuzawa, T. (2007). Working Memory of Numerals in Chimpanzees. *Current Biology, 17*, R1004-R1005. http://dx.doi.org/10.1016/j.cub.2007.10.027

Krubitzer, L., & Disbrow, E. (2008) The Evolution of Parietal Areas Involved in Hand Use in Primates. In J. Kaas, & E. Gardner (Eds.), *The Senses: A Comprehensive Reference* (Volume 6, Somatosensation, pp. 183-214). London: Elsevier.

Lebel, C., & Beaulieu, C. (2011). Longitudinal development of human brain wiring continues from childhood into adulthood. *The Journal of Neuroscience, 31*(30), 10937-10947.

Lieberman, D. (2011). *The evolution of the human head.* Cambridge, MA: Belknap Press of Harvard University Press.

Madden, A. D., Bryson, J., & Palimi, J. (2006). Information behavior in pre-literate societies. In *New directions in human information behavior* (pp. 33-53). Springer Netherlands.

Manasco, H. (2014). *Introduction to neurogenic communication disorders.* Burlington, MA: Jones & Bartlett Learning.

Martin, C. F., Bhui, R., Bossaerts, P., Matsuzawa, T., & Camerer, C. (2014). Chimpanzee choice rates in competitive games match equilibrium game theory predictions. *Scientific reports, 4*, 1-6.

Miller, E. K., Freedman, D. J., & Wallis, J. D. (2002). The Prefrontal Cortex: Categories, Concepts and Cognition. Philosophical Transactions of the Royal Society of London. *Series B, Biological Sciences, 357*, 1123-1136. http://dx.doi.org/10.1098/rstb.2002.1099

Nardini, M., Bedford, R., & Mareschal, D. (2010). Fusion of visual cues is not mandatory in children. *Proceedings of the National Academy of Sciences, 107*(39), 17041-17046. http://dx.doi.org/10.1073/pnas. 1001699107

Purves, D. (2008). Neuroscience. Sunderland, MA: Sinauer.

Rincon, P. (2004). *Human brain began evolving early.* Retrieved from http://news.bbc.co.uk/2/hi/science/nature/ 3496549.stm

Searleman, A. (2007). Is there such a thing as a photographic memory? And if so, can it be learned?. *Scientific American(Nature America).* Retrieved 10 July 2013.

Sloutsky, V. M., & Fisher, A. V. (2004). When learning and development decrease memory: Evidence against category-based induction. *Psychological Science, 15*, 553-558.

Smith, K. (2007). *Cognitive Psychology: Mind and Brain* (pp. 21, 194-199, 349). Upper Saddle River, NJ: Prentice Hall.

Tallinen, T., Chung, J. Y., Biggins, J. S., & Mahadevan, L. (2014). Gyrification from constrained cortical expansion. *Proceedings of the National Academy of Sciences, 111*(35), 12667-12672.

Wahl, E. (2013). *Unthink: Rediscover Your Creative Genius.* Crown Business.

13

Leptin Gene Expression in Rabbits During Pregnancy and Fetal Life

Doaa Kirat[1], Nora E. Abdel Hamid[1], Wafaa E. Mohamed[1], Mohamed Hamada[1] & Shimaa I. Shalaby[1]

[1] Department of Physiology, Faculty of Veterinary Medicine, Zagazig University, Zagazig, Egypt

Correspondence: Doaa Kirat, Department of Physiology, Faculty of Veterinary Medicine, Zagazig University, Zagazig, Egypt. E-mail: doaakirat@hotmail.com

Abstract

Leptin may act as the critical link between adipose tissue and reproduction. Although considerable progress has been achieved in understanding the reproductive actions of leptin, much work is needed for understanding its physiological role. Till now, no data has been published about the distribution and expression levels of leptin in the rabbit maternal adipose tissue, placenta, and various fetal rabbit tissues during pregnancy and postpartum. Our results indicated that circulating leptin levels in rabbit serum during pregnancy were significantly higher than in postpartum and non pregnant rabbits. Furthermore, leptin showed positive correlations with body weight and estrogen and negative correlation with progesterone in pregnant and postpartum rabbits. RT-PCR verified the presence of RNAs encoding leptin in the rabbit maternal perirenal adipose tissue, placenta, and several fetal tissues; including brain, liver, adipose tissue, and bone. The relative abundance of leptin RNA in rabbit maternal adipose was significantly higher at 20th day of pregnancy than that of non pregnant rabbits, while it was significantly decreased at 2nd day after parturition. No significant changes in the placental leptin RNA levels were noticed in pregnant rabbits at 10th, 20th, and 30th day of pregnancy. The relative abundance of fetal leptin transcripts at day 30th of pregnancy was in the order of liver> bone ≥ adipose tissue > brain. The present study provides new evidence for the distribution and expression levels of leptin in the rabbit maternal and fetal tissues during pregnancy and supports the importance of leptin in reproductive physiology and fetal development.

Keywords: leptin, steroids, pregnancy, fetus, rabbit

1. Introduction

Leptin is a 16-kDa cytokine encoded by the obese gene and primarily secreted by adipocytes (Zhang et al., 1994). Leptin, upon binding to specific receptors in different areas of the hypothalamus, is involved in the control of body weight through its effects on food intake and energy expenditure by negative feedback at the hypothalamic nuclei (Zhang et al., 1994). Leptin plays critical actions in the development (Hoggard et al., 1997) and haemopoiesis (Gainsford et al., 1996). Giving that leptin is a pleiotropic and ubiquitous molecule; it has been implicated in several key points of the mammalian reproductive functions such as ovulation (Cunningham et al., 1999), and pregnancy (Mounzih et al., 1998). In particular, leptin may act as the critical link between adipose tissue and the reproductive system, indicating whether adequate energy reserves are present for normal reproductive function. Although considerable progress has been achieved in understanding the reproductive actions of leptin over the past few years, much work is needed before we can arrive at a complete understanding of its physiological role, if any.

In fact, leptin has been identified in the adult human (Butte et al., 1997) and rat (Kawai et al., 1997), mouse (Tomimatsu et al., 1997).

In rabbits, leptin receptor (Ob-R) has been detected in endometrial cells (Gonzalez & Leavis, 2003) and the ovary at day 9 of pseudo-pregnancy by immunohistochemistry and Western blot analysis that supports a permissive luteolytic role for leptin in rabbit Corpus luteum (Zerani et al., 2004). Recently, Koch et al. (2013) identified the leptin gene cloning and expression in the rabbit mammary epithelial cells during pregnancy and lactation. Up till now, no data has been published concerning the expression and distribution of leptin in the rabbit adipose tissue and placenta as well as the different fetal rabbit tissues.

Pregnancy is a hyper metabolic state in which a great increase in maternal body fat and weight occurs and associated with relevant endocrine changes. In addition, energy needs are increased during pregnancy. These increased energy needs may be met through partitioning of nutrients for energy utilization which is under hormonal control. Therefore, in order to verify to what degree pregnancy influences leptin in rabbits, we intended to measure the leptin hormone in the blood and estimate its in vivo expression level in the maternal adipose tissue and placenta at different stages of pregnancy in rabbits. Moreover, to assess whether leptin is involved in the rabbit fetal growth and development, we aimed to investigate the expression level and distribution of RNAs encoding leptin in the 10-day post-coitus rabbit fetus and in various fetal tissues using reverse transcription-PCR.

Moreover, to ascertain the relationship between serum leptin levels and related variables (body weight and steroids in rabbits), the present study also aimed to gain a comprehensive view of leptin correlations with body weight, estrogen, and progesterone in pregnant and non pregnant control rabbits.

2. Materials and Methods

2.1 Animals and Experimental Designs

A total of 30 sexually mature healthy New Zealand White female rabbits weigh between 2600-3200 g and of ~ 6 months old were used. Rabbits were housed in individual stainless steel cages provided with automatic drinkers. All rabbits were raised on balanced diet (20% protein, 2.62% fat, 11% fiber, and 2600 kcal|kg energy). Rabbits were housed in an animal room which maintained on a 12-h light-dark cycle at an ambient temperature of 25 ± 2 °C and 35-50% humidity for two weeks before starting the experiment for adaptation.

After quarantine period, rabbits were distributed into two groups; the control group (6 rabbits) which represent the non pregnant rabbits and the experimental group (24 rabbits) that were introduced into the male's cages. The mated ratio is identically one male one female (for mating on a 1:1 basis). The female was considered mated when the sperms were found in the vaginal washing and that day was designed the first day of pregnancy. It was expected that a female would give birth 30 days following an observed copulatory episode. These mated rabbits were divided equally into 4 groups (6 rabbits each) which designated at pregnancy day 10, 20, 30, and after parturition. The experimental protocol used in the present study was approved by the Laboratory Animal Control Guidelines of Faculty of Veterinary Medicine, Zagazig University.

2.2 Blood and Tissue Samples

Blood samples were collected during sacrifying the rabbits in each group into sterile test tubes without anticoagulant. For serum collection, samples were kept at room temperature for 30 min, and then centrifuged at 3000 rpm for 15 min. The serum samples were stored at -20 °C until used for hormone analysis.

After scarifying on days 10, 20, and 30 of pregnancy as well as days 2 and 7 after parturition, tissue samples from the maternal perirenal adipose tissue and placenta as well as different fetal tissues such as (brain, liver, bone, and perirenal adipose tissue) were immediately collected from the rabbits and washed in ice-cold 0.9% (w/v) NaCl. These tissue samples were immediately snap-frozen in liquid nitrogen and subsequently stored at -80°C until use for RT-PCR analysis.

2.3 Hormonal Analysis

Serum leptin was measured using DRG Leptin ELISA kit (DRG Instruments GbH, Germany) according to the manufacturer protocol.

Serum estradiol II (E_2) level was determined using electrochemiluminescence immunoassay (ECLIA) kit (Catalog Number: 03000079190, Roche Diagnostics, North America) and measured by cobas e 411 immunoassay analyzer according to the manufacturer instructions.

Serum progesterone level was determined using electrochemiluminescence immunoassay (ECLIA) kit (Catalog Number: 12145383160, Roche Diagnostics, North America) and measured by cobas e 411 analyzer according to the manufacturer instructions.

2.4 Molecular Analysis

Leptin gene expression was determined in the female and fetal rabbit's tissues by using the semi-quantitative RT-PCR.

2.4.1 Total RNA Isolation

RNA was isolated from the tissue samples by using Easy-RED™ Total RNA Extraction Kit (iNtRON Cat. No. 17063) following the manufacturer instructions.

2.4.2 Reverse Transcription

Total RNA (5 ug per sample) was reverse transcribed into cDNA using Revert Aid h minus reverse transcriptase (Fermentas, #EP0451, Thermofisher Scientific, European Union) and a mix of oligo (dT) (0.5 ug/reaction, 4 ul 5x reaction buffer, 0.5 ug RNase inhibitor, and 2 ul deoxynucleotide triphosphate (dNTP) mix) in a 12.5 ul total reaction volume at 42°C for 60 min. To terminate the reaction, the tubes were heated at 70°C for 10 min and then stored at -80°C.

2.4.3 Polymerase Chain Reaction (PCR)

The specific oligonucleotide primers used for PCR to amplify the leptin gene were designed using primer 3.0 software (http://www-genome.wi.mit.edu/cgi-bin/primer/primer3_www.cgi) based on the published leptin sequences of rabbits (GenBank accession no. 001163069.1) (Zhao & Wu, 2005). The forward primer was 5'-GTCGTCGGTTTGGACTTCATC-3' and the reverse primer was 5'-CGGAGGTTCTCCAGGTCGTTG-3'. As a control, PCR amplification with a pair of GAPDH primers based on published sequences with GenBank accession number 001082253.1 (The forward primer 5'-GGAGCCAAAAGGGTCATC-3' and reverse primer 5'-CCAGTGAGTTTCCCGTTC-3') were used. The oligonucleotide primers were synthesized by Metabion International AG, Planegg/Martinsried, Germany.

The isolated cDNA were amplified using Maximo Taq DNA Polymerase 2X-preMix kit (GeneOn GbH, Deutschland, Germany) following the manufacturer protocol. In brief, the PCR was carried out in a reaction volume of 50 μl, containing 3.0 ul cDNA template (50 ng), 25 ul 2x Taq master mix, 0.5 ul (10 μM) forward primer, 0.5 ul (10 μM) reverse primer, and 21 ul sterile nuclease free water. The thermal cycling parameters were an initial denaturation at 94°C for 3 min, 38 cycles of amplification in a Techne, TC-3000 thermal cycler (Bibby Scientific Ltd, Staffordshire, UK) were performed under the following conditions: 94°C for 1 min, 55–60° C for 1 min, and 72°C for 80 sec followed by a final extension at 72°C for 10 min.

The PCR products (146 and 346 bp for leptin and GAPDH, respectively) were electrophoresed on 1.5% agarose gel, stained by ethidiumbromide, and visualized under UV light. The amplified RT-PCR products were cloned into pSTBlue-1 Accep-Tor Vector (Novagen, Darmstadt, Germany) followed by sequencing with the BigDye Terminator v3.1 Cycle Sequencing kit (Applied BioSystems, Foster City, Calif., USA), according to the manufacturer's instructions, in an ABI Prism 3100 automated sequencer (Applied Biosystems). Homology searches of the cDNA sequences were carried out against previously identified genes by using the Basic Local Alignment Search Tool (BLASTx) program (http://www.ncbi.nlm.nih.gov/blast/ Blast.cgi) of the GenBank database (National Center for Biotechnology Information, Washington, D.C., USA).

The relative abundance of leptin RNA was normalized with respect GABDH. Densitometric analysis for the leptin bands was quantified using ImageJ gel analysis program (NIH, Bethseda, MD, USA). (http://rsb.info.nih.gov/ij/index.html).

2.5 Statistical Analysis

Results are expressed as means ± S.E.M. All statistical analyses were performed using one-way analysis of variance (ANOVA) (SPSS 17.0; SPSS, Inc., Chicago, IL, USA) or Duncan's test. The diagrams were drawn by GraphPad Prism 5.0 (GraphPad Software, San Diego, CA, USA). Differences were considered significant at P values ≤0.05. Pearsons correlation (simple linear correlation) was used to measure the correlation values and relationship between two variables (SPSS 17.0; SPSS, Inc., Chicago, IL, USA). Positive correlation is represented by the value +1.00, no correlation by 0.00, and negative correlation by -1.00.

3. Results and Discussion

The present study, for the first time, estimated the serum leptin levels in the non-pregnant, pregnant, and lactating rabbits (Figure 1). Additionally, to the best of our knowledge, this is the first report to describe the existence of leptin transcripts in the adipose and placental tissues of rabbits during different pregnancy stages (Figures 4 and 5). Our quantitative RT-PCR data confirmed that leptin expression level in the maternal perirenal adipose tissue during lactation was lower than in pregnancy but higher than non-pregnant rabbits (Figure 4). Our findings, which were similar to those obtained in other species, support a direct role of leptin in the regulation of several reproductive functions. Additionally, the present study demonstrates, for the first time, that leptin gene expression was present in a number of tissues in the fetal rabbits (Figure 6). This expression of leptin in the rabbit fetal tissues suggests that the leptin may be involved in the growth and development of the fetus. Furthermore, the study proved that serum leptin level was correlated positively to the body weight (Figure 3) and estrogen (Figure 8), and negatively to progesterone (Figure 10) along the stages examined.

3.1 Serum Leptin Levels Throughout Pregnancy and After Parturition in Rabbits

Figure (1) showed the significant increase in circulating serum leptin level (2.40±0.08 ng/ml and 2.32±0.04 ng/ml) at 20th and 30th day of pregnancy in rabbits, respectively when compared to the non-pregnant control (1.88 ± 0.04ng/ml) group. Interestingly, after parturition, the serum leptin level in rabbits was significantly lower, as detected on the 2nd day post-partum (2.06±0.02 ng/ml) than that was detected on the 20th and 30th day of pregnancy in rabbits (2.40±0.08 ng/ml and 2.32±0.04 ng/ml, respectively). Whereas, the serum leptin level after parturition (2.06±0.02 ng/ml) was significantly ($P < 0.05$) higher than that was detected in the non-pregnant rabbits and on the 10th day of pregnancy (1.88 ± 0.04 ng/ml and 1.76±0.02 ng/ml, respectively). Additionally, it is apparent from the data in Figure (1) that there was no significant changes in serum leptin levels between non pregnant control rabbits (1.88 ± 0.04 ng/ml) and at the 10th day of pregnancy (1.76 ±0.02 ng/ml), also there was no significant changes in serum leptin levels between the pregnancy day 20 (2.40±0.08 ng/ml) and 30 (2.32±0.04 ng/ml).

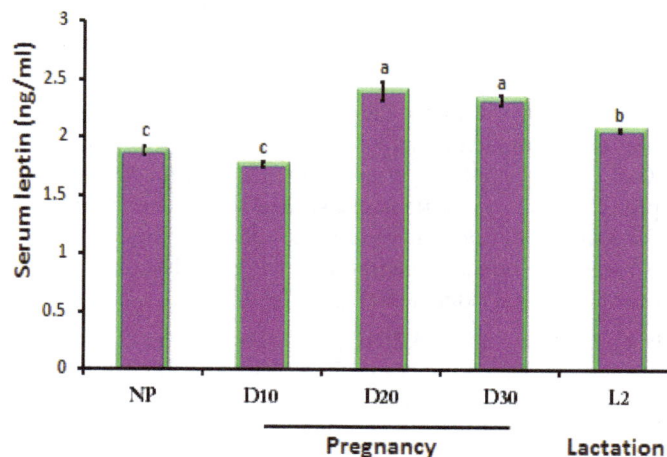

Figure 1. Changes in circulating leptin concentrations during pregnancy and lactation in rabbits. Leptin concentrations (ng/ml) were measured in the serum samples obtained from rabbits at 10th, 20th, 30th day of pregnancy (D10, D20, and D30, respectively), lactation day 2 (L2), and non pregnant (NP). Different letters denote significant differences (P≤0.05). Data are expressed as means ± S.E.M. of six animals each group

Increased leptin level appears to be a ubiquitous feature of pregnant mammals. Most reports indicate that there were substantial increase occurs in the rat from midgestation, and this is followed by a clear pre-partum decline (Kawai et al., 1997; Amico et al., 1998; Seeber et al., 2002). Additionally, the maternal plasma leptin concentrations remain relatively stable during the first half of rodent pregnancy but then increase dramatically from midgestation in the mouse (Gavrilova et al., 1997; Tomimatsu et al., 1997) and humans (Butte et al., 1997). The high levels of maternal leptin serum throughout pregnancy and its decline drastically postpartum suggests a functional importance of leptin during pregnancy. One study reported, however, that pregnancy induced increases in plasma leptin are not evident if compared to the non pregnant control rats (Terada et al., 1998).

The rise in plasma leptin from mid pregnancy appears due to the combined effects of increased plasma leptin binding activity (Gavrilova et al., 1997; Seeber et al., 2002) and enhanced maternal adipocyte expression of leptin mRNA (Kawai et al., 1997; Tomimatsu et al., 1997). Since leptin is cleared by the kidney, at least partially by filtration (Cumin et al., 1996), binding of the leptin to the binding protein should decrease its clearance. Thus hyperleptinemia during pregnancy is probably due to reduced renal clearance of bound leptin. Nonetheless, the prepartum decline in plasma leptin levels in the rat may reflect reduced fat mass in late pregnancy (Herrera et al., 2000).

3.2 Correlation Between Leptin Level & Body Weight During Pregnancy and Lactation in Rabbits

Gestational weight gain is a unique and complex biological phenomenon that supports the functions of growth and development of the fetus (Gunderson & Abrams, 2000). Consistent with this concept, the results of the present study (Figure 2) revealed that the body weight in rabbits was significantly increase (P≤0.05) at 30th day of pregnancy (3150±220.22 g) in comparison to rabbits at 10th day of pregnancy (2700±141.42 g) and the non-pregnant control group (2675± 79.84 g). Moreover, there was significant increase in the body weight of rabbits at 20th day of pregnancy (2875±58.09 g) in comparison to the non-pregnant control group (2675± 79.84

g). Besides, (Figure 2) showed increase in the body weight of pregnant rabbits at 30th day of pregnancy (3150±220.22 g) more than that of the 20th (2875±58.09 g) day of pregnancy but such increase was not significant (P>0.05). Whereas, there was significant decrease (P≤0.05) in body weight of rabbits at 2 days postpartum (2656±213.69 g) compared to the rabbits at 30th day of pregnancy (3150±220.22 g) (Figure 2).

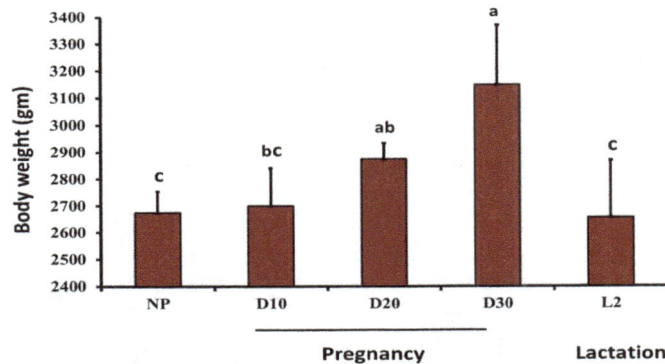

Figure 2. Body weight of rabbits during pregnancy and lactation. Body weights were estimated at 10th, 20th, 30th day of pregnancy (D10, D20, and D30, respectively), lactation day 2 (L2), and non pregnant (NP) rabbits. Different letters denote significant differences (P≤0.05). Data are expressed as means ± S.E.M. of six animals each group

Pregnancy is the only normal physiologic process where body weight increases by ~20 percent and additional fat stores for lactation are deposited during the pregnancy period (Ueland & Ueland, 1986). Gestational weight gain has been reported to be the primary and most important determinant of weight change from preconception to postpartum (postpartum weight change). Postpartum weight change is made up of the sum of gestational weight gain and the post-delivery weight loss (Gunderson & Abrams, 2000).

Moreover, our study estimated the correlation between serum leptin level and body weight during pregnancy and postpartum in rabbits. The scatter graph (bivariate plot; Figure 3) gives a clear visual representation of the strength and direction of correlation between leptin and body weight. As seen in (Figure 3), the serum leptin was positively correlated with body weight (0.670; P≤0.05). This data was in harmony and confirmed those reported in human and rodent (Considine et al., 1996; Butte et al., 1997). Maffei et al. (1995) revealed that leptin may play an important role in regulating body weight by signaling the size of the adipose tissue mass.

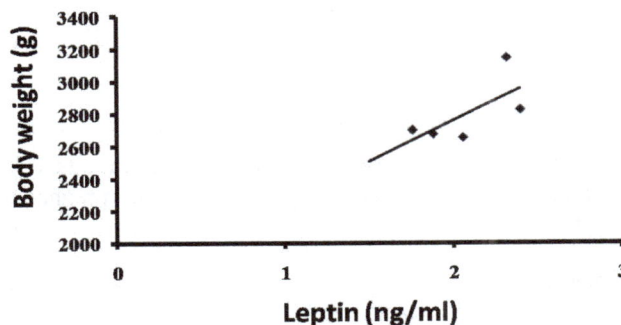

Figure 3. Correlation between serum circulating leptin levels and body weights in rabbits during pregnancy and lactation

In all mammals thus far examined, serum leptin levels increase during pregnancy and then decline shortly before or after parturition (Butte et al., 1997). The elevated maternal concentrations of leptin are not associated with decreased food intake or increased metabolic rate, as might be expected from known actions of leptin in non-pregnant mammals (Mounzih et al., 1998). Leptin resistance is a state in which there is an excess of serum leptin but the body does not effectively respond to these increased levels by reducing food intake or body weight (Frederich et al., 1995). This suggests that a pregnancy-associated state of partial leptin resistance may exist within the hypothalamus. Recently, resistance to the central effects of leptin during pregnancy and its effect on food intake has been verified in mice (Ladyman et al., 2012). Grattan et al. (2007) have demonstrated that intra cerebro-ventricular leptin is unable to suppress food intake in pregnant rats, as it does in non-pregnant animals. However, whether the suggested purpose of such resistance to leptin during pregnancy is to preferentially transport nutrients and/or promote fetal growth is not fully recognized. Tessier et al. (2013) has been reported that leptin resistance in healthy pregnancy seems to be central and beneficial for mobilizing energy stores to support adequate fetal growth. The central leptin resistance may act as a compensatory mechanism to meet the developing fetal energy needs. In later stages of a healthy pregnancy, central leptin resistance occurs to allow increased nutrient availability for the fetus (Tessier et al., 2013). These changes are physiologically appropriate, providing increased energy reserves to meet the high metabolic demands of fetal development and lactation.

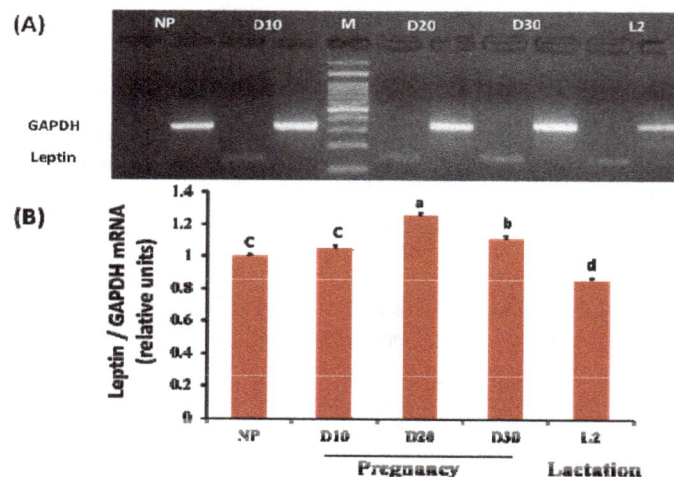

Figure 4. RT-PCR analysis of leptin in the rabbit maternal perirenal adipose tissue. (A) representative RT-PCR of leptin RNA expression in perirenal adipose tissue of rabbits during pregnancy and lactation. RNA from perirenal adipose tissue of non pregnant (NP), pregnant day 10, 20, and 30 (D10, D20, and D30, respectively), and lactating day 2 (L2) rabbits, were reverse transcribed and amplified using the PCR method and specific primers for leptin. Eight microliters of the PCR reaction were run on 1.5% agarose gel. A 100 base pair (bp) DNA ladder 100 bp DNA Ladder (Cat. No. M-214; Jena Bioscience GbH, Germany) was included as a reference for fragment size. GAPDH was used as a loading control. (B) Bar graph for the relative expression of leptin after normalization to GAPDH using the ImageJ program. Different letters denote significant differences (P≤0.05). Data are expressed as means ± S.E.M. of six animals each group

3.3 Expression of Leptin RNA in the Adipose Tissue of Non Pregnant, Pregnant, & Postpartum Rabbits

To examine the leptin expression and its levels in perirenal adipose tissue of rabbits, RT-PCR was used to detect the RNA of leptin in perirenal adipose tissue of non pregnant, pregnant (pregnancy day 10, 20, 30), and parturated (post-partum day 2) rabbits (Figure 4). As demonstrated in figure (Figure 4), leptin was expressed in all adipose tissue samples. The relative abundance of leptin RNA were significantly increased (P ≤ 0.05) in rabbit maternal adipose at 10th, 20th and 30th day of pregnancy as compared to the control non pregnant group (Figure 4). The highest expression level was observed at 20th day of pregnancy and the lowest was noticed at the 10th day of pregnancy (Figure 4). However, leptin RNA was significantly decreased at 2nd day after parturition as compared to the control group (P ≤ 0.05) (Figure 4). The elevated circulating leptin levels during pregnancy in rabbits, suggested that maternal leptin may play a role in maintenance of pregnancy and preparation for parturition and lactation.

Leptin is secreted by adipose cells, and its circulating concentration normally correlates with body adiposity (Considine et al., 1996). Leptin RNA is expressed in mice adipocytes, as shown using in situ hybridization, cell fractionation, and immunohistochemistry (Maffei et al., 1995). The plasma levels of leptin are highly correlated with adipose tissue mass and fall in both humans and mice after weight loss (Maffei et al., 1995).

Kronfeld-Schor et al. (2000) directly demonstrated that leptin secretion rates from mouse adipose tissue in vitro are decreased during early pregnancy and up-regulated during late pregnancy and lactation and these changes in leptin secretion rates in vitro paralleled those of circulating leptin in vivo during pregnancy. On the other hand, the increase in circulating leptin levels during mid to late pregnancy does not consistently correlate with adiposity in rats (Kawai et al., 1997; Amico et al., 1998), mice (Gavrilova et al., 1997; Tomimatsu et al., 1997), and humans (Butte et al., 1997). A clear increase in adipose leptin mRNA, however, has been observed during pregnancy in mice (Tomimatsu et al., 1997), suggesting that adipose tissue may be a major source of elevated serum leptin during pregnancy.

3.4 Expression of Leptin RNA in the Placenta During Pregnancy in Rabbits

The non adipose sources of leptin also might involve in the changes of circulating leptin during pregnancy. Several works have shown that in addition to adipose tissue, leptin is present in substantial amounts in the placentas of several species. Most available information describing placental leptin has been obtained from studies of rodents and humans. Our RT-PCR data proved the presence of RNA transcripts for leptin in the placenta of the pregnant rabbits (Figure 5). Even though no significant change in the placental leptin RNA expression levels was noticed among pregnant rabbits at 10th (1.0±0.045), 20th (0.91±0.058), and 30th (0.99±0.040) day of pregnancy (Figure 5).

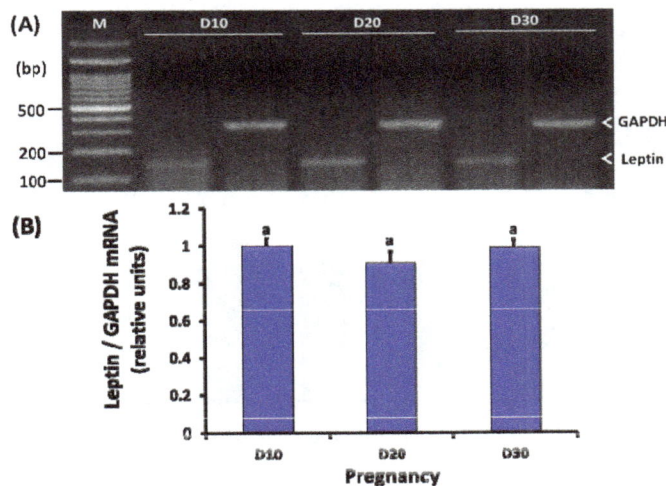

Figure 5. RT-PCR analysis of leptin in the rabbit maternal placenta. (A) representative RT-PCR of leptin RNA expression in planentas of rabbits during pregnancy and lactation. RNA from perirenal adipose tissue of non pregnant (NP), pregnant day 10, 20, and 30 (D10, D20, and D30, respectively), and lactating day 2 (L2) rabbits, were reverse transcribed and amplified using the PCR method and specific primers for leptin. Eight microliters of the PCR reaction were run on 1.5% agarose gel. A 100 base pair (bp) DNA ladder 100 bp DNA Ladder (Cat. No. M-214; Jena Bioscience GbH, Germany) was included as a reference for fragment size. GAPDH was used as a loading control. (B) Bar graph for the relative expression of leptin after normalization to GAPDH using the ImageJ program. Different letters denote significant differences (P≤0.05). Data are expressed as means ± S.E.M. of six animals each group

The human placenta contains substantial amounts of leptin mRNA and protein (Bi et al., 1997). Comparable amounts of immunoreactive leptin occur in the rodent placenta, although expression of rodent placental leptin mRNA appears to be much lower than in humans (Hoggard et al., 1997; Kawai et al., 1997; Tomimatsu et al., 1997; Terada et al., 1998). Importantly, the rodent placenta appears to contribute little, if any, leptin to the maternal circulation (Kawai et al., 1997; Hoggard et al., 2000). In agreement with our results (Figures 1 and 5), Gavrilova et al. (1997) have reported a dramatic increase of circulating leptin levels during mouse gestation with, however, no detectable increase in placental leptin production. On the other hand, Amico et al. (1998) revealed that as gestation advanced in rats, the serum leptin concentrations increased significantly and that was concomitant to a significant increase of placental leptin production. The benefit of placental leptin to the fetus may be for growth and angiogenesis, as leptin is considered an important growth factor in intrauterine and neonatal development (Hassink et al., 1997).

On the contrary with the suggestion that placenta provides a major contribution to circulating leptin during pregnancy, is that the plasma leptin levels do not return to baseline until several days after parturition and expulsion of the placenta (Gavrilova et al., 1997). One possible explanation for this is that adipose tissue continues to secrete leptin for a brief time after parturition (Kronfeld-Schor et al., 2000).

3.5 Distribution and Expression Levels of Leptin RNA in the Various Fetal Tissues of Rabbits

To date, the only indications that leptin may have a biologic function in the fetus is based on data obtained in mice and rats showing the expression of leptin receptor mRNA in several fetal tissues including cartilage, bone, lung, kidney, testes, and hypothalamus (Hoggard et al., 1997). To assess whether leptin is involved in the rabbit fetal growth and development, we have examined the distribution and relative expression levels of RNAs encoding leptin in various fetal tissues during rabbit pregnancy using RT-PCR analysis (Figure 6).

In the present study, RT-PCR verified the existence of RNAs encoding leptin in the whole rabbit fetus at day 10 of pregnancy and also in the brain, liver, perirenal adipose tissue, and bone of the rabbit fetus at 30th day of pregnancy (Figure 6). Additionally, the relative abundance of leptin transcripts, as measured by the densitometric analysis, showed significant differences (P≤ 0.05) among the different regions of the fetal rabbit tissues (Figure 6). The leptin was most abundantly in the liver, at intermediate levels in adipose tissue and bone, and low level in the brain of fetuses on day 30th of pregnancy (Figure 6).

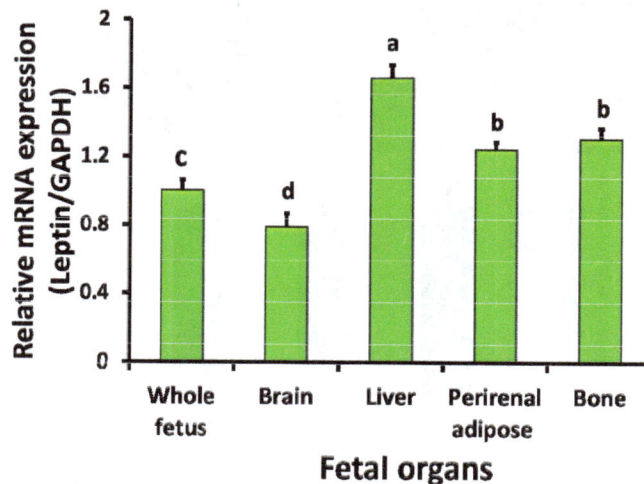

Figure 6. Detection of leptin expression in various rabbit fetal tissues by RT-PCR analysis. RNA from whole rabbit fetus at 10th day of pregnancy as well as the rabbit brain, liver, perirenal adipose, and bone of fetus at 30th day of pregnancy, were reverse transcribed and amplified using the PCR method and specific primers for leptin. The bar graph represents the relative expression of leptin after normalization to GAPDH using the ImageJ program. Different letters denote significant differences (P≤0.05). Data are expressed as means ± S.E.M. of six animals each group

On the contrary, the study by Hoggard et al. (1997) revealed that no leptin receptor (OB-R) gene expression either on the mRNA or protein level was identified in the heart, kidney, liver, adrenal, or pancreatic primordium of the 14.5-day postcoitus murine fetus. High levels of RNA expression for leptin and its receptor, however, were seen in the cartilage/bone, hair follicles, and lung as well as the leptomeninges and choroid plexus of the brain of the 14.5-day postcoitus murine fetus, suggesting a possible function of leptin as an autocrine or paracrine regulator in the fetus (Hoggard et al., 1997).

The expression of leptin in the rabbit fetal tissues (Figure 6) suggests that leptin may be involved in the growth and development of the fetus with one possible function being a fetal growth factor or a signal to the fetus of maternal energy status. Alternatively, fetal leptin could provide a signal to the mother of fetal growth and development. This implies a role for leptin in fetal bone and/or cartilage development that may be linked to its influence on hematopoiesis in the adult (Gainsford et al., 1996). The importance of SNAT (System A sodium dependant neutral amino acid transport) in fetal growth regulation has been well demonstrated (Jansson et al., 2003). Leptin has been shown to enhance SNAT activity, suggesting a role for leptin as a mediator of amino acid delivery to the fetus via the placenta (Jansson et al., 2003).

3.5 Correlation Between Leptin Level and Steroids During Pregnancy and Lactation in Rabbits

The "big two" of pregnancy hormones; estrogen and progesterone play vital roles during pregnancy, including triggering fetal development and common pregnancy symptoms. Estrogen and progesterone are two of the primary female sex hormones. The concentrations of estradiol (E_2) and progesterone in the serum of rabbits during the course of pregnancy have been determined by electrochemiluminescence immunoassay (ECLIA).

3.5.1 Estrogen

Estrogen influences various aspects of placental function and fetal development in humans and primates, and plays roles in the regulation of the onset of parturition, placental steroideogenesis, release of neuropeptides, release of glycoproteins, and leptin secretion (Chardonnens et al., 1999).

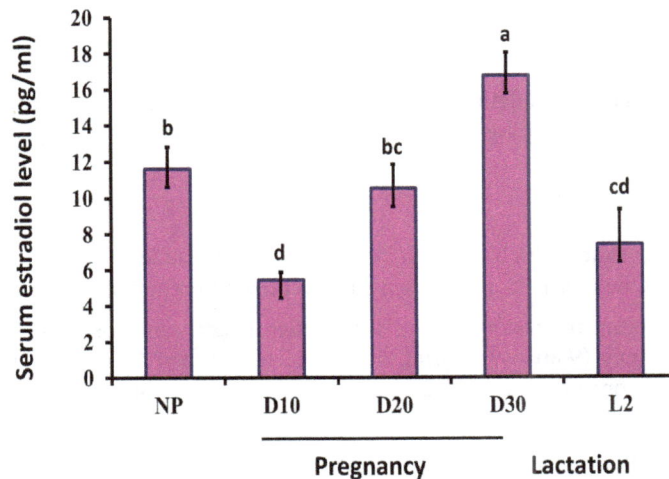

Figure 7. Serum circulating estradiol II (E_2) levels in rabbits during pregnancy and lactation. Estradiol II (E_2) concentrations (pg/ml) were measured in the serum samples obtained from rabbits at 10th, 20th, 30th day of pregnancy (D10, D20, and D30, respectively), lactation day 2 (L2), and non pregnant (NP). Different letters denote significant differences (P≤0.05). Data are expressed as means ± S.E.M. of six animals each group

Estradiol is present throughout pregnancy with a tendency to increase towards the end of gestation (Figure 7). Figure (7) revealed that the significant highest level of serum estradiol (E_2) was seen at 30th day of pregnancy (16.77±1.26 pg/ml). On the other hand, significant decrease (P≤ 0.05) in serum E_2 level was observed at 10th day of pregnancy and at 2nd day after parturition (5.46±0.42 pg/ml and 7.40±1.95 pg/ml, respectively) compared to its level in the non pregnant rabbits (11.62 ± 1.20 pg/ml) as well as in rabbits at 20th and 30th of pregnancy (10.53±1.30 pg/ml and16.77±1.26 pg/ml, respectively) (Figure 7).

Furthermore, there was a non-significant change in the serum estradiol level between rabbits at 20th of pregnancy (10.53±1.30 pg/ml) and the non pregnant control group (11.62±1.20 pg/ml). Besides, there was a non-significant change in the serum estradiol level between rabbits at 10th of pregnancy (5.46±0.42 pg/ml) and rabbits at 2nd day after pregnancy (7.40±1.95 pg/ml) (Figure 7).

During pregnancy, the level of E_2 increases as a result of placental production. This increase of E_2 contributes to leptin production. Our results, as shown in Figure (8), detected a positive correlation between both serum leptin and estradiol levels in rabbits during pregnancy and lactation (0.621; P≤0.05). This means that when estradiol level increases in value, the leptin level also increases in value and vice versa.

It has been found that leptin pulsatility is positively and strongly correlated with LH and estrogen levels in normal cycling women (Licinio et al., 1998). Moreover, estrogen sensitizes leptin signaling (Ainslie et al., 2001).

Ainslie et al. (2001) has showed that estrogen deficiency causes central leptin insensitivity and increased hypothalamic NPY. Furthermore, chronic estrogen withdrawal, as in ovariectomy in rodents and postmenopause in humans, causes leptin resistance, whereas estrogen replacement solves this problem (Ainslie et al., 2001). Estradiol administration enhances the expression of leptin mRNA transcripts and protein secretion by adipocytes, both in vitro and in vivo. Similarly, leptin expression in isolated rat adipocytes is inhibited by an estrogen receptor antagonist, and diminution of leptin expression in white adipose tissue in rats following ovariectomy is reverted with E_2 administration (Henson & Castracane, 2006).

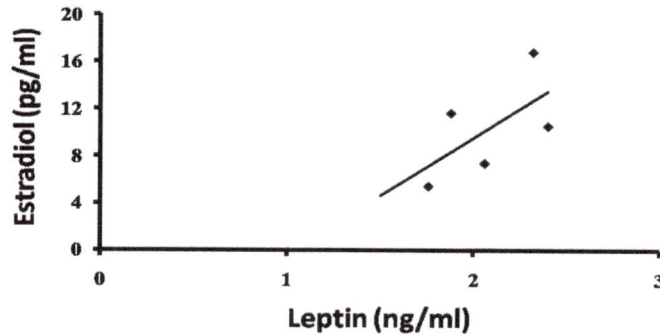

Figure 8. Correlation between serum circulating leptin levels and serum circulating estrogen levels in rabbits during pregnancy and lactation

3.5.2 Progesterone

Progesterone is produced by the ovary which is essential for the maintenance of pregnancy for 30-32 days in rabbits and is crucial for keeping the uterus in a quiescent state to prevent premature onset of labor.

The present study showed that serum progesterone levels were significantly increased (P≤0.05) in the rabbits at 10th, 20th, and 30th day of pregnancy (9.86±1.93 ng/ml, 5.25±0.53 ng/ml, and 4.10±1.51 ng/ml, respectively) in comparison to the non-pregnant control group and 2 days after parturition (2.40 ± 0.73 ng/ml, and 1.58± 0.05 ng/ml, respectively) (Figure 9).

Moreover, we found that the serum progesterone level was significantly increases (P≤0.05) in rabbits at 10th day of pregnancy (9.86±1.93 ng/ml) when compared to the other groups (Figure 9). Whereas, no significant difference was observed in the serum progesterone level at 20th day (5.25±0.53 ng/ml) and 30th day of pregnancy (4.10±1.51 ng/ml) in rabbits (Figure 9).

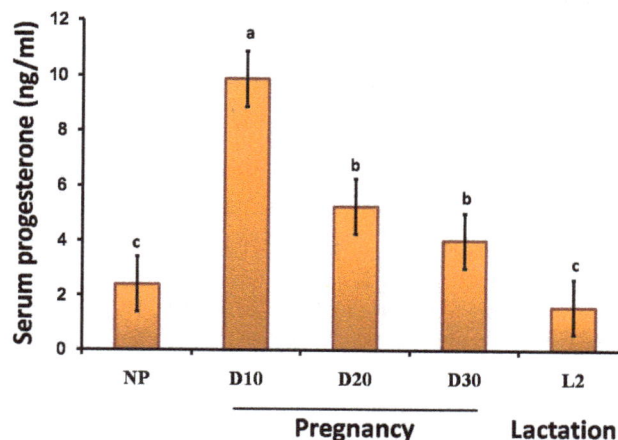

Figure 9. Serum circulating progesterone levels in rabbits during pregnancy and lactation. Progesterone concentrations (ng/ml) were measured in the serum samples obtained from rabbits at 10th, 20th, 30th day of pregnancy (D10, D20, and D30, respectively), lactation day 2 (L2), and non pregnant (NP). Different letters denote significant differences (P≤0.05). Data are expressed as means ± S.E.M. of six animals each group

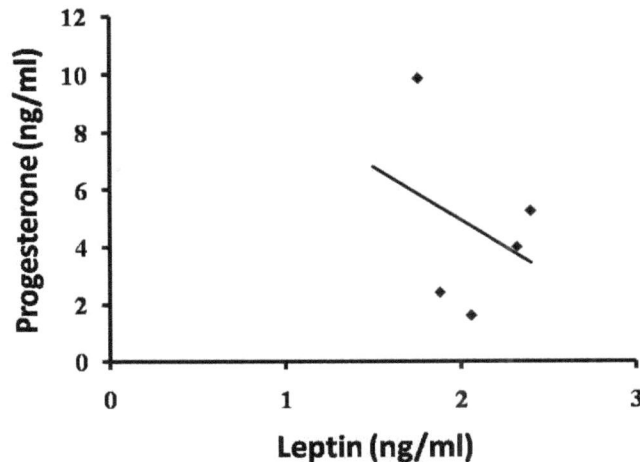

Figure 10. Correlation between serum circulating leptin levels and serum circulating progesterone levels in rabbits

In late pregnancy when placental progesterone production is at its height, progesterone has been reported to inhibit leptin secretion by human placental cells in culture (Coya et al., 2005). On the other hand, leptin always led to an inhibitory effect on progesterone secretion (Cameo et al., 2003).

The negative correlation found between serum leptin and progesterone levels in rabbits (-0.33; P≤0.05) (Figure 10) means that during pregnancy and lactation when the leptin level increases in value, the progesterone level decreases in value. Our findings are in agreement with an inhibitory effect of leptin on progesterone secretion found in other tissues, such as rat granulosa cells (Spicer et al., 2000) and human term placental trophoblast cells (Cameo et al., 2003).

4. Conclusion

To the best of our knowledge this is the first report on the distribution and expression levels of leptin in rabbit maternal adipose tissue and placenta as well as in different fetal tissues including; brain, liver, bone, and perirenal adipose tissue during pregnancy. Moreover, our study proved that leptin has positive correlations with body weight and estrogen and negative correlation with progesterone in pregnant and postpartum rabbits. Results of our study imply a fundamental role of leptin in the maintenance of pregnancy and preparation for lactation. The present data strongly reinforce the idea that circulating leptin levels may provide a growth-promoting signal for fetal development during late pregnancy. Overall, the present work supports the importance of leptin in reproductive physiology and fetal development. Although great progress has been made in the last few years towards understanding the physiological roles of leptin, several fundamental questions still need to be answered.

References

Ainslie, D. A., Morris, M. J., Wittert, G., Turnbull H., Proietto J., & Thorburn, A. W. (2001). Estrogen deficiency causes central leptin insensitivity and increased hypothalamic neuropeptide Y. *Int J Obes Relat Metab Disord, 25*, 1680–1688.

Amico, J. A., Thomas, A., Crowley, R. S., & Burmeister, L. A. (1998). Concentrations of leptin in the serum of pregnant, lactating, and cycling rats and of leptin messenger ribonucleic acid in rat placental tissue. *Life Sci, 63*, 1387–1395.

Bi, S., Gavrilova, O., Gong, D. W., Mason, M. M., & Reitman, M. (1997). Identification of a placental enhancer for the human leptin gene. *J Biol Chem, 272*(48), 30583-30588.

Butte, N. F., Hopkinson, J. M., & Nicolson, M. A. (1997). Leptin in human reproduction: serum leptin levels in pregnant and lactating women. *J Clin Endocrinol Metab, 82*, 585–589.

Cameo, P., Bischof, P., & Calvo, J. C. (2003). Effect of leptin on progesterone, human chorionic gonadotropin, and interleukin-6 secretion by human term trophoblast cells in culture. *Biol Reprod, 68*(2), 472-477. http://dx.doi.org/ 10.1095/biolreprod.102.006122

Chardonnens, D., Cameo, P., Aubert, M. L., Pralong, F. P., Islami, D, Campana, A, ... Bischof, P. (1999). Modulation of human cytotrophoblastic leptin secretion by interleukin-1alpha and 17beta-oestradiol and it effect on HCG secretion. *Mol Hum Reprod, 5*, 1077–1082. http://dx.doi.org/ 10.1093/molehr/5.11.1077

Considine, R. V., Sinha, M. K., Heiman, M. L., Kriauciunas, A., Stephens, T. W., Nyce, M. R., ... Caro, J. F. (1996). Serum immunoreactive-leptin concentrations in normal-weight and obese humans. *N Engl J Med, 334*(5), 292-295. http://dx.doi.org/10.1056/NEJM199602013340503

Coya, R., Martul, P., Algorta, J., Aniel-Quiroga, M. A., Busturia, M. A., & Señarís, R. (2005). Progesterone and human placental lactogen inhibit leptin secretion on cultured trophoblast cells from human placentas at term. Gynecol. *Endocrinol, 21*(1), 27-32. http://dx.doi.org/10.1080/09513590500099305

Cumin, F., Baum, H. P., & Levens, N. (1996). Leptin is cleared from the circulation primarily by the kidney. *Int J Obes Relat Metab Disord, 20*(12), 1120-1126.

Cunningham, M. J., Clifton, D. K., & Steiner, R. A. (1999). Leptin's Actions on the Reproductive Axis: Perspectives and Mechanisms. *Biology of reproduction, 60*, 216–222. http://dx.doi.org/10.1095/ biolreprod60.2.216

Gainsford, T., Willson, T. A., Metcalf, D., Handman, E., McFarlane, C., Ng, A., ... Hilton, D. (1996). Leptin can induce proliferation, differentiation, and functional activation of hemopoietic cells. *J Proc Natl Acad Sci USA, 93*, 14564–14568.

Gavrilova, O., Barr, V., Marcus-Samuels, B., & Reitman, M. (1997). Hyperleptinemia of pregnancy associated with the appearance of a circulating form of the leptin receptor. *J Biol Chem, 272*, 30546–30551. http://dx.doi.org/10.1074/jbc.272.48.30546

Gonzalez, R. R., & Leavis, P. C. (2003). A peptide derived from the human leptin molecule is a potent inhibitor of the leptin receptor function in rabbit endometrial cells. *Endocrine, 21*(2), 185-95. http://dx.doi.org/10.1385/ENDO:21:2:185

Grattan, D. R., Ladyman, S. R., & Augustine, R. A. (2007). Hormonal induction of leptin resistance during pregnancy. *Physiol Behav, 91*(4), 366-374. http://dx.doi.org/10.1016/j.physbeh.2007.04.005

Gunderson, E. P., & Abrams, B. (2000). Epidemiology of gestational weight gain and body weight changes after pregnancy. *Epidemiol Rev, 22*(2), 261-274.

Hassink, S. G., deLancey, E., Sheslow, D. V., Smith-Kirwin, S. M., O'Connor, D. M., Considine, R. V., ... Funanage, V. L. (1997). Placental leptin: an important new growth factor in intrauterine and neonatal development? *Pediatrics, 100*, E1–E6. http://dx.doi.org/10.1542/peds.100.1.e1

Henson, M. C., & Castracane, V. D. (2006). Leptin in pregnancy: an update. *Biol Reprod, 74*(2), 218-29. http://dx.doi.org/10.1095/biolreprod.105.045120

Herrera, E., Lasunción, M. A., Huerta, L., & Martín-Hidalgo, A. (2000). Plasma leptin levels in rat mother and offspring during pregnancy and lactation. *Biol Neonate, 78*(4), 315-20. http://dx.doi.org/10.1159/ 000014286

Hoggard, N., Hunter, L., Duncan, J. S., Williams, L. M., Trayhurn, P., & Mercer, J. G. (1997). Leptin and leptin receptor mRNA and protein expression in the murine fetus and placenta. *Proc Natl Acad Sci U S A, 94*, 11073–11078. http://dx.doi.org/10.1073/pnas.94.20.11073

Jansson, N., Greenwood, S. L., Johansson, B. R., Powell, T. L., & Jansson, T. (2003). Leptin stimulates the activity of the system A amino acid transporter in human placental villous fragents. *J Clin Endocrinol Metab, 88*, 1205-1211. http://dx.doi.org/http://dx.doi.org/10.1210/jc.2002-021332

Kawai, M., Yamaguchi, M., Murakami, T., Shima, K., Murata, Y., & Kishi, K. (1997). The placenta is not the main source of leptin production in pregnant rat: gestational profile of leptin in plasma and adipose tissues. *Biochem Biophys Res Commun, 240*, 798–802. http://dx.doi.org/10.1006/bbrc.1997.7750

Koch, E., Hue-Beauvais, C., Galio, L., Solomon, G., Gertler, A., Révillon, F., ... Charlier, M. (2013). Leptin gene in rabbit: cloning and expression in mammary epithelial cells during pregnancy and lactation. *Physiol Genomics, 45*(15), 645-652. http://dx.doi.org/10.1152/physiolgenomics.00020.2013

Kronfeld-Schor, N., Zhao, J., Silvia, B. A., Bicer, E., Mathews, P. T., Urban, R., ... Widmaier E. P. (2000). Steroid-dependent up-regulation of adipose leptin secretion in vitro during pregnancy in mice. *Biol Reprod, 63*, 274–280. http://dx.doi.org/10.1095/biolreprod63.1.274

Ladyman, S. R., Fieldwick, D. M., & Grattan, D. R. (2012). Suppression of leptin-induced hypothalamic JAK/STAT signalling and feeding response during pregnancy in the mouse. *Reproduction, 144*, 83-90. http://dx.doi.org/10.1530/REP-12-0112

Licinio J., Negrao A. B., Mantzoros, C., Kaklamani, V., Wong, M. L., Bongiorno, P. B., ... Gold, P. W. (1998). Synchronicity of frequently sampled, 24-h concentrations of circulating leptin, luteinizing hormone, and estradiol in healthy women. *Proc Natl Acad Sci USA, 95*, 2541–2546.

Maffei, M., Halaas, J., Ravussin, E., Pratley, R. E., Lee, G. H., Zhang, Y., ... Friedman, J. M. (1995). Leptin levels in human and rodent: measurement of plasma leptin and ob RNA in obese and weight-reduced subjects. *Nat Med, 1*(11), 1155-1161.

Mounzih, K., Qiu, J., Ewart-Toland, A., & Chehab, F. F. (1998). Leptin is not necessary for gestation and parturition but regulates maternal nutrition via a leptin resistance state. *Endocrinology, 139*(12), 5259-5262.

Seeber, R. M., Smith, J. T., & Waddell, B. J. (2002). Plasma leptin-binding activity and hypothalamic leptin receptor expression during pregnancy and lactation in the rat. *Biol Reprod, 66*(6), 1762-1767. http://dx.doi.org/10.1095/biolreprod66.6.1762

Spicer, L. J., Chamberlain, C. S., & Francisco, C. C. (2000). Ovarian action of leptin: effects on insulin-like growth factor-I-stimulated function of granulosa and thecal cells. *Endocrine, 12*, 53-59. http://dx.doi.org/10.1385/ENDO:12:1:53

Terada, Y., Yamakawa, K., Sugaya, A., & Toyoda, N. (1998). Serum leptin levels do not rise during pregnancy in age-matched rats. *Biochem Biophys Res Commun, 253*(3), 841-844. http://dx.doi.org/10.1006/bbrc.1998.9861

Tessier, D. R., Ferraro, Z. M., & Gruslin, A. (2013). Role of leptin in pregnancy: consequences of maternal obesity. *Placenta, 34*(3), 205-211. http://dx.doi.org/http://dx.doi.org/10.1016/j.placenta.2012.11.035

Tomimatsu, T., Yamaguchi, M., Murakami, T., Ogura, K., Sakata, M., Mitsuda, N., ... Murata, Y. (1997). Increase of mouse leptin production by adipose tissue after midpregnancy: gestational profile of serum leptin concentration. *Biochem Biophys Res Commun, 240*, 213-215. http://dx.doi.org/10.1006/bbrc.1997.7638

Ueland, K., & Ueland, F. R. (1986). Physiologic adaptations to pregnancy. In R. A. Knuppel, & J. E. Drukker (Eds.), *High-risk pregnancy, a team approach* (pp. 148-172). Philadelphia, PA: Saunders.

Zerani, M., Boiti, C., Zampini, D., Brecchia, G., Dall'Aglio, C., Ceccarelli, P., & Gobbetti, A. (2004). Ob receptor in rabbit ovary and leptin in vitro regulation of corpora lutea. *J Endocrinol, 183*(2), 279-288. http://dx.doi.org/10.1677/joe.1.05507

Zhang, Y., Proenca, R., Maffei, M., Barone, M., Leopold, L., & Friedman, J. M. (1994). Positional cloning of the mouse obese gene and its human homologue. *Nature, 372*, 425- 432. http://dx.doi.org/10.1038/372425a0

Zhao, S. P., & Wu, Z. H. (2005). Atorvastatin reduces serum leptin concentration in hypercholesterolemic rabbits. *Clin Chim Acta, 360*(1-2), 133-140. http://dx.doi.org/10.1016/j.cccn.2005.04.021

14

Physiological and Ethological Effects of Fluoxetine, A Study Using Ants as Biological Models

Marie-Claire Cammaerts[1] & David Cammaerts

[1] Faculté des Sciences, Université Libre de Bruxelles, Av. F. Roosevelt, Bruxelles, Belgium

Correspondence: Marie-Claire Cammaerts, Faculté des Sciences, DBO, CP 160/12, Université Libre de Bruxelles, 50 Av. F. Roosevelt, 1050, Bruxelles, Belgium. E-mail: mtricot@ulb.ac.be

Abstract

Using ants as biological models, we showed that fluoxetine (the active substance of the most consumed antidepressants) largely affects the individual's physiology and behavior. It increases sinuosity of movement, decreases precision of reaction and response to pheromones, decreases food consumption and brood caring, and induces aggressiveness towards nestmates while decreasing that towards aliens. Under fluoxetine consumption, ants lost their olfaction and their learning ability, having also lower cognitive ability. There is no habituation to, and no dependence on, fluoxetine consumption, which effects vanish in two or two and a half days. Attention should be paid whenever that drug is used, not only for the humans but also for all the organisms living on field and depending on natural water. A future work will examine antidepressants free of fluoxetine, hoping that they could advantageously replace those containing that harmful substance.

Keywords: activity, aggressiveness, cognition, dependence, food consumption, memory

1. Introduction

In some rivers (for instance, in River St Laurent, Canada), fishes appeared to present aberrant behavior. Fifty trouts maintained in St Laurent water mixed to that coming from Montreal city presented, after three months, small amounts of antidepressants in their liver, brain and muscles, as well as some decrease of their brain synaptic activity (Sauvé, 2011). Similar observations were made in Wisconsin, USA, on minnows living in water contaminated by antidepressants (Klaper, 2013). Males often stayed under stones, no longer approached females, took more time for capturing preys, and under larger amounts of drugs, aggressed congeners, and even killed females. Also, females laid fewer eggs, and young minnows behaved as being anxious. Such aberrant behavior could be explained by changes observed in the young minnows' brains (Klaper, 2013). Apart this, many other drugs used by humans presently contaminate natural water (Lecomte, 2014).

Antidepressants are present in natural water because they are presently largely consumed by humans, are eliminated by the kidneys, and then transported into natural water via sewerage systems. There exist four kinds of such drugs. The 'IMAO' inhibit the monoamine oxidize and are now given only in hospitals, thus in limited amounts. The 'ATC' are tricycle ones, and inhibit several neurotransmitters. The 'ISRS' inhibit the recapture of serotonin, while the 'IRSNa' inhibit the recapture of serotonin and noradrenalin (Antidépresseur, ND). Acting on the recapture of neurotransmitters, they may have unwanted effects on humans. Effectively, several studies revealed adverse effects of antidepressants: let us cite, among others, the works of Parent (2011), Simon (2002), Cipriani et al. (2009), and Lane (2009). However, not all potential adverse effects have been examined for the humans as well as for other living organisms, and, concerning the humans, it is possible that not all the obtained results have been revealed.

All these information leaded us to presume that some danger could exist for living organisms, especially aquatic ones, when they are exposed to those substances. Aquatic macroinvertabrates could be exposed to antidepressants through the release of these substances with untreated or inefficiently teated waste water. Aquatic macroinvertebrates play such an important role in the ecology of water bodies, occupying a variety of ecological niche and performing a huge variety of behaviors, so that any modifications of these behaviors could lead to significant changes in the trophic relationship between organisms of the aquatic fauna. Complex equipment is required for studying the behavior of aquatic invertebrates in laboratory conditions. So, we use ants as biological models to study the effects of antidepressants on animals' behavior and physiology.

Presently, the ISRS are the most commonly consumed antidepressants, and their most commonly used active substance is fluoxetine. We examine thus, in the present work, some effects of that substance, and will study in a future work those of an 'ATC' and an 'IRSNa', each time using ants as biological models.

As most of the biological processes are similar for all animals, including humans (i.e. genetics, metabolism, nervous cells functioning), a lot of invertebrates and vertebrates can be used as models for studying biological questions (Kolb & Whishaw, 2002; Whener & Gehring, 1999; Russell & Burch, 2014). Invertebrates are more and more used as biological models because they offer scientists many advantages, among others a short life cycle, a simple anatomy, and being available in large numbers (Wolf & Heberlein, 2003, Sovik & Barron, 2013). Some species became largely used as biological models, for instance, the flatworm *Dendrocelium lacteum*, the nematode worm *Caenorhabdotes elegans*, the mollusk *Aplysia californica*, the beetle *Tribolim castaneum*, the fruit fly *Drosophila melanogaster*, and the domestic bee *Apis mellifera*. Among the invertebrates, insects, especially social hymenoptera and among them, bees, are advantageously used as biological models (André, Wirtz, & Das, 2008; Abramson, Wells, & Janko, 2007).

Ants also could be used. Colonies containing thousands of ants can easily be maintained in laboratories, at low cost and very conveniently, throughout the entire year. Ants are also among the most evolved organisms as for their morphology, their physiology, their social organization and their behavior. They are among the most morphologically elaborate hymenoptera, having indeed a unique resting position of their labium, mandibles and maxilla (Keller, 2011), as well as a lot of glands emitting numerous, efficient compounds (Billen & Morgan, 1998). Their societies are highly organized with a strong division of labor, an age-based polyethism and a social regulation (Holldobler & Wilson, 1990). Their behavior is highly developed: they care for their brood, build sophisticated nests, chemically mark the inside of their nest, and, differently, their nest entrances, their nest surroundings and their foraging area (Passera & Aron, 2005). They generally use an alarm signal, a trail pheromone, and a recruitment signal (Passera & Aron, 2005); they are able to navigate using memorized visual and olfactory cues (Cammaerts, 2012, and references therein); they efficiently recruit nestmates where, when and as long as it is necessary (Passera, 2006), and, finally, they clean their nest and provide their area with cemeteries (Keller & Gordon, 2006). So, according to the complexity of their society and their behavior, it looks reasonable to use ants as biological models for studying physiological and ethological effects of neuronal active substances.

We have largely studied the ant's species belonging to the genus *Myrmica*, and above all *Myrmica sabuleti* Meinert 1861. We know its ecological characteristics, eye morphology, visual perception, navigation system, visual and olfactory conditioning capabilities, and recruitment strategy (Rachidi, Cammaerts, & Debeir, 2008; Cammaerts, 2008; Cammaerts & Rachidi, 2009; Cammaerts, Rachidi & Cammaerts, 2011). The ontogenesis of cognitive abilities of *Myrmica* species, including *M. sabuleti*, has also been approached (Cammaerts & Gosset, 2014b, Cammaerts, 2013a, 2013b, 2014a, 2014b). Studies on the impact of age, activity and diet on *M. ruginodis*' conditioning capability (Cammaerts & Gosset, 2014a) leaded to presume that *M. sabuleti* could be a good biological model. This was confirmed by the study of the effects of caffeine, theophylline, cocaine, and atropine (Cammaerts, Rachidi & Gosset, 2014b), of nicotine (Cammaerts, Gosset, & Rachidi, 2014a), and of morphine and quinine on *M. sabuleti* (Cammaerts & Cammaerts R., 2014). The results of these studies brought information and precision about effects of the alkaloids on humans. We thus have compiled a sufficient amount of information about the behavior and physiology of that species to be able to conduct a similar analysis about potential adverse physiological and ethological effects generated by fluoxetine consumption.

2. Experimental Planning

The 19 following behavioral and physiological traits were assessed on two colonies of *M. sabuleti* before and after they consumed fluoxetine. A third colony was used to provide 'alien workers'. Most of these traits had previously similarly been examined while studying effects of alkaloids and drugs (see references at the end of the 'Introduction' section).

1 – the locomotion (and thus the general activity) through the ants' linear and angular speed,

2 – the precision of reaction through the orientation towards a source of their alarm pheromone,

3 – the response to pheromones through the trail following behavior,

4 – the "audacity" through the numbers of ants coming onto a test apparatus,

5 – food consumption through the numbers of ants coming onto meat food,

6 – the tactile sensation (or "pain" perception) through the ants' behavior in an uncomfortable situation,

7 – cognition through the ability in performing a task requiring cognition (moving through chicanes),

8 – the potential aggressiveness against nestmates through ants' behavior in the course of dyadic encountering,

9 – the expected aggressiveness against alien ants through ants' behavior in the course of dyadic encountering,

10 – the caring behavior through the behavior in front of larva removed from the nest,

11 – the visual perception through the distinguishing of two colors,

12 – the olfactory perception through the distinguishing of two odors,

13 – the visual learning ability through the acquisition of a visual conditioning,

14 – the visual memory through the duration of the remembering a learned visual cue,

15 – the olfactory learning ability through the acquisition of an olfactory conditioning,

16 – the olfactory memory through the duration of the remembering a learned olfactory cue,

17 – the habituation to the drug consumption through the speed of movement and the orientation to an alarm signal, seven days after continuous drug consumption,

18 – the dependence on drug consumption through the numbers of ants choosing food containing the drug,

19 – the decrease of the effects of fluoxetine after its consumption end, through the ants' sinuosity of movement and orientation to an alarm signal after that fluoxetine had been removed from the food.

3. Material and Methods

3.1 Collection and Maintenance of the Ants

The study was made on three colonies of *M. sabuleti,* two ones (labeled 1 and 2) having been devoted to the experimental work while the third one only furnished the 'alien workers' required in one experiment. The three colonies were collected in summer 2013 in an abandoned sandstone quarry located at Treignes (Ardenne, Belgium). The ants were nesting under stones, in a field covered with small plants and brushes. The collected colonies were demographically similar, containing about 600 workers, one or two queens and brood at larval and nymphal stages. They were maintained in the laboratory in artificial nests made of one to three glass tubes half-filled with water, with a cotton-plug separating the ants from the water. These glass tubes were deposited in trays (34 cm x 23 cm x 4 cm), the sides of which were covered with talc to prevent the ants from escaping. The trays served as foraging areas, food being delivered into them. The ants were fed with sugar-water provided *ad libitum* in a small glass tube plugged with cotton, and with pieces of *Tenebrio molitor* (Linnaeus 1758) larva provided twice a week on a glass slide. Temperature was maintained between 18°C and 22°C, humidity at about 80%, these conditions remaining constant over the course of the study. Lighting had a constant intensity of 330 lux while caring for the ants, training and testing them. During other time periods, the lighting was dimmed to 110 lux. The ambient electromagnetic field had an intensity of 2-3 $\mu W/m^2$.

3.2 Acquisition of Fluoxetine, Realization of Aqueous Solutions for Ants

Five hundred milligrams of fluoxetine, produced by the manufacturer CERTA, were provided by the pharmacist J. Cardon (1050 Brussels). The product was provided as a white bright powder, at the highest level of purity possible. According to the amount of fluoxetine given to depressive humans, and the quantity of water these humans drink, it could be established that the most appropriate solution of fluoxetine would be 1 mg of fluoxetine into 1,000 mg of water. Therefore, 30 mg of fluoxetine were weighted using a precision balance, and were then dissolved in 30 ml (= 30,000 mg of water) of a saturated solution of brown sugar, the ants' usual liquid food. The concentration in drug of the final solution was thus 30 mg in 30 ml of water, so 1/1,000. This solution was given to the ants, like their usual liquid food, in a small glass tube plugged with cotton, the cotton being refreshed each two days and the entire solution renewed each 14 days. It was checked each day if ants actually consumed the given liquid food containing fluoxetine.

3.3 Orientation, Linear and Angular Speed (1, 2)

Ants' linear and angular speed was assessed for detecting excitation or sleepiness in the animals. This assessment was made on ants freely moving on their foraging area. Ants' orientation towards an isolated congener's head allows measuring the ants' precision of reaction. An isolated worker's head, with widely opened mandibles, is a source of alarm pheromone identical to that of an alarmed worker, in terms of the dimensions of the emitting source (the mandibular glands' opening) and of the quantity of pheromone emitted (Cammaerts-Tricot, 1973). Each time, such assessment was made on ants of the two nests having never consumed fluoxetine, then on ants of these two nests having consumed the drug during two days. For each assessment, the movement of ten ants of each nest (n = 20 ants) was analyzed.

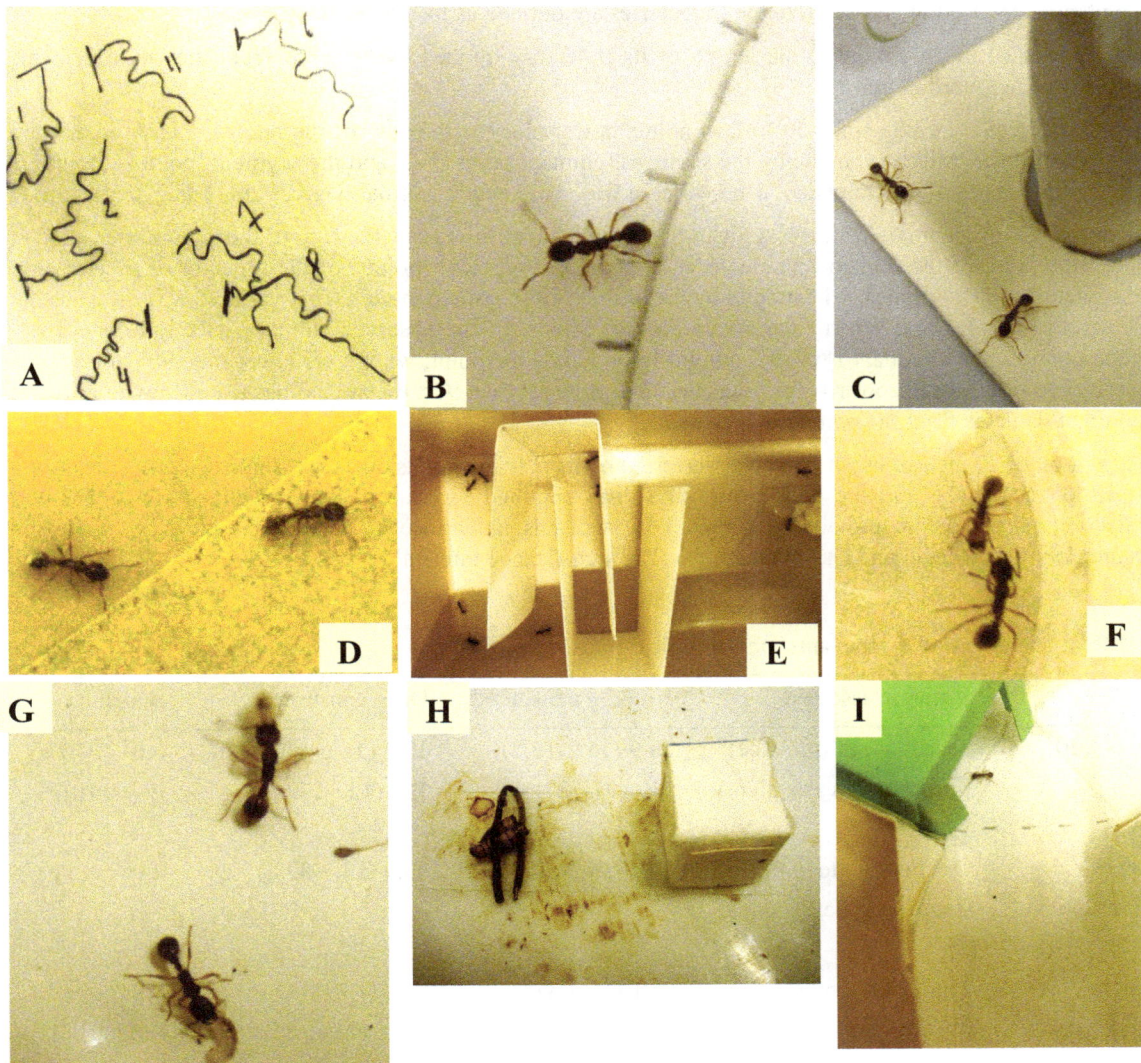

Figure 1. Some views of the experiments

A: ants' trajectories under fluoxetine consumption; the trajectories were sinuous. **B**: ants' trail following behavior under fluoxetine consumption; ants failed in well following the trail. **C**: ants under fluoxetine consumption, and confronted with a risky apparatus; as without this drug, they were not inclined to move on the apparatus. **D**: ants under normal diet, confronted with a very rough substrate; the ant on the right presented difficulties when moving on the rough substrate. **E**: ants under normal diet set in front of chicanes; some of them could find their way through the chicanes, up to a larger area. **F**: two nestmates, under fluoxetine consumption, in front of one another; they opened their mandibles, a rather aggressive behavior. **G**: caring of larvae by ants under normal diet; the ants gently pulled up the larvae and quickly brought them into the nest. **H**: experimental design used for assessing the ants' visual perception, under fluoxetine consumption; a cube with a blue and a yellow face was set four cm from the ants' meat food. **I**: an ant among twenty others, under fluoxetine consumption, tested in a Y apparatus for assessing the ants' visual learning ability

Trajectories were recorded manually, using a water-proof marker pen, on a glass slide placed horizontally 3 cm above the area where the tested individuals were moving. A metronome set at 1 second was used as a timer for assessing the total time of each trajectory. Each trajectory was recorded during 5 to 10 seconds or until the ant reached the stimulus. All the trajectories were then traced, with a water-proof marker pen, onto transparent polyvinyl sheets (Figure 1A) using the glass slide as the reference model, and the polyvinyl sheets were affixed to a PC monitor screen. The trajectories were analyzed using specifically designed software (Cammaerts, Morel, Martino & Warzée, 2012) so that the trajectory parameters could be quantified.

The three parameters used to characterize the trajectories are defined as follows:

The linear speed (V) of an animal is the length of its trajectory divided by the time spent moving along this trajectory. It was measured in mm/s.

The angular speed (S) (i.e. the sinuosity) of an animal's trajectory is the sum of the angles, measured at each successive point of the trajectory, made by the segment 'point i - point i – 1' and the segment 'point i - point i + 1', divided by the length of the trajectory. This variable was measured in angular degrees/cm (= ang. deg. / cm).

The orientation (O) of an animal towards a given point (here an ant's head) is the sum of the angles, measured at each successive point of the registered trajectory, made by the segment 'point i of the trajectory - given point' and the segment 'point i - point i + 1' divided by the number of measured angles. This variable was measured in angular degrees (= ang. deg.). When such a variable (O) equals 0°, the observed animal perfectly orients itself towards the point; when O equals 180°, the animal fully avoids the point; when O is lower than 90°, the animal has a tendency to orient itself towards the point; when O is larger than 90°, the animal has a tendency to move in a direction that deviates from the point.

Each distribution of 20 measurements was characterized by its median and its quartiles (since being not Gaussian; Table 1, table lines 1, 2), and the distribution of values obtained for ants having consumed fluoxetine was statistically compared to that previously obtained for the ants having never consumed that drug, using the non-parametric χ^2 test (Siegel and Castellan, 1989). The significance threshold was set to $\alpha = 0.05$.

Table 1. Effect of fluoxetine on five ants' physiological traits

Traits	Variable assessed	No drug consumed	Drug consumed	Statistics
Activity n = 20	Linear speed (mm/sec)	14.9 (14.2-16.8)	13.1 (11.6-14.6)	0.01<P<0.02
	Sinuosity (ang.deg./cm)	113 (97-129)	149 (129-172)	P<0.001
Precision of reaction n=20	Orientation to an alarm signal (ang.deg.)	34.5 (29.9-50.7)	69.9 (53.7-82.5)	0.001<P<0.01
Response to pheromones n = 40	Trail following behavior (n° of arcs walked)	C: 1 (1 - 1.3) T:12.0 (7.0-16.3)	C: 1 (1 - 1) T:3.5 (2.0-6.3)	NS P<0.001
"Audacity" n = 20	N° of ants on a tower (mean, extremes)	1.35 (0 - 2)	1.62 (0 - 4)	P = 0.0858
Food consumption n = 20	N° of ants eating meat (mean, extremes)	2.2 (1 - 4)	0.46 (0 - 2)	P = 0.000002

The traits, the variables, and the experimental methods are detailed in the text. Fluoxetine increased the ants' sinuosity of movement, decreased their precision of reaction, their response to pheromones, and their food consumption; it did not impact the ants' audacity.

3.4 Trail Following Behavior (3)

This behavior was assessed for examining the ants' response to a pheromone. The trail pheromone of *Myrmica* ants is produced by the workers' poison gland. Ten of these glands were isolated in 0.5 ml (500µl) hexane and stored for 15 min at -25 °C. To perform one experiment, 0.05 ml (50µl) of the solution was deposited, using a metallic normograph pen, on a circle (R = 5 cm) pencil drawn on a piece of white paper and divided into 10 angular degrees arcs (= ang. deg.). One minute after being prepared, the piece of paper with the artificial trail was placed in the ants' foraging area. When an ant came into contact with the trail, its movement was observed (Figure 1B). Its response was assessed by the number of arcs of 10 angular degrees it walked without departing from the trail, even if it turned back while walking on the trail. If an ant turned back when coming in front of the trail, its response was assessed as "zero arc walked"; when an ant crossed the trail without following it, its response equaled "one walked arc". Before testing the ants on a trail, they were observed on a "blank" circumference imbibed with 50μl of pure hexane, and the control numbers of walked arcs were so obtained (Table 1, table line 3, C = control, T = test). On such experimental trails, *Myrmica* workers do not deposit their trail pheromone because they do so only after having found food or a new nest site. Each time, these manipulations were made firstly on ants having never consumed fluoxetine, then on the same colony having consumed this drug for 3 days. For each control and test experiment, 20 individuals of each two used colonies were observed (n = 40). Each distribution of values was characterized by its median and its quartiles (since being not Gaussian; Table 1, table line 3). The distributions of values obtained for ants having consumed fluoxetine were compared to the corresponding ones obtained for ants having never consumed this substance, by using the non parametric χ^2 test (Siegel & Castellan, 1989).

3.5 Ants' "Audacity" (4)

Before the ants consumed fluoxetine, and three days after they had consumed that drug, a cylindrical tower built in strong white paper (Steinbach ®, height = 4 cm; diameter = 1.5 cm) was set on the ants' foraging area (Figure 1C), and the ants present on it, at any place, were counted 10 times, in the course of 10 min. The mean and the extreme values of the obtained values were established each time and the two series of values were compared using the non parametric Mann-Whitney U test (Siegel & Castellan, 1989; Table 1, table line 4).

3.6 Ants' Food Consumption (5)

Before the ants consumed fluoxetine, and after they had consumed that drug for three days, the workers present on the meat food (pieces of *T. molitor* larva) at a time ants must be fed were counted 10 times in the course of 10 min. The numbers obtained for the two kinds of food intake (with no drug, then with drug) were statistically compared using the Mann-Whitney U test (same reference as above), and the mean as well as the extreme values of the recorded numbers were established (Table 1, table line 5).

3.7 Ants' Tactile Sensation (Presumed 'Pain Sensation') (6)

It was tempted to assess this physiological trait by setting ants in an experimental apparatus made of a small tray (15 cm x 7 cm x 4.5 cm) into which a piece (3 cm x 11 cm) of rough emery paper (number 280) was duly folded (11 cm: 2 cm + 7 cm + 2 cm) and tied to the bottom and the border of the tray, so dividing the tray in three zones: a small initial smooth zone (3 cm long), a zone (3 cm long) on which ants' walking should be uncomfortable, and a large smooth zone (9 cm long) for inciting the ants crossing the uncomfortable zone. Two such apparatus were used, one for each used colony. The ants were tested before they consumed fluoxetine, then five days after they had continuously received that drug. Each time, 12 ants were set, all together, at the same time, in the small initial zone. The ants present in each of the three zones of the apparatus were counted after 0, 2, 4, 6, 8, 10 min, and the linear as well as the angular speed of 12 ants for each two tested colonies (so n = 24) moving on the rough paper were assessed using the method briefly explained in point 3.3 (Figure 1D; Table 2, table line 1). The numbers of ants obtained for ants having consumed fluoxetine were statistically compared to those previously obtained for ants of the same colonies having never ingested that drug, using the non parametric Wilcoxon test (Siegel & Castellan, 1989), while the linear and angular speed of such ants were statistically compared using the non parametric χ^2 test (same reference as above).

3.8 An Ants' Cognitive Ability Requiring No Memory (7)

This ability was assessed on ants of the two used nests first while these nests did not received fluoxetine, then seven days after they had continuously fluoxetine in their liquid food. The assessment was made using an adequate experimental apparatus schematically presented in the figure 3 of Cammaerts et al. (2014b) and here shown in Figure 1 E. This apparatus consisted in a small tray (15 cm x 7 cm x 4.5 cm) inside of which two pieces of white extra strong paper (Steinbach ®, 12 cm x 4.5 cm), duly twice folded, were inserted in order to create a way with four chicanes between a narrow (too narrow for 15 ants) initial space (initial loggia) and a larger area (free loggia). Two such experimental apparatus were built and used, each one, for one of the two nests. Each time, for each nest and each feeding situation, 15 ants were collected from their colony and set all together, at the same time, in the initial loggia of the apparatus, and those located in this loggia as well as in the free loggia were counted after 0, 5, 10, 15 and 20 min (Table 2, table line 2). The numbers obtained for ants consuming fluoxetine were statistically compared to those previously obtained for ants having never received this substance using the non parametric Wilcoxon test (Siegel & Castellan, 1989).

Table 2. Effect of fluoxetine on five other ants' physiological traits

Trait examined	Variable assessed	No drug consumed						Drug consumed					
Tactile perception	time:	t= 0	2	4	6	8	10'	t= 0	2	4	6	8	10'
	# ants in the initial zone	24	20	17	16	18	16	24	18	19	17	16	16
	in the rough zone	0	2	4	6	3	4	0	6	2	3	4	4
	in the large zone	0	2	3	2	3	4	0	0	3	4	1	4
	linear speed, mm/sec	4.1 (3.1 – 4.2)						4.3 (3.4 – 4.7)					
	sinuosity, ang.deg./cm	295 (268 – 338)						324 (231 – 346)					
Cognitive ability requiring no memory	time:	t= 0	5	10	15	20'		t= 0	5	10	15	20'	
	# ants in the small loggia	24	21	17	15	13		29	24	20	19	15	
	in the free loggia	0	2	2	2	5		0	0	0	0	0	
Aggressiveness	levels of aggressiveness	*vs* congeners	*vs* alien ants					*vs* congeners	*vs* alien ants				
	0	72	0					3	9				
	1	27	49					72	60				
	2	0	62					67	85				
	3	0	56					5	19				
	4	0	39					0	3				
Brood caring	time:	t= 0	2	4	6	8	10'	t= 0	2	4	6	8	10'
	# of 10 larvae not re-entered	10	8	6	4	1	1	10	9	7	6	6	5

Fluoxetine appeared to have no effect on the ants' tactile perception (they went on having difficulties in crossing a rough bottom), but to affect their cognitive ability (fewer ants reached the free zone, through chicanes), their aggressiveness towards congeners (which was increased) and towards alien ants (which was decreased), as well as their brood caring behavior (which became of poor quality).

3.9 Ants' Aggressiveness Towards Congeners or Alien Workers (8, 9)

This trait was quantified before the ants consumed fluoxetine, then eight days after they had continuously consumed that drug. Ants' potential aggressiveness towards nestmates as well as expected aggressiveness towards alien workers (i.e. belonging to another colony) was assessed in the course of dyadic encounters of five ants of each of the two used colonies, the encountering being conducted in a small glass (base diameter = 3 cm, top diameter = 4 cm, height = 5 cm), the borders of which had been slightly covered with talc. Each time (in total ten encounters with nestmates and ten encounters with alien workers, each time before then under fluoxetine consumption), the ant of the presently used colony was observed for 3 minutes and its meetings with the other ant was characterized by the numbers of times it did nothing (level 0 of aggressiveness), it touched the other ant with its antennae (level 1), it opened its mandibles in front of the other ant (level 2; Figure 1F), it gripped and/or pulled the other ant (level 3), and it tried to sting or stung the other ant (level 4). The numbers recorded for each two used colonies were added (Table 2, table line 3), and the results obtained for ants consuming fluoxetine were compared to those obtained for ants having never consumed that drug, using the non parametric χ^2 test (Siegel & Castellan, 1989).

3.10 Ants' Caring Behavior (10)

This trait was examined, for the two used colonies, before the ants consumed fluoxetine, then ten days after they had consumed that drug. Each time, a few larva were removed from the inside of the nest and deposited in front of the nest tube entrance. Five of them were carefully observed, as well as the ants' behavior in front of the larva (Figure 1G). The numbers of these five larva still remaining out of the nest were counted after 0, 2, 4, 6, 8, 10 minutes, and the numbers recorded for each two colonies were added (Table 2, table line 4). The results obtained for ants consuming fluoxetine were compared to those similarly obtained for ants having never consumed that drug using the non parametric Wilcoxon test (Siegel & Castellan, 1989).

3.11 Ants' Visual Perception (11)

For examining this sensorial ability, ants from each colony were trained with their own training apparatus and then tested using another similar outfit so that, again, each colony had its own test apparatus. The apparatus consisted of a glass slide (2.6 cm X 7.6 cm) with a cube (2 cm X 2 cm X 2 cm) made of extra strong white paper (Steinbach ®) placed at an end. On one face of the cube was a blue square (1.5 cm x 1.5 cm) and on the opposite face a yellow identical square. The cube was positioned on the glass slide so that the blue cue appeared to the left and the yellow cue to the right of the glass slide. The squared cues were cut from strong colored paper (Canson ®) the colors of which had previously been analyzed for their wavelengths reflection (Cammaerts, 2007, Cammaerts & Cammaerts, 2009). During training, a *T. molitor* larvae cut in two pieces was tied to the end of the glass slide where there was no cube, so at 4 cm from the cube (Figure 1H). In this way, the reward was located 4 cm to the right of the blue visual cue, as well as 4 cm to the left of the yellow visual cue. During the tests, no meat was placed on the apparatus.

Four tests were performed six days, as well as nine days, after the ants were exposed to (and so could see) the adequate apparatus on their foraging area, the apparatus being provided with a cube free of any colored cues (first test), or a cube with a blue square (second test), or a cube with a yellow square (third test), or a cube with a blue and a yellow cues (fourth test). Each time, the ants present on the glass slide were counted 10 times, for each of the two colonies, and the mean value of the 20 counts was calculated (Table 3). The four kinds of counts were compared to one another using the non parametric test of Wilcoxon (Siegel & Castellan, 1989). If the ants coming onto the glass slide in the presence of one, or the other, or the two colored cues were more numerous than those coming in the presence of a cube free of any cue, then the ants had been able to visually perceive the colored cues (Cammaerts & Rachidi, 2009, Cammaerts et al., 2011).

3.12 Ants' Olfactory Perception (12)

For examining this sensorial ability, ants from each two colonies were trained by way of an own experimental apparatus. Afterward they were tested using another, similar apparatus, each colony having also its own test apparatus. The apparatus consisted of a piece of extra strong white paper (Steinbach ®, 12 cm X 6 cm) orthogonally folded lengthwise to present a horizontal and a vertical part. A small glass tube (length: 7 cm; diameter: 1 cm) was inserted into a hole (diameter: 1.2 cm) cut in the middle of the vertical part very close to the base. The glass tube was placed in the foraging area with the opening in the middle of the apparatus. A schema of such an apparatus is given in Cammaerts and Rachidi (2009). During training, the glass tube was filled with sugared water (the reward) and closed with a cotton plug, while pieces of thyme and pieces of estragon were deposited on a glass slide cover slip (2.2 cm x 2.2 cm) located respectively on the left and on the right horizontal ends of the experimental apparatus. In this way, the reward (sugared water, renewed when necessary) was located 4 cm to the right of the thyme and 4 cm to the left of the estragon. Another glass slide cover slip was located in the middle of the apparatus in front of the opening of the sugared water glass tube. The experimental apparatus used for testing was free of odorous plants, or provided with thyme, or with estragon, or with both cues depending on the experiments, and the glass tube was empty (i.e., no reward was given during tests), but closed with a cotton plug to prevent entry.

Four tests were so performed six days, as well as nine days, after the ants were exposed to (and so could perceive) the two odors, together with their sugar food, on their foraging area, using the four here above cited experimental designs. During each series of tests, the ants were counted 10 times, for each of the two used colonies, on the entire area of the apparatus (free of any plants), or on the left-half area of the apparatus (provided with pieces of thyme), or on the right-half area of the apparatus (provided with pieces of estragon), or on the entire area of the apparatus (provided with pieces of the two odorous plants). Each time, the mean value of the 20 counts was calculated (Table 3), while the four kinds of counts were compared to one another using the non parametric test of Wilcoxon (Siegel & Castellan, 1989). If the ants coming onto the apparatus in the presence of one, or the other, or the two olfactory cues were more numerous than those coming in the absence of such cue, then the ants had been able to olfactory perceive the odorous cues (Cammaerts & Rachidi, 2009, Cammaerts et al., 2011).

Table 3. Visual and olfactory perception of ants under fluoxetine consumption

Visual perception	no cue	blue cue	yellow cue	the two cues
after 6 days	0.25	1.80	1.00	1.20
after 9 days	0.40	0.85	1.10	0.85

Olfactory perception	no odor	thyme	estragon	the two odors
after 6 days	0.60	0.70	0.65	1.00
after 9 days	0.30	0.25	0.35	0.55

After having been in presence of two visual cues and two odorous ones, for six, then nine days, ants were tested in front of each of these cues. They reacted to the visual cues but not significantly to the olfactory ones. More experimental and statistical details are given in the text.

3.13 Ants' Visual and Olfactory Operant Conditioning Ability and Memory (13, 14, 15, 16)

Briefly, at a given time, either a green hollow cube or pieces of dried shallots were set above the pieces of *T. molitor* larva, this time tied to the supporting piece of glass. The ants so underwent, either visual or olfactory operant conditioning. Each time, tests were performed, in the course of time, while the ants were expected acquiring conditioning, then, after having removed the green cube or the pieces of shallots, while the ants were expected to partly lose their conditioning.

In detail, ants were collectively visually trained to a hollow green cube constructed of strong paper (Canson ®) according to the instructions given in Cammaerts and Nemeghaire (2012) and set over the meat food which served as a reward. The color has been analyzed to determine its wavelengths reflection (Cammaerts, 2007). Only the ceiling of each cube was filled, this allowing ants entering the cube. Choosing the green cube was considered as giving the 'correct' choice when ants were tested as explained below. The ants were olfactory conditioned by setting pieces of dried shallots aside the tied pieces of *T. molitor* larva. Choosing the pieces of shallots was considered as giving the 'correct' choice when ants were tested as explained below.

Table 4. Effects of fluoxetine an ants' visual and olfactory learning and memory

Studied function	no fluoxetine	+ fluoxetine	statistics
Visual learning ability n= 20 + 20 = 40	* C: 61/59 50 ** 7 hrs 11/9 55% 24 hrs 12/8 60% 30 hrs 12/8 60% 48 hrs 13/7 65% 55 hrs 13/7 65% 72 hrs 14/6 70% 79hrs 13/7 65%	7 hrs 11/9 55% 24 hrs 11/9 55% 30 hrs 11/9 55% 48 hrs 10/10 50% 55 hrs 10/10 50% 72 hrs 10/10 50% 79 hrs 10/1 50%	P = 0.016
Visual memory n= 20 + 20 = 40	** 7 hrs 13/7 65% 24 hrs 15/5 75% 30 hrs 14/6 70% 48 hrs 14/6 70% 55 hrs 14/6 70% 72 hrs 14/6 70% 79 hrs 14/6 70% 96 hrs 14/6 70%	no memory	
Olfactory learning ability n = 20 + 20 = 40	* C: 61/59 50% ** 7 hrs 11/9 55% 24 hrs 13/7 65% 30 hrs 14/6 70% 48 hrs 15/5 75% 55 hrs 16/4 80% 72 hrs 16/4 80% 79hrs 16/4 80%	7 hrs 9/11 45% 24 hrs 9/11 45% 30 hrs 10/10 50% 48 hrs 9/11 45% 55 hrs 9/11 45% 72 hrs 9/11 45%	P = 0.016
Olfactory memory N= 20 + 20 = 40	** 7 hrs 13/7 65% 24 hrs 12/8 60% 30 hrs 12/8 60% 48 hrs 11/9 55% 55 hrs 11/9 55%	no memory	

Ants were trained to a green hollow cube or to pieces of shallots set above or aside the meat food. Then, tests were made in the course of time: the numbers of ants giving the correct and the wrong response were counted, and the percentage of correct responses for the ant population was determined. The percentages obtained for ants consuming fluoxetine were compared to those previously obtained (by * Cammaerts et al., 2011, ** Cammaerts, Gosset, Rachidi, 2014b) for ants having never consumed this drug, using the non parametric Wilcoxon test (Siegel and Castellan, 1989). These results are graphically presented in Figure 2.

Ants were individually tested in a Y-shaped apparatus (Figure 1I) constructed of strong white paper according to the instructions given in Cammaerts et al. (2011), and set in a small tray (30 cm x 15 cm x 4 cm), apart from the experimental colony's tray. Each colony had its own testing device. The apparatus had its own bottom and the sides were slightly covered with talc to prevent the ants from escaping. In the Y-apparatus, the ants deposited no trail since they were not rewarded. However, they could utilize other chemical secretions as traces. As a precaution, the floor of each Y-apparatus was changed between tests. The Y-apparatus was provided with either a green cube, or pieces of dried shallots, in one or the other branch. Half of the tests were conducted with the cube, or the odorous plant, in the left branch and the other half with the cube, or the odorous plant, in the right branch of the Y maze, and this was randomly chosen. Control experiments had previously been made on never conditioned ants and on trained ants of colonies having never received fluoxetine (Table 4: * from Cammaerts et al., 2011: ** from Cammaerts et al., 2014b). This must be done because, once an animal is conditioned to a given stimulus, it becomes no longer naïve for such an experiment. It was so impossible to perform, on the same ants, conditioning without then with fluoxetine in the ants' food. The only solution was thus to use previous results obtained in the course of identical experiments made on very similar colonies never fed with any drug.

To conduct a test on a colony, 10 workers - randomly chosen from the workers of that colony - were transferred one by one onto the area at the entrance of the Y-apparatus. Each transferred ant was observed until it turned either to the left or to the right in the Y-tube, and its choice was recorded. Only the first choice of the ant was recorded and this only when the ant was beyond a pencil drawn thin line indicating the entrance of a branch (Figure 1I). Afterwards, the ant was removed and transferred into a polyacetate cup, in which the border was covered with talc, until 10 ants were so tested, this avoiding testing the same ant twice. All the tested ants were then placed back on their foraging area. For each experiment, the numbers of ants, among n = 10 + 10 = 20, which turned towards the "correct" green cube or pieces of shallots, or went to the "wrong" empty branch of the Y were recorded. The percentage of correct responses for the tested ant population was so established (Table 4, Figure 2). The results obtained for ants that have consumed fluoxetine were compared to previous results obtained for ants that had never consumed that substance, using the non parametric Wilcoxon test (Siegel & Castellan, 1989). The value of N, T, and P, according to the nomenclature given in the here above reference, are defined in the results section.

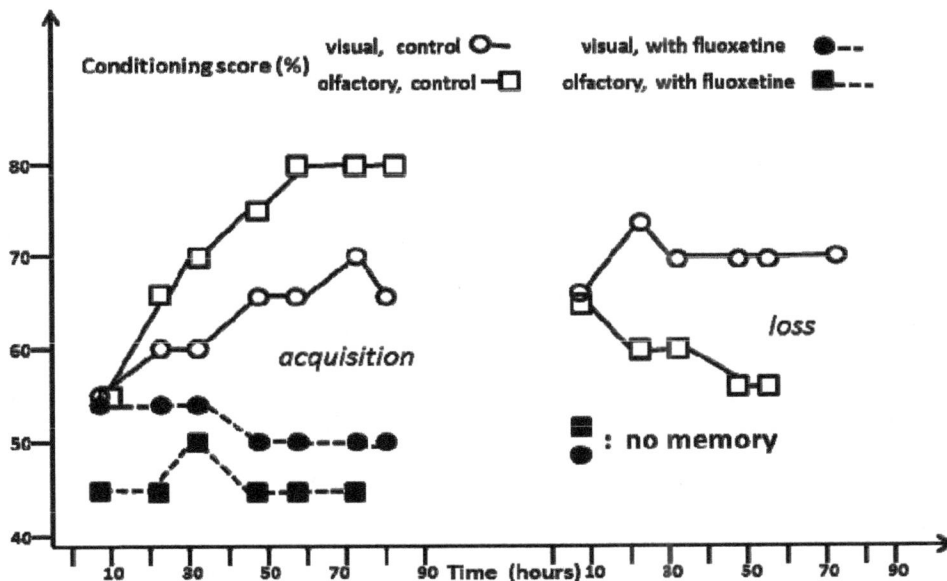

Figure 2. Ants' conditioning to a visual and an odorous cue

Numerical results and statistical evaluation are given in Table 4. The legend is identical to that of this table. Since under fluoxetine consumption, ants could not acquire conditioning, their memory could not be assessed (they behaved just as if they had no visual and no olfactory memory).

3.14 Ants' Habituation to the Drug Consumption (17)

Fourteen days after the ants had continuously consumed fluoxetine, their linear and angular speed, as well as their orientation towards an isolated worker's head, were assessed (Table 5, table line 1). The results were compared to the control ones and to those obtained after two days of drug consumption using the non parametric χ^2 test (Siegel & Castellan, 1989).

Table 5. Habituation to, and dependence on, fluoxetine consumption, as well as decrease in the course of time of the effects of this drug

Effect studied	Variable assessed	Numerical results	
habituation	linear speed (mm/sec)	12.3 (10.6-13.4)	vs control: P<0.001; vs two days: NS
	angular speed (ang.deg/cm)	176 (168-194)	vs control: P<0.001; vs two days: NS
	orientation towards an alarm signal (ang.deg.)	73.9 (62.2-93.1)	vs control: P<0.001; vs two days: NS
dependence (12 counts)	choices between sugar water *and* sugar water + fluoxetine	nest 1: 15 ants *and* 7 ants → 68.18% *and* 31.81% nest 2: 36 ants *and* 25 ants → 59.01 *and* 40.98%	
duration of effects	angular speed (ang. deg./cm) *followed by* orientation towards an alarm signal (ang.deg.) control: 113 *and* 34.5	T 0hrs 182 (151 − 193) 86.7 (74.0 - 97.4) T 4hrs 168 (151 − 212) 83.8 (76.1 − 93 0) T 8hrs 157 (139 − 166) 74 9 (62.6 - 98.4) T 12.5hrs 148 (131 − 168) 70.6 (67 3 - 81.7) T 20.5hrs 140 (114 − 147) 67.3 (60.4 - 92.1) T 24.5hrs 135 (113 − 151) 63.9 (54.9 - 78.5) T 30.5hrs 130 (114 − 148) 60.6 (42.9 − 78.2) T 37hrs 123 (87 − 149) 49.5 (28.2 − 57.3) T 45hrs 119 (116 − 136) 46.5 (40.0 − 58.5) T 50hrs 41.0 (33.6 − 51.3) T 58hrs 34.9 (23.6 − 50.4)	

Experimental methods and more statistical analysis are given in the text. Fluoxetine leaded to no habituation, and to no dependence. Its effects on locomotion (and so general activity) ended in about two days, and on olfaction and precision of reaction in about two and a half days.

3.15 Ants' Dependence on the Drug Consumption (18)

After the ants had continuously consumed fluoxetine during five days, an experiment was performed for examining if they acquired some dependence on the consumed drug. Fifteen ants of each two used colonies were transferred into a small tray (15 cm × 7 cm × 5 cm), the borders of which had been covered with talc and in which two tubes (h = 2.5cm, diam. = 0.5 cm) were laid, one containing sugar water, the other sugar water and fluoxetine (at the concentration 1/1,000), each tube being plugged with cotton. In one of the trays, the tube containing the drug was located on the right; in the other tray, it was located on the left. Photographs of such an experimental design can be seen in Cammaerts et al., 2014 a, b, and in Cammaerts & Cammaerts R., 2014. The ants drinking each liquid food were counted 12 times, the mean values being then established for each kind of food (Tables 5, table line 2). They were statistically compared to the values expected if ants randomly went drinking each kind of food, using the non parametric goodness of fit χ^2 test (Siegel & castellan, 1989).

3.16 Decrease of the Effects of the Drug, After Its Consumption Ended (19)

Three weeks after that the ants had continuously consumed fluoxetine, the liquid food containing the drug was removed from the ants' tray and replaced by sugar water free of any drug. This change was made at a given recorded time. After that, the ants' angular speed and their orientation towards an isolated worker's head were assessed after successive given time periods (Tables 5, table line 3; Figure 3). The results revealed the decrease of the effects of the drug, on ants. Their statistical significance could be estimated via the non parametric χ^2 test (Siegel & Castellan, 1989).

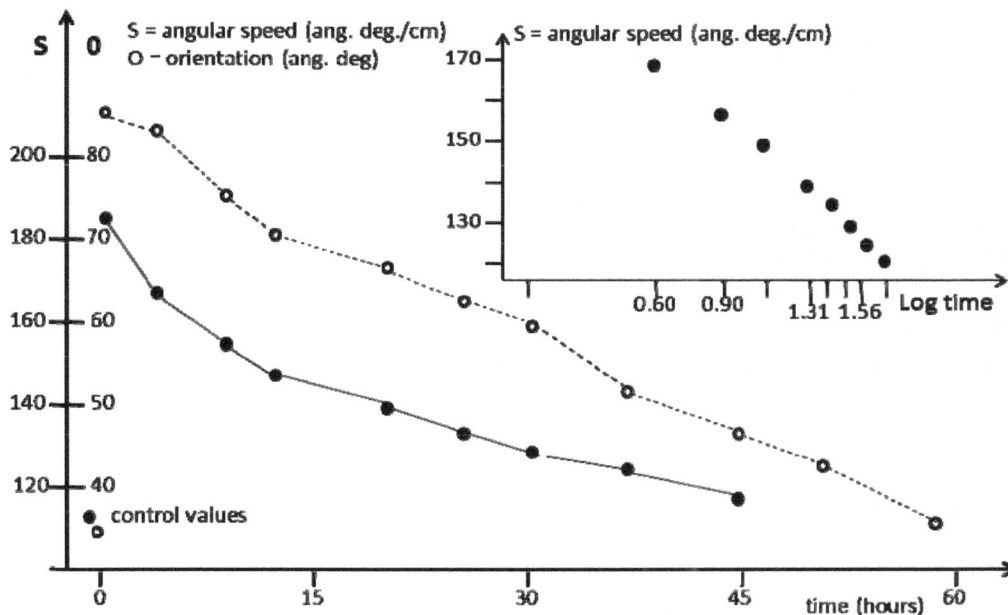

Figure 3. Decrease of the effects of fluoxetine on sinuosity (S) and on precision of reaction, in the course of time

The first effect (black circles) vanished in about two days [S = F (Log t)], while the second one (empty circles) ended after about two and a half days (O = F t). Experimental and statistical details are given in the text.

4. Results

4.1 Locomotion, General Activity (1)

These traits were affected by fluoxetine consumption.

While under normal diet, ants moved rather rapidly (median = 14.9 mm/sec) and with a moderate sinuosity (median = 113 ang. deg./cm), after having consumed fluoxetine for two days, they moved more sinuously (median = 149 ang.deg./cm) and consequently somewhat more slowly (median = 13.1 mm/sec) (Figure 1A). These results were significant (Table 1, table line 1; linear speed: $\chi^2 = 8.6$, df = 2, $0.01 < P < 0.02$; angular speed: $\chi^2 = 17.58$, df = 2, $P < 0.001$).

The ants' general activity appeared to have changed. As soon as 24 hrs after having consumed fluoxetine, several workers seemed excited, they often turned back on their way, they presented shivering and sometimes unusual movements of the legs. In front of congeners, they stopped, as in alert, often opened their mandibles, and sometimes roughly contacted them. Few ants were observed on the food sites while many ones acted as guarding at the entrances of the nest tubes.

4.2 Orientation Towards a Source of Alarm Pheromone (= Precision of Reaction) (2)

Under normal diet, ants very well oriented themselves towards a punctual source of alarm pheromone (median = 34.5 ang. deg.). After having consumed fluoxetine for two days, they badly did so (median = 69.4 ang. deg.): they moved towards a wrong direction, or they went near then over the alarm signal without succeeding in approaching it as they usually do. Such a result was significant (Table 1, table line 2; $\chi^2 = 11.63$, df = 2, $0.001 < P < 0.01$).

4.3 Trail Following Behavior (= Response to Pheromones) (3)

On a blank circumference, either ants under normal diet or those consuming fluoxetine did not move along the traced line free of pheromone. Ants under normal diet very well followed a circular trail imbibed with one poison gland extract (median = 12.0 arcs of 10°), a value similar to previous obtained ones (Cammaerts et al., 2014 a, b; Cammaerts & Cammaerts R., 2014). After having consumed fluoxetine for three days, ants very badly followed such a circular trail (median = 3.5 arcs). When reaching the trail, they seemed to have difficulties in perceiving the pheromone; they often followed it along only a few arcs, and sometimes only crossed the trail without following it. Sometimes, they stopped, turned, and moved away from the trail (Figure 1B). Such a result was highly significant (Table 1, table line 3; $\chi^2 = 38.63$, df = 2, $P < 0.001$).

4.4 "Audacity" (4)

Such a trait (assessed through the number of ants coming onto a test apparatus, Figure 1C) was only slightly and not statistically affected by fluoxetine consumption (1.62 *vs* 1.35, Table 1, table line 4; U = 136, Z = -1.72, P = 0.0858). Once more, during this experiment, ants were seen stopping in front of congeners and opening then their mandibles.

4.5 Food Consumption (5)

This trait was affected by fluoxetine consumption. Under normal diet, meanly 2.2 ants were present on the meat site, while after having consumed fluoxetine for three days, only meanly 0.46 ants were observed on that site. Such a result was statistically significant (Table 1, table line 5; U = 24, Z = 4.75, P = 0.000002).

4.6 Tactile Sensation (Pain Perception) (6)

This trait was not affected by fluoxetine consumption (Table 2; line 1).

The numbers of ants under fluoxetine consumption remaining in the initial zone, the numbers of ants crossing the rough zone, and of ants reaching the large zone were similar and statistically not different from the corresponding numbers of ants having never consumed fluoxetine. Each time, only few ants could cross the rough zone. The statistical results were: initial zone: N = 4, T = -6, P = 0.438; rough zone: N = 4, T = +7, P = 0.313; large zone: N = 3, T = -4, P = 0.375. Also, the locomotion of ants consuming fluoxetine, on the rough bottom, was similar and statistically not different from that of ants having never consumed fluoxetine, on the same substrate. Each time, ants moved very slowly, cautiously, presenting a very large sinuosity, and appearing to be uncomfortable (Figure 1D). The control and test median values, and the statistical results were the following ones: linear speed: 4.1 *vs* 4.3 mm/sec, χ^2= 0, df = 1, P = 1.00; angular speed: 295 *vs* 324 ang. deg./cm, χ^2= 0.11, df = 1, P $\approx$ 0.95. Thus, fluoxetine did not decrease the ants' potential "pain" perception, or at least did not affect the tactile sensation when walking.

4.7 Cognitive Ability Requiring No Memory (7)

This trait assessed before ants consumed fluoxetine, then seven days after they had continuously consumed that drug, appeared to be affected by fluoxetine consumption (Table 2, table line 2). Indeed, among the 30 ants set in the initial small loggia, under normal diet 13 ants were still there after 20 min. while five ones were moving in the large free area (Figure 1E), and under fluoxetine consumption, 15 ants were still in the initial loggia while no any ant could reach the free large loggia. This result was significant: initial loggia: N = 5, T = 15, P = 0.031; free loggia: N = 4, T = 10, P = 0.063.

4.8 Aggressiveness Against Nestmates and Alien Workers (8, 9)

These two traits appeared to be somewhat affected by fluoxetine consumption (Table 2, table line 3).

Concerning potential aggressiveness towards nestmates, ants under normal diet never aggressed their congeners during the ten artificial dyadic encounters, while ants consuming fluoxetine were rather aggressive in identical circumstances (Figure 1F). Indeed, these last ants presented in total 67 gripping behavior and 5 stinging ones. This result was statistically significant (χ^2 = 163.6, df = 2, P < 0.001).

Concerning expected aggressiveness towards alien ants of the same species, ants under normal diet immediately (without delay) attacked the alien ant in the course of each of the ten artificial dyadic encounters performed. In total, 62 mandibles openings, 56 gripping behaviors, and 39 stinging ones were observed. On the contrary, ants under fluoxetine consumption experimented in identical circumstances differed in time their aggressive behavior, sometimes moved away from the alien ant, and exhibited in total 69 non aggressive contacts, 85 mandibles openings, 19 gripping behaviors, and only 3 stinging ones. This result was statistically significant (χ^2 = 50.46, df = 2, P < 0.001). In fact, under fluoxetine consumption, the ants' behavior in front of an alien worker varied from escaping to slight aggressiveness.

Thus, under fluoxetine behavior, ants became somewhat aggressive towards nestmates, but less aggressive towards alien workers. However, they were still a little (and largely less than under normal diet) more aggressive against alien ants than against congeners (χ^2 = 10.61, df = 2, 0.001 < P < 0.01).

4.9 Caring Behavior (10)

This trait was affected by fluoxetine consumption (Table 2, table line 4).

Under normal diet, all the five larva of nest 1 removed from their nest were re-entered by workers in six minutes, and four larva among five were re-entered, in ten minutes, by workers of nest 2 (Figure 1G).

Under fluoxetine consumption, after 10 minutes, two larva among five for nest 1, and three larva among five for nest 2 were still outside of the nest tubes, at about one cm of the nest entrances. Such a difference between the two experiments was statistically significant (N = 5, T = 15, P = 0.031).

During the last experiment, the ants' behavior was carefully observed. In general, when approaching a larvae, an ant under fluoxetine consumption did not take the larvae with its mandibles, or took it then dropped it, or took the larvae but moved onto the foraging area instead of towards a nest entrance, or moved towards the entrance, holding the larvae, but failed in correctly entering the nest. Finally, with the help of several workers, the larva was re-entered, in successive right and wrong steps. Also, several nymphs present in nest 1 appeared to have been eaten by workers. Some eluvia of nymphs were also present in nest 1 while under normal diet such eluvia are not observed inside a nest (Cammaerts & Gosset, 2014b).

So, brood caring was of poor quality in ant population having fluoxetine in their liquid sugared food.

4.10 Visual and Olfactory Perception (11, 12)

Myrmica sabuleti workers are known for having a rather good visual perception (at least color perception (Cammaerts, 2007) and an excellent olfactory perception, using primarily odors for traveling (Cammaerts & Rachidi, 2009).

Results of the tests performed six, then nine days after ants, under fluoxetine consumption, have had the possibility to perceive two colors (blue and yellow) and two odors (thyme and estragon), were perfectly in agreement (Table 3).

As for the ants' visual perception (Figure 1H), the number of ants reacting to the blue cue as well as to the yellow one was statistically higher than those of ants reacting to no color at all (first test: blue: N = 10, T = 55, P = 0.001; yellow: N = 9, T = 45, P = 0.002; second test: blue: N = 7, T = 28, P = 0.008; yellow: N = 10, T = 55, P = 0.001). The ants better reacted to the blue cue during the first test and to the yellow cue during the second test because the experimental apparatus was differently oriented after the first test, the yellow cue fronting then the nest entrances. Of course, the ants were also more numerous in the presence of the two cues than in the absence of color: first test: N = 10, T = 55, P = 0.001; second test: N = 9, T = 36, P = 0.064). Moreover, for each of the two tests, the number of ants reacting to the two cues was lower than the sum of those of ants reacting to each of the two cues, and approached each of these two last numbers (first test: 1.20 < 1.80 + 1.00; second test: 0.85 < 1.10 + 0.85). This allowed concluding that same ants have memorized, so clearly perceived, each of the two cues.

As for the ants' olfactory perception, the number of ants reacting to thyme, to estragon, or to the two odors was not statistically different from that of ants reacting in the absence of any odor. The statistical results were the following ones: first test: thyme: N = 2, NS; estragon: N = 3, NS; the two odors: N = 9, T = 29, P = 0.248; second test: thyme: N = 5, T = -4, P > 0.50; estragon: N = 6, T = 5, P > 0.50; the two odors: N = 8, T = 7, P > 0.53). Moreover, for each of the two tests, the ants reacting in the presence of the two odors were more numerous than those reacting to each of these odors and approached the sum of the two last numbers. More precisely, for the first test, 1.00 > 0.70, 1.00 > 0.65 and 1.00 ≈ 0.70 + 0.65, while for the second test, 0.55 > 0.25, 0.55 > 0.35 and 0.55 ≈ 0.25 + 0.35. This allowed to presume that ants have remembered (so clearly perceived) only one of the two odors, and not the two ones. Such a lack of (or at least some deficiency in) olfactory perception is not at all usual for the studied species M. sabuleti. Such a physiological effect may explain several ethological ones such as aggressiveness, lower response to pheromones, brood caring of poor quality.

4.11 Visual and Olfactory Learning and Memory (13, 14, 15, 16)

Even after 79 hours of training, ants under fluoxetine consumption could not acquire visual conditioning (Table 4, upper part; Figure 2, black circles). Tested ants often came towards the green hollow cube, then stopped, opened their mandibles, and went away from the cube. Only half the tested ants moved under the green cube (Figure 1I). Fluoxetine so impacted the insects' visual conditioning capability.

In the same way, and even to a slightly stronger extent, ants under fluoxetine consumption could not acquire olfactory conditioning (Table 4, lower part; Figure 2, black squares). They behaved just as if they really did not know at all the presented odor. This confirmed the lack of olfactory perception, and the incapability of ants to acquire conditioning (pieces of shallots could have been seen), under fluoxetine consumption.

4.12 Habituation to Fluoxetine Consumption (17)

Two assessments were made for examining such a habituation (Table 5, upper part).

The ants' locomotion was assessed two days, and then 14 days after the ants had continuously consumed fluoxetine. The impact of the drug on the ants' sinuosity of movement (and consequently on linear speed) still remained after 14 days of drug consumption. Indeed, the obtained median values were 176 ang. deg./cm (12.3

mm/sec) after 14 days, while after two days they equaled 149 ang. deg/cm (13.1 mm:sec), and during the control experiment, 113 ang. deg./cm (14.9 mm/sec). Such a result was statistically significant: 14 days vs two days: linear speed: $\chi^2 = 2.74$, df = 2, $0.20 < P < 0.30$; angular speed: $\chi^2 = 1.13$, df = 1, $0.20 < P < 0.30$; 14 days vs control: linear speed: $\chi^2 = 17.33$, df = 2, $P < 0.001$; angular speed: $\chi^2 = 36.19$, df = 1, $P < 0.001$.

The ants' orientation towards a source of alarm pheromone was assessed after two days, then after 14 days of fluoxetine consumption. The impact of the drug on the ants' precision of response remained in the course of time: the median value was 73.9 ang. deg. after 14 days while it equaled 69.9 ang. deg. after two days, and 34.5 ang. deg. during the control experiment. Such a result was statistically significant: 14 days vs two days: $\chi^2 = 2.09$, df = 2, $0.30 < P < 0.50$; 14 days vs control: $\chi^2 = 16.53$, df = 2, $P < 0.001$.

We can thus conclude that individuals presented no habituation to the effects of fluoxetine, at least during the duration of the experiment (one month in total).

4.13 Dependence on Fluoxetine Consumption (18)

Results obtained for nest 1 and nest 2 were in agreement (Table 5, middle part). Fifteen ants of nest 1 were counted on the food free of fluoxetine while 7 ones were on food containing fluoxetine. For nest 2, the corresponding counts were 36 and 25 respectively. So, in total 51 ants were counted, during 24 counts, on the food free of fluoxetine; the mean number of ants on such a food equaled thus 2.13. In the same way, 32 ants were counted, during 24 counts, on the food containing fluoxetine; the mean number of ants choosing that food equaled thus 1.33. These results can also be expressed in percentages: $51/83 = 61.45\%$ ants were seen on food free of fluoxetine while $32/83 = 38.55\%$ were seen on food containing fluoxetine. A non parametric goodness of fit χ^2 test gave $\chi^2 = 4.35$, df = 1, $0.02 < P < 0.05$. Ants thus did not present any dependence on fluoxetine consumption, even preferring food free of that drug, this last result being not highly but yet significant.

4.14 Decrease of Fluoxetine Effects, in the Course of Time (19)

The drug very slowly lost its effects. The ants' behavior perfectly reflected the obtained numerical results. They moved less and less sinuously, in the course of time, being at the same time less and less aggressive towards congeners, until they moved as usual about 45 hours after they ceased to consume fluoxetine. They went on having difficulties for perceiving the presented alarm signal and for orienting themselves towards it during about 30 hours after their consumption of fluoxetine ended, then they progressively did so more easily, and finally, 58 hours after they ceased consuming fluoxetine, they behave normally. The effect of the drug on olfactory perception and precision of response lasted so a somewhat longer time period (about 2½ days) than its effect on the locomotion (about 2 days) (Table 5, lower part, Figure 3). However, statistically, the ants' sinuosity of movement no longer differed from the control one after 30½ hrs (30½ hrs: $\chi^2 = 2.89$, df = 2, NS; 37 hrs: $\chi^2 = 0.91$, df = 2, NS), and the ants' orientation no longer differed from the control one after 45 hrs (45 hrs: $\chi^2 = 3.75$, df = 2, $0.10 < P < 0.20$; 50 hrs: $\chi^2 = 0.48$, df = 2, NS). It also appeared that the effect on olfaction and precision of reaction linearly vanished with time (= t) (O = 0.83 t), while the effect on sinuosity (so on general activity) vanished with the logarithm of time (S = F (Log t); S linearly decreased with the logarithm of t: Figure 3).

5. Discussion

Working on *Myrmica sabuleti* as a biological model, we found that fluoxetine changed the general activity of the ants, increasing their sinuosity of movement, decreased their precision of response and their response to pheromones, and drastically decreased their food consumption. Fluoxetine did not affect their tactile perception, but decreased their cognition. This substance also induced aggressiveness towards nestmates while reducing that towards aliens. Under fluoxetine consumption, ants less cared of their brood (and even eat nymphs), kept intact their visual perception but lost their olfactory one, and could not acquire visual or olfactory conditioning. Ants presented no habituation to, and no dependence on, fluoxetine consumption. After fluoxetine consumption ended, the effect of the drug on the ants' locomotion vanished in about two days, while that on the ants' olfaction and precision of response vanished in two and a half days.

The present work on the effects of fluoxetine on ants is in agreement with the conclusions of Sauvé (2011) and Klaper (2013): that drug affects the physiology and the behavior of animals, and may so affect any aquatic organism living in contaminated water. We think, among others, to excitation, aggressiveness, loss of olfaction, reduced food consumption, loss of remembering, these effects being rather long lasting. The potency of the here pointed out effects of fluoxetine should also be taken into account when this drug is given to humans, even if conditions of exposure (dose, duration, ...) as well as human physiology and neural processes could differ from those of ants. However, some of these effects are presently presumed for humans or known but not divulgated and really not enough considered especially during long lasting treatments (Antidépresseur, ND). A future work on ants will examine effects of antidepressants free of fluoxetine, for checking if these effects (somewhat reported by Brodin, Fick, Jonsson & Klaminder, 2013 and Thomas, Joshi & Klaper, 2012) are weaker than those of antidepressants containing fluoxetine, and if physical dependence to them occurs. Let us recall that, in the course of the present experimentation, ants never develop habituation to or dependence on fluoxetine consumption.

Punctual points of the present work should now be discussed.

We used two experimental colonies and, generally, 20 – 40 individuals for each experiment. The samples are so rather small but, in fact, not too small for ethological studies, and we used non parametric statistics for evaluating the results. Nnumerical results were in accordance with our general impression resulting from direct observations of the ants. For instance, fewer food consumed was observed every days and not only during the experiment. Aggressiveness between congeners was evident, with many widely mandibles openings; less care of the brood was obvious and, after the experimental period of two months, there was nearly no more living brood in the nests. Also, during the experiment relative to potential dependence, ants obviously avoided the drug, and one day after fluoxetine was removed from the ants' food, many ants came onto the sugar food free of the drug, indicating that they only scarcely drunk the sugar food containing fluoxetine. This allows us to suspect that if an animal, including humans, present dependence on fluoxetine consumption, it is only a psychological one, and not a physical one.

For all the studied traits, our results are in agreement with what has been observed on fishes, and is actually observed in humans (see references in the Introduction section). Here, we found that ants consuming fluoxetine lost their olfaction, or have a reduced one. We so propose to examine, on fishes, rats, mice, and humans, if, under fluoxetine consumption, they also present some decrease of their olfactory perception. In general, attention should be paid to the more and more numerous and abundant drugs presently consumed, and consequently existing in the natural water. Before allowing the consumption of any drug, experiments should be made on several living organisms, and results largely communicated. The present work, as previous ones (references in the Introduction section), shows that ants can advantageously be used as biological models.

Acknowledgments

We are very grateful to Dr R. Cammaerts for having made the Mann-Whitney tests, for his patience while experimenting, and for his help in writing the manuscript.

References

Abramson, C. I., Wells, H., & Janko, B. (2007). A social insect model for the study of ethanol induced behavior: the honey bee. In R. Yoshida (Ed.), *Trends in Alcohol Abuse and Alcoholism Research* (pp. 197-218). Nova Sciences Publishers, Inc.

Andre, R. G., Wirtz, R. A., & Das, Y. T. (1989). Insect Models for Biomedical Research. In A. D. Woodhead (Ed.), *Nonmammalian Animal Models for Biomedical Research* (November 13, 2008). Boca Raton, FL: CRC Press.

Antidépresseur. (ND). In Wikipedia, the free encyclopedia. Retrieved from http://fr.wikipedia.org/wiki/Antidépresseur

Antidépresseurs. (2014). In *Psycom, information et sante mentale*. Retrieved from http://www.psycom.org/Medicaments/Antidepresseurs

Billen, J., & Morgan, E. D. (1998). Pheromone communication in social insects - sources and secretions. In R. K. Vander Meer, M. D. Breed, K. E. Espelie, & M. L. Winston (Eds.), *Pheromone Communication in Social Insects: Ants, Wasps, Bees, and Termites* (pp. 3-33). Boulder, Oxford: Westview Press.

Brodin, T., Fick, J., Jonsson, M., & Klaminder, J. (2013). Dilute Concentrations of a Psychiatric Drug Alter Behavior of Fish from Natural Populations. *Science, 339, n° 6121*, 814-815. http://dx.doi.org/10.1126/science.1226850

Cammaerts, M.-C. (2004). Some characteristics of the visual perception of the ant Myrmica sabuleti. *Physiological Entomology, 29*, 472-482. http://dx.doi.org/10.1111/j.0307-6962.2004.00419.x

Cammaerts, M.-C. (2007). Colour vision in the ant *Myrmica sabuleti* MEINERT, 1861 (Hymenoptera: Formicidae). *Myrmecological News, 10*, 41-50.

Cammaerts, M.-C. (2008). Visual discrimination of cues differing as for their number of elements, their shape or their orientation, by the ant *Myrmica sabuleti. Biologia, 63*, 1169-1180. http://dx.doi.org/10.2478/s11756-008-0172-2

Cammaerts, M.-C. (2012). Navigation system of the ant *Myrmica rubra* (Hymenoptera, Formicidae). *Myrmecological News, 16*, 111-121.

Cammaerts, M.-C. (2013a). Ants' learning of nest entrance characteristics (Hymenoptera, Formicidae). *Bulletin of Entomological Research*, 6. http://dx.doi.org/10.1017/S0007485313000436

Cammaerts, M.-C. (2013b). Learning of trail following behaviour by young *Myrmica rubra* workers (Hymenoptera, Formicidae). *ISRN Entomology*, Article ID 792891, 2013b, 6 pages.

Cammaerts, M.-C. (2014). Learning of foraging area specific marking odor by ants (Hymenoptera, Formicidae). *Trends in Entomology, 10*, 11-19.

Cammaerts, M. C. (2014). Performance of the species-typical alarm response in young workers of the ant Myrmica sabuleti is induced by interactions with mature workers. *Journal of Insect Sciences, 14*(1). http://dx.doi.org/10.1093/jisesa/ieu096

Cammaerts, M.-C., & Cammaerts, D. (2009). Light thresholds for colour vision in the workers of the ant *Myrmica sabuleti (Hymenoptera: Formicidae). Belgian Journal of Zoology, 138*, 40-49.

Cammaerts, M. C., & Cammaerts, D. (2014). Comparative outlook over physiological and ecological characteristics of three closely-related Myrmica species. *Biologia, 69*(8), 1051-1058. http://dx.doi.org/10.2478/s11756-014 -0399-z

Cammaerts, M.-C., & Cammaerts, R. (2014). Physiological and ethological effects of morphine and quinine, using ants as biological models. *Journal of Pharmaceutical Biology, 4,* 43-58.

Cammaerts, M.-C., & Gosset, G. (2014a). Impact of age, activity and diet on the conditioning performance in the ant *Myrmica ruginodis* used as a biological model. *International Journal of Biology, 6*(2), 10-20. http://dx.doi.org/ 10.5539/ijb.v6n2p10

Cammaerts, M.-C., & Gosset, G. (2014b). Ontogenesis of visual and olfactory kin recognition, in the ant *Myrmmica sabuleti* (Hymenoptera, Formicidae). *Annales de la Société Entomologique de France, 50,* 1-19. http://dx.doi.org/10.1080/0003792271.2014.981406

Cammaerts, M.-C., & Nemeghaire, S. (2012). Why do workers of *Myrmica ruginodis* (Hymenoptera, Formicidae) navigate by relying mainly on their vision? *Bulletin de la Société Royale Belge d'Entomologie, 148*, 42-52.

Cammaerts, M.-C., & Rachidi, Z. (2009). Olfactive conditioning and use of visual and odorous elements for movement in the ant *Myrmica sabuleti* (Hymenoptera, Formicidae). *Myrmecological news, 12*, 117-127.

Cammaerts, M.C., Gosset, G., & Rachidi, Z. (2014a). Some physiological and ethological effects of nicotine; studies on the ant *Myrmica sabuleti* as a biological model. *International Journal of Biology, 6*, 64-81.

Cammaerts, M.-C., Morel, F., Martino, F., & Warzée, N. (2012b). An easy and cheap software-based method to assess two-dimensional trajectories parameters. *Belgian Journal of Zoology, 142*, 145-151.

Cammaerts, M.-C., Rachidi, Z., & Cammaerts, D. (2011). Collective operant conditioning and circadian rhythms in the ant *Myrmica sabuleti* (Hymenoptera, Formicidae). *Bulletin de la Société Royale Belge d'Entomologie, 14*7, 142-154.

Cammaerts, M.-C., Rachidi, Z., & Gosset, G. (2014b). Physiological and ethological effects of caffeine, theophylline, cocaine and atropine; study using the ant *Myrmica sabuleti* (Hymenoptera, Formicidae) as a biological model. *International Journal of Biology, 3*, 64-84.

Cammaerts, M-C., & Cammaerts, R. (1980). Food recruitment strategies of the ants *Myrmica sabuleti* and *Myrmica ruginodis. Behavioural Processes, 5,* 251-270. http://dx.doi.org/10.1016/0376-6357(80)90006-6

Cammaerts-Tricot, M-C. (1973). Phéromone agrégeant les ouvrières de *Myrmica rubra. Journal of Insect Physiology, 19*, 1299-1315. http://dx.doi.org/10.1016/0022-1910(73)90213-8

Cipriani, A., Furukawa, T. A., Salanti, G., Geddes, J. R., Higgins, J. P., Churchill, R., ... Barbui, C. (2009). Comparative efficacy and acceptability of 12 new-generation antidepressants: a multiple-treatments meta-analysis. *Lancet., 373*(9665), 746-58. http://dx.doi.org/10.1016/S0140-6736(09)60046-5

Hölldobler, B., & Wilson, E. O. (1990). *The ants.* Springer-Verlag Berlin: Harvard University Press. http://dx.doi.org/10.1007/978-3 -662-10306-7

Keller, L., & Gordon, E. (2006). *La vie des fourmis* (p. 204). Odile Jacob, Paris.

Keller, R. A. (2011). A phylogenetic analysis of ant morphology (Hymenoptera: Formicidae) with special reference to the Poneromorph subfamilies. *Bulletin of the American Museum of Natural History, 355*, 99. http://dx.doi.org/10.1206/355.1

Klaper R. (2013). Les poissons vivant dans des eaux contaminées par des antidépresseurs deviennent anxieux, anti-sociaux et peuvent développer des tendances meurtrières, confirme une étude réalisée par des chercheurs de l'université de Milwaukee-Wisconsin (USA). Retrieved from http://egora12.rssing.com/chan-15648280/all_p1.html

Kolb, B., & Whishaw, I. Q. (2002). Neuroscience & cognition: cerveau et comportement. Eds Worth Publishers, New York, Basing Stoke, 635pp.

Lane, C. (2009). Comment la psychiatrie et l'industrie pharmaceutique ont médicalisé nos émotions. *Flammarion*.

Lecomte, E. (2014). *7 choses à savoir sur la présence de médicaments dans l'eau*. Retrieved from http://www.sciencesetavenir.fr/nature-environnement/20140829.OBS7600/7-choses-a-savoir-sur-la-presence-de-medicaments-dans-l-eau.html

Parent, H. (2011). Analyse de la responsabilité pénale des personnes faisant l'usage d'antidépresseurs et ayant commis des infractions criminelles en cours de traitement ou de servage. *Éditions Revue de Droit de l'Université de Sherbrooke*.

Passera, L. (2006). La véritable histoire des fourmis. Librairie Fayard (p. 340).

Passera, L., & Aron, S. (2005). *Les fourmis: comportement, organisation sociale et évolution*. Les Presses Scientifiques du CNRC, Ottawa, Canada, 480 pp.

Rachidi, Z., Cammaerts, M.-C. & Debeir, O. (2008). Morphometric study of the eye of three species of Myrmica (Formicidae). *Belgian Journal of Entomology, 10*, 81-91.

Russell, W. M. S., & Burch, R. L. (2014). The Principles of Humane Experimental Technique. Johns Hopkins University.

Sauvé, S. (2011). *Poissons sous antidépresseurs*. Retrieved from http://www.quebecscience.qc.ca/actualites/Poissons-sous-antidepresseurs

Siegel, S., & Castellan, N. J. (1989). *Nonparametric statistics for the behavioural sciences*. Singapore: McGraw-Hill Book Company.

Simon, G. E. (2002). Evidence review: efficacy and effectiveness of antidepressant treatment in primary care. *General Hospital Psychiatry, 24*(4), 213-24. http://dx.doi.org/10.1016/S0163-8343(02)00198-6

Søvik, E., & Barron, A. B. (2013). Invertebrate models in addiction research. *Brain Behavior and Evolution, 82*, 153-165. http://dx.doi.org/10.1159/000355506

Thomas, M. A., Joshi, P. P., & Klaper, R. D. (2012). Gene-class analysis of expression patterns induced by psychoactive pharmaceutical exposure in fathead minnow (Pimephales promelas) indicates induction of neuronal systems, *Comparative Biochemistry and Physiology, part C, Toxicology and Pharmacology, 155*(1), 109-120. http://dx.doi.org/10.1016/j.cbpc.2011.05.014

Wehner, R., & Gehring, W. (1999). Biologie et physiologie animales. Eds. De Boek Université, Thieme Verlag, Paris, Bruxelles, 844 pp.

Wolf, F. W., & Heberlein, U. (2003). Invertebrate models of drug abuse. *Journal of Neurobiology, 54*, 161-178. http://dx.doi.org/10.1002/neu.10166

15

Antioxidant and Free Radical Scavenging Capacity of Crude and Refined Oil Extracted From *Azadirachta indica* A. Juss.

Sunday O. Okoh[1,2], Ade O. Oyewole[1], Ruth O. Ishola[1], Adenike D. Odusote[1], Omobola O. Okoh[2], Chima C. Igwe[1] & Gloria N. Elemo[3]

[1] Department of Chemical Fibre and Environmental Technology, FIIRO, Lagos, Nigeria

[2] Department of Chemistry, University of Fort hare, Eastern Cape, South Africa

[3] Department of Food Technology, FIIRO, Lagos, Nigeria

Correspondence: Sunday O. Okoh, Department of Chemical, Fibre and Environmental Technology, FIIRO, Lagos, Nigeria. E-mail: Sunday.okoh@fiiro.gov. ng, sunnyokoh2003@yahoo.com

Abstract

Naturally nutritive and non-nutritive occurring antioxidants have been proven potent and safe for management of variety of diseases. This study investigated the antioxidant and free radical scavenging capacity of crude and refined *Azadirachta indica* (Neem tree) oil. The neem crude oil (NCO) was extracted from the seeds by mechanical press and degummed. The neem oil was de-pigmented with activated charcoal, and fractionated with silica gel in a capillary column. The ability of the oils to act as hydrogen/electrons donor were determined *in-vitro* using 2, 2-dipphenyl-1-picrylhydrazyl (DPPH), 2, 2-azinobis - (3-ethylbenzothiazolin - 6-sulfonic acid) diammonium salt (ABTS), lipid peroxyl (LP) and nitric oxide (NO) radicals scavenging assays, at different extract concentrations (0.1, 0.2, 0.3 and 0.4 mg/mL). The IC_{50} of the NCO oil (1.50 ± 0.10 mg/mL) showed that antioxidant activity is comparable to vitamin C and β-carotene (1.60 ± 0.10 and 1.27 ± 0.12 mg/mL respectively) in scavenging DPPH radical. The crude neem oil exhibited superior activity against NO radical, than the refined oil and vitamin C. Generally, in the four antioxidant assays, a significant correlation existed between concentrations of the oils and percentage inhibition of the four different radicals. GC/MS analyses identified monounsaturated and saturated fatty acids, aldehydes and pentanethiol as the major compounds in the oils, these may account for their antioxidant capacity.

Keywords: *Azadirachta indica*, antioxidant, DPPH, nitric oxide radicals, unsaturated fatty acids

1. Introduction

Natural antioxidant has been proven safe and potent in reduction or inhibition of free radicals which play vital role in damaging various cellular macromolecules (Morten, Mygind, & Rikke, 2012; Camila et al., 2013). The generation of free radicals such as superoxide $[O_2^\bullet]$ hydroxyl $[OH^\bullet]$, peroxyl $[RO_2^\bullet]$, hydroperoxyl $[HO_2^\bullet]$, alkoxyl $[RO^\bullet]$, nitric oxide $[NO^\bullet]$ and lipid peroxyl $[LOO^\bullet]$ beyond the antioxidant capacity of a biological system gives rise to oxidative stress (Murray, Rodriguez, Frontera, & Mulet, 2004). Oxidative stress has been implicated in the pathogenesis of a variety of inflammatory diseases such as arthritis, stroke, intestinal ischemia, acquired immunodeficiency syndrome (AIDS), hypertension, neurologic diseases e.g., multiple sclerosis, Alzheimer's disease, diabetes, cancers, and atherosclerosis (Sachdev, & Davies, 2008; Mahmood, Soheila, & Saeid, 2008; Saikat, Chakraborty, Sridhar, Reddy, & Biplab, 2010; Mimica-Dukic, Dušan, Slavenko, Dragana, & Branka, 2010; Okoh, Asekun, Familoni, & Afolayan, 2014). Plants contain antioxidant compounds that function as free radical scavengers, reducing agents and quenchers of singlet oxygen formation. There have been increasing suggestions and demand for use of natural antioxidant as potential substitute for the synthetic ones (Nweze, & Okafor, 2010; Tuttolomondo et al., 2013). The side effects of the currently available synthetic drugs pose some limitation and studies have shown that synthetic antioxidants unlike the natural analog cannot be recycled or re-used by the organism once they have donated their electron (Wang, Wu, Zu, & Fu, 2008). Consequently, they become harmful metabolic byproducts that increase, rather than decrease the total load of oxidative stress (Miller et al., 2005). One of the current trends in green chemistry is the utilization of plant extracts as drugs in the form of food confectionaries, disinfectants, insecticides, nutraceuticals, herbal drinks, fragrance, and supplement to prevent diseases or for health management (Morten, Mygind, & Rikke, 2012). *Azadirachta indica* (Neem tree) is used in traditional medicine in many West African countries for the treatment of various human diseases. The leaf serves as roof in traditional houses to repel mosquitoes and aqueous extracts of the leaves are used as anti-diabetic agent and malaria treatment (Anaso & Anaso, 2001). *A. indica* seed and leaf extracts are currently used in some countries especially in India as active ingredient in the manufacture of neem tea, cosmetics, insect repellents and herbal dentifrice (Okujagu, 2009). The extracts from Asian neem tree were reported to contain the active compound azadirachtin, nimbinin and nimbandiol, as well as several natural steroids (Sadeghian, & Mortazaienezhad, 2007). The anti-malarial activity of neem has been attributed to azadirachtin, which is a limonoid compound (Mak-Mensah & Firempong, 2011). The leaf, flower, bark and root extracts of neem tree have been reported to possess antioxidant activity (Subapriya, & Nagini, 2005; Ghimeray, Jin, Ghimire, & Cho, 2009; Nahak & Sahu, 2010; Nahak, & Sahu, 2011). One of the strong antioxidant compounds documented in the neem crude seed oil is catechin (Anokwuru, Ajibaye, & Adesuyi, 2011). Neem oil was also reported to exhibit high percentage inhibition against DPPH radical (Ghimeray, Jin, Ghimire, & Cho, 2009).

The crude neem oil is however very bitter and further study may be necessary to eliminate or reduce the bitter taste for wider applications (Okoh, Ahmed, Okoh, & Igwe, 2014).There is also scanty information on the refined neem oil as well as the comparative study of the antioxidant capacity of the crude and refined (de-bittered) seed oils in variety of radicals quenching assays. Therefore, this present research was conducted to investigate the *in-vitro* antioxidant capacity and free radical scavenging activity of the crude seed-oil extracted from *A. indica* as well as the refined oil. This research is part of an ongoing study aimed at discovering oil seed with bioactive or industrial potentials either as nutraceuticals, cosmetics, soap and industrial chemicals that are indigenous to Nigeria.

2. Materials and Methods

2.1 Chemicals Used

Potassium persulfate (PPS), 2, 2-dipphenyl-1-picrylhydrazyl (DPPH) and 2, 2-azinobis-(3-ethylbenzothiazolin-6-sulfonic acid) diammonium salt (ABTS) were purchased from Sigma - Aldrich St Louis, USA). Methanol was purchased from Fluka Chemicals (Buchs, Switzerland). All other chemicals used were analytical grade.

2.2 Plant Materials and Oil Refining

The neem crude oil was obtained from National Research Institute for Chemical Technology; Zaria (NARICT) was extracted from the seeds by mechanical press. The neem crude oil (NCO) (200 mL) was degummed by previously as described method (Asuquo, Etim, Ukpong, & Etuk, 2012). It was then refined to reduce the oil crude pigment with 10 % activated charcoal at 65°C. The oil after cooling was decanted, centrifuged at 500 rpm for 10 min and filtered to obtain the first refined neem oil (RNO1). Fifty percent of the RNO1 was subsequently fractionated using silica gel 5.4% in a 200 x1.47 mm column to obtain colourless de-bittered fraction of refined neem oil (RNO2).

2.3 Gas Chromatography/Mass Spectrometry (GC/MS) and Identification of Compounds

The GC/MS analyses of the oils were conducted on a Hewlett- Packed HP 5973 mass spectrometer interfaced with an HP 6890 gas chromatograph. The following column and temperature conditions were used; initial temperature 70°C, equilibration time 3.00 min, ramp 4°C/min, final temperature 240°C; inlet: splitless, initial temperature 220°C, pressure 8.27 psi, purge flow 30 mL/min, purge time 0.20 min, helium gas; column: capillary, 30 m x 0.25 mm i.d; 0.25 μm, film thickness 0.7 ml/min, average velocity 32 cm/sec; MS: EI method at 70 eV.

The oils compounds were identified by matching their mass spectra data with those of authentic standards held in the computer library (Wiley 275, New York) and by comparing the calculated retention indices with those in literature. The percentage composition was calculated from summation of the peak areas of the total oil composition (Asekun, Okoh, Familoni, & Afolayan, 2013).

2.4 Antioxidant Activity

The antioxidant activities were examined by DPPH radical scavenging test, ABTS free radical decolonization assay, nitric oxide and lipid peroxyl radicals scavenging assays.

2.2.1 DPPH Assay

The 2, 2 – diphenylpicrylhydrazyl radicals (DPPH) test of the oil was carried out as described by previous method (Okoh, Asekun, Familoni, & Afolayan, 2011) with minor change (oils concentrations prepared in ethyl acetate). A solution of 0.135 mm DPPH in methanol was prepared and 1.0 ml of this solution was mixed with 1.0 ml of the oil prepared in ethyl acetate containing 1.0 - 4.0 mg/mL of the oil and standard drugs (vitamin C and β carotene). The reaction mixture was vortexed thoroughly and left in the dark at room temperature for 30 min. The absorbance of the mixture was measured spectrophotometrically at 517 nm. Ethyl acetate was used as blank and all measurements were performed in triplicate. The ability of the oil to scavenge DPPH radical was calculated as % inhibition using the following equation.

$$\% \text{ inhibition} = \{(\text{Abs}_{control} - \text{Abs}_{sample})\} / (\text{Abs control}) \times 100$$

Where Abs $_{control}$ is the absorbance of the DPPH radical + methanol; Abs $_{sample}$ is the absorbance of DPPH radical + oil or standard.

2.4.2 ABTS Assay

In the 2, 2-azinobis-(3-ethylbenzothiazolin - 6-sulfonic acid) diammonium salt (ABTS) free radicals assay, the method of Witayapan and co-workers (Witayapan, Sombat, & Siriporn, 2007) was adopted with minor changes (ABTS stock solution diluted in methanol). The pre-formed radical monocation of ABTS is generated by oxidation of ABTS solution (7 mm) with 2.45 mM potassium persulfate solution in equal amount. The mixture was allowed to react for 12 h in the dark at room temperature. The resulting solution (1 ml) was diluted in 60 ml of ethyl acetate to obtain an absorbance of 0.706 ± 0.001 at 734 nm using spectrophotometer. The ABTS radical cation solution (1 ml) was added to series of oil solutions and standards drugs (vitamin C and β -carotene) of different concentrations 0.1 - 0.4 mg/mL, prepared by diluting with ethyl acetate. Ethyl acetate without oil sample was used as blank. The absorbance, after 7 min were measured spectrophotometrically at 734 nm. All measurements were carried out in triplicate. All measurements were carried out in triplicates. The percentage inhibitions of ABTS radical by the oils were calculated using the equation as described in the DPPH assay.

2.4.3 Lipid Peroxidation Assay

The Thiobarbituric acid-reactive species (TBARS) assay as described by previous method (Badmus, Odunola, & Obuotor, 2010) was used to measure the lipid peroxide formed, using egg-yolk homogenates as lipid-rich media with a minor change (methanol used for dilution of oils instead water). Egg homogenate (0.5 mL, 1 0 % in distilled water, v/v) and the oils of different concentrations 1.0 - 4.0 mg/mL prepared by diluting with ethyl acetate were mixed in test tubes and the volume was made up to1 ml, by adding methanol. Finally, 0.05 ml FeSO$_4$ (0.07 M) was added to the above mixture and incubated for 30 min, to induce lipid peroxidation. Thereafter, 1.5 ml of 20 % acetic acid (pH adjusted to 3.5 with NaOH) and 1.5 ml of 0.8 % Thiobarbituric acid (w/v) (prepared in 1.1 % sodium dodecyl sulphate) and 0.05 ml 20 % trichloroacetic acid were added, the test tubes were vortexed and then heated in a boiling water bath for 60 min. After cooling, 5.0 ml of 1-butanol was added to each tube and centrifuged at 3000 rpm for 10 min. The absorbance of the organic upper layer was measured at 532 nm. For the blank 0.1 ml of ethyl acetate was used in place of the oil. The percentage inhibitions of lipid peroxyl radical by the oils were calculated using the equation as described in the DPPH assay.

2.4.4 Nitric Oxide Assay

The method described by Makhija et al., (Makhija, Aswatha, Shreedhara, Vijay, & Devkar, 2011) was adopted. Nitric oxide radicals were generated from sodium nitroprusside solution at physiological pH. Sodium nitroprusside (1.0 mL of 10 mm) was mixed with 1.0 mL of oil in different concentrations 1.0 - 4.0 mg/mL in phosphate buffer (pH 7.4). The mixture was incubated at 25°C for 150 min. To 1.0 ml of the incubated solution, 1.0 ml of Griess' reagent (1 % sulphanilamide, 2 % o-phosphoric acid and 0.1% napthyl ethylene diamine dihydrochloride) was added. Absorbance was read at 546 nm and percentage inhibition of nitric oxide radical by the oil was calculated using the equation as described in DPPH assay. All measurements were run in triplicates and mean values were calculated.

2.5 Statistics

Data were calculated as means $\pm$ SD. Pearson's correction analysis (SPSS15.0 for windows, SPSS Inc) was used to test form the significance of the relationship between the concentration and percentage inhibition. Significant difference was considered at a level of $P < 0.05$.

3. Results and Discussion

3.1 Antioxidant Activity of the Oils

The neem crude (NCO) and refined neem oils (RNO1) and (RNO2) with activated charcoal and the colourless de-bittered fraction respectively were examined for radicals scavenging and antioxidant activities using four different assay methods. The percentage inhibitions of the oils were concentration dependent. The percentage inhibitions for DPPH assay are given in Figure 1. At all concentrations (0.1 – 0.4 mg/mL) the crude neem oil showed stronger DPPH radicals scavenging activity than the refined oils. Interestingly, the oils activity at 0.2 mg/mL and 0.4 mg/mL were far above average and demonstrated comparable scavenging activity to vitamin C. Unlike in DPPH assay, the oils were less active in scavenging ABTS radicals, while vitamin C and carotene activities were superior to the three oils at all concentrations. However the crude neem oil scavenging activity at 0.4 mg/mL was above average (Figure 2).

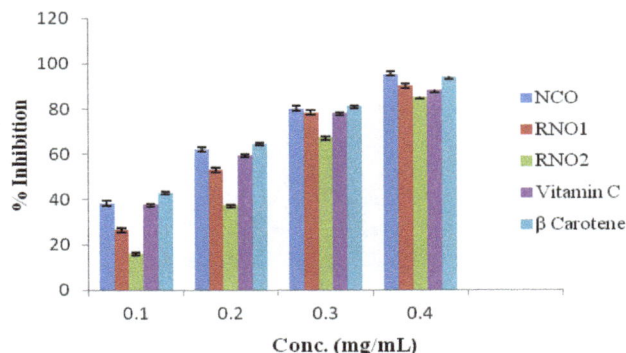

Figure 1. Antioxidant activity of crude and refined Neem oils on DPPH radicals

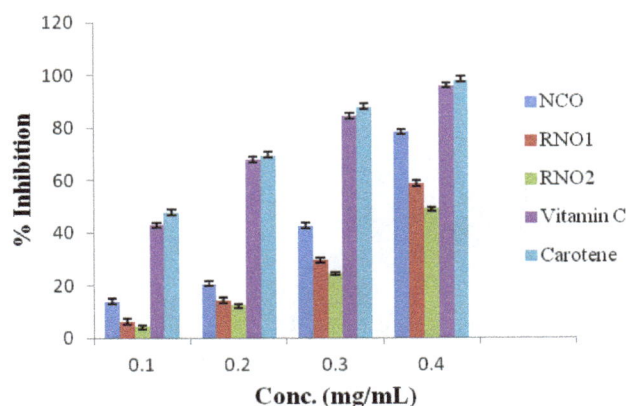

Figure 2. Antioxidant Effects of Crude and Refined Neem oils on ABTS radicals

The lipid peroxide radicals scavenging activity of the neem oils results at concentrations (0.1-0.4 mg/mL) is shown in Figure 3. The crude neem oil demonstrated more potent lipid peroxide radicals scavenging activity than the refined neem oil and two commercial antioxidants (vitamin C and β-carotene). At all concentrations (0.1- 0.4 mg/mL), the crude oil displayed stronger lipid peroxide radicals scavenging activity than the refined oil while there is no significant difference ($P < 0.05$) in peroxide radicals scavenging activity between the crude neem oil and the two commercial antioxidants.

The radical scavenging activity of the crude and refined neem oils at various concentrations against nitric oxide (NO) radicals generated from sodium nitroprusside solution are presented in Figure 4. The percentages of inhibition of the NO radicals by crude neem oil were not significantly different ($p < 0.05$) from the crude neem oil refined with activated charcoal (RNO1), vitamin C and β-carotene at 0.1- 0.3 mg/mL. Notable is the significance difference at a higher concentration (0.4 mg/mL), as the crude neem oil nitric oxide radicals scavenging activity was superior, while RNO1 and RNO2 activities were comparable to vitamin C and β-carotene.

Previous *in-vitro* antioxidant studies have noted that DPPH test does not discriminate between types of radical species, but gives rough estimates of radicals quenching ability (Sachdev & Davies, 2008). Therefore, we studied quantitative and qualitative antioxidant capacity of the oils for suspected antioxidant activity using four radical quenching assays. In the series of the *in vitro* tests the crude and refined oils of *A. indica* (neem) seeds exhibited significant antioxidant activity by acting as donators of proton or electron in the DPPH, ABTS assays and possessed strong hydroxyl, lipid peroxide (LP) and nitric oxide (NO) radical scavenging capacities. In DPPH, LP and NO assays, the crude and refined neem oil (RNO1) demonstrated similar radicals scavenging activity to the two positive controls (vitamin C and β carotene). These results are presented in Figures 1–4, while the IC$_{50}$ values for the oils are presented in Table 1.

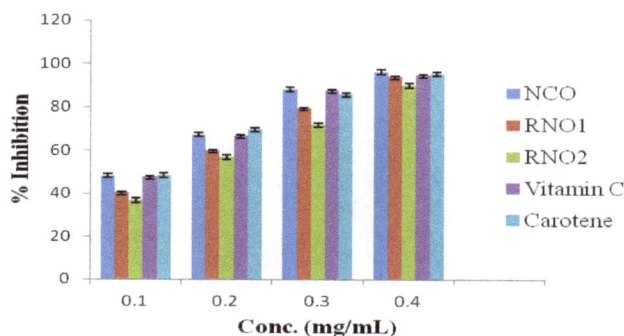

Figure 3. Antioxidant Effects of Crude and Refined Neem oils on Lipid peroxyl radicals

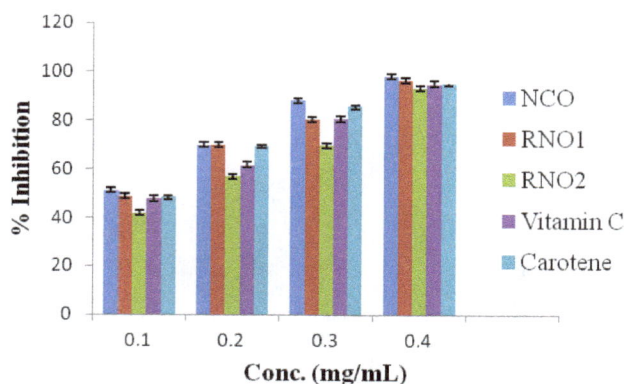

Figure 4. Antioxidant Effects of the Neem oils on Nitric oxide radicals

3.2 Antioxidant Capacity of the Oils

The crude neem oil (NCO) and refined neem oils (RNO1 & RNO2) were able to reduce the stable purple DPPH radicals to a yellow-coloured DPPH-H, reaching 50 % reduction, with IC_{50} values 1.50 ± 0.10, 1.94 ± 0.11 and 2.40 ± 0.10 mg/mL respectively (Table 1).

Table 1. Antioxidant capacity of neem crude and refined neem oils (mg/mL)

Activity	Crude Neem Oil	Refined Neem Oil		Commercial Antioxidants (Positive Controls)	
	NCO (IC_{50})	RNO1 (IC_{50})	RNO2 (IC_{50})	Vitamin C (IC_{50})	β-Carotene (IC_{50})
DPPH$^\bullet$	1.50 ± 0.10	1.94 ± 0.11	2.40 ± 0.10	1.60 ± 0.13	1.27 ± 0.12
ABTS$^{+\bullet}$	3.01 ± 0.20	3.81 ± 0.12	4.36 ± 0.11	1.2 ± 0.14	0.97 ± 0.30
LP$^\bullet$	0.98 ± 0.01	1.48 ± 0.22	1.70 ± 0.04	1.02 ± 0.04	0.90 ± 0.11
NO$^\bullet$	0.82 ± 0.02	0.94 ± 0.04	1.57 ± 0.13	1.18 ± 0.12	0.93 ± 0.20

CNO: crude neem oil, RNO1 = refined neem oil with activated charcoal neem oil, RNO2 = refined neem oil with silica gel. Values are mean $\pm$ SD, n = 3. The IC_{50} (mg/mL) was obtained from standard curve for each oil and positive controls. The lower IC_{50} the higher antioxidant capacity. Significant difference was considered at a level of P < 0.05.

The antioxidant capacities of the three oils in the DPPH radicals assay were comparable to the commercial antioxidants (vitamin C and β-carotene). The three oils antioxidant capacities were similar and less effective in scavenging ABTS radicals than the DPPH radicals with IC_{50} values 3.01 ± 0.20, 3.81 ± 0.12 and 4.36 ± 0.11 mg/mL respectively, while the vitamin C and carotene antioxidant capacities were superior to the oils (Table 2). These results corroborated previous reports that the complexity of oils, polarity and chemical properties may lead to varying bioactivity depending on the method adopted (Guerrini et al., 2009).

The crude neem oil, vitamin C and β-carotene exhibited excellent antioxidant capacity against lipid peroxide radicals with IC_{50} values 0.98 ± 0.01, 1.02 ± 0.04 and 0.90 ± 0.11 mg/mL respectively. However, the refined oils (RNO1 & RNO2) antioxidant capacities (1.48 ± 0.22 and 1.70 ± 0.04 mg/mL) were lower than crude neem oil. In the nitric oxide antioxidant experiment, refined neem oils (RNO1& RNO2) and vitamin C exhibited similar antioxidant capacity and effectively reduced the generated NO radicals with significant IC_{50} of 0.94 ± 0.04, 1.57 ± 0.13 and 1.18 ± 0.12 mg /mL respectively. It is noteworthy that the crude neem oil displayed slightly higher antioxidant capacity than β-carotene and vitamin C. Factors like stereo-selectivity of the radicals or the solubility of the oil in the different testing systems have also been reported to affect the capacity of oils in quenching different radicals [30]. Studies have also shown that some antioxidants are good electron donor (active antioxidants) in DPPH assay are inactive in ABTS assay vice- versa (Wang, Wu, Zu & Fu, 2008). Therefore, the ability of the crude and refined neem oils to scavenge different free radicals in different systems is remarkable. This indicates that the oils may be useful for preventing radical related pathological damage, especially breakdown of biomolecules and DNA by LP and NO radicals that may lead to arteriosclerosis, carcinogenesis and inflammation (Valko, Leibfritz, Moncol, Mazur, & Telser, 2007).]. Several studies have shown that chronic expression of NO radicals is associated with inflammation conditions including juvenile diabetes, multiple sclerosis, arthritis, ulcerative colitis (Sathyavathi, Suchetha, Vijay, Ullal, & Praveen, 2012; Fini, Johnson, Stenmark & Wright, 2013).

3.3 Composition of the Oil Extracted

The GC/MS quantitative and qualitative analyses of the crude and refined neem oils in our previous report (Okoh, Ahmed, Okoh, & Igwe, 2014), revealed the presence 9 - octadecenoic (oleic) acid (30.42 %), 6 -hexadecenoic acid (11.02 %), palmitic acid (21.51 %), Stearic acid (17.75 %) and methyl octadecanoate (16.88 %) as the dominant compounds. Other important compounds in the crude and the refined oils (NCO, RNO1 and RNO2) were 2-pentanethiol (1.30-5.28 %), 3, 13-octadecedienol (2.50 - 2.80 %), 2-dodecenal (0.23-3.59 %) and trans-2-decenal (0.42-2.71 %). In the refined oils (RNO1 and RNO2), the 2-pentanethiol content was significantly reduced to 1.30 and 1.37 % respectively. The high content of the monounsaturated fatty acids (oleic, 6 -hexadecenoic acids) and substantial 2-pentanethiol identified in theses oils could be responsible for the strong antioxidant capacity displayed. These results suggest possible synergistic effects of the compounds (Peana, Marzocco, & Popolo, 2006; Luh, Wong, & El-shimi, 2007). The findings in this study are in agreement with previous oil studies, that some fatty acids that are mono and polyunsaturated are good natural antioxidant and anti-inflammatory agents (Cockbain, Toogood, & Hull, 2013).

4. Conclusion

This study shows that besides the traditional uses of the plant extract, the crude and refined oils extracted from *A. indica* seeds have good antioxidant potential, and could probably replace synthetic antioxidants in further studies.

Acknowledgments

The authors are grateful to Management of Federal Institute of Industrial Research Oshodi (FIIRO) Lagos, Nigeria and Govan Mbeki Research and Development Centre, University of Fort Hare, Eastern Cape, South Africa for supporting this research work.

Conflicts of Interest

The authors declare no conflict of interest.

References

Anaso, A. B., & Anaso, C. E. (2001). Formulation of Neem Based Biopectiticides in Nigeria. Challenges before Research and Development Institutions. *UNIDO Report*, 135-139.

Anokwuru, C. P., Ajibaye, O., & Adesuyi, A. (2011). Comparative antioxidant activity of water extract of *Azadiractha indica* stem bark and *Telfairia occidentalis* leaf. *Current Research Journal of Biological, Sciences* 3(4), 430-434.

Asekun, O. T., Okoh, S. O., Familoni, O. B., & Afolayan, A. J. (2013). Chemical Profiles and Antioxidant activity of essential oils extracted from the leaf and stem of *Parkia biglobosa* (Jacq) Benth. *Research Journal of Medicinal Plants, 7*, 82-91. http://dx.doi.org/10.3923/rjmp.2013.82.91

Asuquo, J. E., Etim, E. E, Ukpong, I. U., & Etuk, S. E. (2012). Extraction, Characterisation and fatty acid profile of *pogaoleosa* oil. *Intl. J. Mod. Anal. Sep. Science, 1*, 23-30.

Badmus, J. A., Odunola, O. A., & Obuotor, E. (2010). Phytochemical and *in vitro* antioxidant potentials of *Holarr-hena floribunda* leaf. *Afr. J. Biotechnology, 9*, 340-346.

Camila, C. S., Mirian, S. S., Vanine, G. M., Luciana, M. C., Antonia, A. C., Guilherme, A. L., & Reinaldo, N. A.(2013). Antinociceptive and Antioxidant Activities of Phytol *in vivo* and *in vitro* Models. *Neuroscience Journal,* Article ID 949452.

Cockbain, A. J., Toogood, G. J., & Hull, M. A. (2013). Omega-3 polyunsaturated acids for treatment and prevention of colorectal cancer: Recent Advances in basic science. *Gut ,16*, 135-149.

Fini, M. A., Johnson, R. J., Stenmark, K. R., & Wright, R. M. (2013). Hypertension, nitrate-nitrite, and xanthine oxidoreductase catalyzed nitric oxide generation: Pros and Cons. *Hypertension, 62*, e9. *floribunda* leaf. *Afr. J. Biotechnology, 9*, 340-346.

Ghimeray, A. K., Jin, C., Ghimire, B. K., & Cho, D. H. (2009). Antioxidant activity and quantitative estimation of azadirachtin and nimbin in *Azadirachta Indica* A. Juss grown in foothills of Nepal, *Journal of Biotechnology, 8*(23), 3084-3091.

Guerrini, A., Sacchetti, G., Rossi, D., Paganetto, G., Muzzoli, M., & Andreotti, E. (2009). Bioactivities of *Piper aduncum* and *Piper obliquum* essential oils from Eastern Ecuador. *Environ. Toxicology Pharmacology, 27*, 39-48. http://dx.doi.org/10.1016/j.etap.2008.08.002

Luh, S., Wong, S., & El-shimi, E. (2007). Effect of processing on some chemical constituents of *Pistachio* nuts. *J. Food Quality, 5*, 33-41. http://dx.doi.org/10.1111/j.1745-4557.1982.tb00954.x

Mahmood, R. M., Soheila, M., & Saeid, A. (2008). Radical scavenging and reducing power of *Salvia mirzayanii* subfractions. *Molecules, 13*, 2804-2813. http://dx.doi.org/10.3390/molecules13112804

Makhija, I. K., Aswatha, H. N., Shreedhara, C. S., Vijay, K. S., & Devkar, R. (2011). *In vitro* antioxidant studies of *Sitopaladi Churna*, a polyherbal Ayurvedic formulation. *Free Radical Antioxidant, 1*, 37-41. http://dx.doi.org/10.5530/ax.2011.2.8

Mak-Mensah, E. E., & Firempong, C. K. (2011). Chemical characteristics of toilet soap prepared from neem *(Azadirachta indica)* seed oil *Asian J. Plant Sci. Res., 1*(4), 1-7.

Miller, E. R., Pastor-Barriuso, R., Dalal, D., Riemersma, R. A., Appel, L. J., & Guallar, E. (2005). Vitamin E su pplementation may increase all-cause mortality. *Ann. Intern. Med., 1*, 37-46. http://dx.doi.org/10.7326/0003-4819-142-1-200501040-00110

Mimica-Dukic, N., Dušan, B., Slavenko, M., Dragana, V. G., & Branka, O. D. (2010). Essential oil of *Myrtus communis* L. as a potential antioxidant and antimutagenic agents. *Molecules, 15*, 2759-2770. http://dx.doi.org/10.3390/molecules15042759

Moncol, J., Mazur, M., & Telser, J. (2007). Free radicals and antioxidants in normal physiological functions and human disease. *Int. J. Biochem. Cell Biology, 39*, 44-84. http://dx.doi.org/10.1016/j.biocel.2006.07.001

Morten, H., Mygind, T., L. & Rikke, M. (2012). Essential Oils in Food Preservation: Mode of Action, Synergies, and Interactions with Food Matrix Components. *Front Microbiology, 3*, 12. http://dx.doi.orrg/10.3389/fmicb.2012.00012

Murray, A. P., Rodriguez, M. A., Frontera, M. A., & Mulet, M. C. (2004). Antioxidant metabolites from *limonium brasiliene, Naturforsch, 59*, 477-480.

Nahak, G., & Sahu , R. K.. (2010). Antioxidant activity in bark and roots of Neem (*azadirachta indica*) and Mahaneem. *Continental J. Pharmaceutical Sciences, 4*, 28-34.

Nahak, G., & Sahu, R. K. (2011). Evaluation of antioxidant activity of flower and seed oil of *Azadirachta indica A. juss, Journal of Applied and natural Science, 3*(1), 78-81.

Nweze, E. I., & Okafor, J. I. (2010). Activities of a wide range of medicinal plants and essential oil against *Scedospaorium* isolates. *Am. Eurasian J. Sci. Res., 5*, 161-169.

Okoh, S. O., Ahmed, S. A., Okoh, O. O., & Igwe, C. C. (2014). GC/MS Constituents and Physicochemical Properties of Crude and Refined *Azadirachta indica* Seed Oils. *Pittsburgh Conference, Symposium and Exhibition on analytical Chemistry & Applied Spectroscopy*, March 2-6, 2014, Chicago, Illinois, USA. http://dx.doi.org/10.3390/antiox3020278

Okoh, S. O., Asekun, O. T., Familoni O. B., & Afolayan A. J. (2014). Antioxidant and Free Radical Scavenging Capacity of Seed and Shell Essential Oils Extracted from *Abrus precatorius, Antioxidant, 3*, 278-287.

Okoh, S. O., Asekun, O. T., Familoni, O. B., & Afolayan, A. J. (2011). Composition and Antioxidant Activities of leaf and root volatile oils of *Morinda lucida. J. Natural Product Communications, 6*(10), 1537-1541.

Okujagu, T. F. (2009). Medicinal plants of Nigeria from North East zone of Nigeria. *Natural Medicine Development agency* (NNMDA), Lagos. 109. Nigeria.

Pasquale, A., & Saija, A. (2013). Biomolecular characterization of wild sicilia oregano: Phytochemical screening of essential oils and extracts, and evaluation of their antioxidant activities. *Chem. Biodiversity, 10*, 411-433. http://dx.doi.org/10.1002/cbdv.201200219

Peana, A., Marzocco, S., & Popolo, A. (2006). Linalool inhibits *in vitro* NO formation: Probable involvement in the antinociceptive activity of this monoterpene compound. *Life Science, 78*, 719-723. http://dx.doi.org/10.1016/j.lfs.2005.05.065

Reinaldo, N. A. (2013). Antinociceptive and Antioxidant Activities of Phytol *in vivo* and *in vitro* Models. *Neuroscience Journal,* Article ID 949452.

Sachdev, S., & Davies, K. (2008). Production, detection, and adaptive responses to free radicals in exercise. *Free Radical Biology and Medicine, 44*, 215-223. http://dx.doi.org/10.1016/j.freeradbiomed.2007.07.019

Sadeghian, M. M., & Mortazaienezhad, F. (2007). Investigation of Compounds from *Azadirachta indica* (Neem) *Asian Journal of Plant Sciences, 6*, 444-445. http://dx.doi.org/10.3923/ajps.2007.444.445

Saikat, S., Chakraborty, R., Sridhar, C.Y., Reddy, S.R., & Biplab, D. (2010). Free radicals, antioxidants, diseases and phytomedicine: Current status and future prospect. *Int. J. Pharm. Sci. Rev. Res., 3*, 91-100.

Sathyavathi, A., Suchetha, N., Vijay, R., Ullal, D., & Praveen, A. (2012). Status of Phosphodiesterase, Nirtic oxide and Arginase levels in hypo and hyperthyroidism. *Int. J. Res. Pharm. Biomed. Science, 3*, 541-544.

Subapriya, R., & Nagini, S. (2005), *Curr. Med. Chem. Anticancer Agents, 5*(2), 149-156. http://dx.doi.org/10.2174/1568011053174828

Tuttolomondo, T., La Bella, S., Licata, M., Virga, G., Leto, C., Saija, A., ... & Ruberto, G. (2013). Biomolecular characterization of wild Sicilian oregano: phytochemical screening of essential oils and extracts, and evaluation of their antioxidant activities. *Chemistry & biodiversity, 10*(3), 411-433. http://dx.doi.org/10.1002/cbdv.201200219

Wang, W., Wu, N., Zu, G., & Fu, Y (2008). Antioxidant activity of *Rosmarinus officinalis* essential oil compared to its main components. *Food Chem., 108*, 1019-1022. http://dx.doi.org/10.1016/j.foodchem.2007.11.046

Witayapan, N., Sombat, C., & Siriporn, O. (2007). Antioxidant and antimicrobial activities of *Hyptis suaveolens* essential oil. *Scientia Pharmaceutica, 75*, 35-46. http://dx.doi.org/10.3797/scipharm.2007.75.35

16

In vitro and *In vivo* Anti-Diabetic Activity of Extracts From *Actinidia kolomikta*

Xi Yuan[1], Xuansheng Hu[1], Yu Liu[1], Hongyi Sun[1], Zhenya Zhang[1] & Delin Cheng[2]

[1] Graduate School of Life and Environmental Sciences, University of Tsukuba, Ibaraki, Japan

[2] Taisei Kogyo Co. Ltd., Tokyo, Japan

Correspondence: Zhenya Zhang, Graduate School of Life and Environmental Sciences, University of Tsukuba, Ibaraki 305-8577, Japan. E-mail: zhang.zhenya.fu@u.tsukuba.ac.jp

Abstract

Diabetes is a chronic disease, which occurs when the pancreas not produces enough insulin, or when the body cannot effectively utilise the insulin it produced. The anti-diabetic activities of both roots and leaves extracted by water and ethanol from *Actinidia kolomikta* (*A. kolomikta*) have been corroborated by this research by *in vitro* and *in vivo* tests. In the alpha-glucosidase inhibitory activity test, ethanol extract of roots showed the best inhibitory activity (74.2%, 6 mg/ml). Aqueous extracts from leaves prevented the increase in blood glucose level without causing a hypoglycemic state in the oral glucose tolerance test (120 min, 0.8 mg/g, 85.2 mg/dl) which performed *in vivo* and the maximum effect of the extract showed in 60 and 90 min after the administration of glucose. In the long term anti-diabetic test, the activities of regulation for the blood indicators demonstrated ethanol extracts of the roots from *A. kolomikta* has the effect to prevent and regulate the issue dysfunctions cause by complications of diabetes. The results reveal that roots and leaves of *A. kolomikta*, as forestry waste before, possess a high potential for the development of novel anti-diabetic drugs.

Keywords: *Actinidia kolomikta*, anti-diabetic activity, alpha-glucosidase

1. Introduction

Diabetes mellitus, which was considered as a disease of minor significance to world health, now is taking its place as one of the main threats to human health in the 21st century (Zimmet, 2000). The number of people afflicted with diabetes experienced an explosive increase in the past 20 years worldwide (Amos et al., 1997; King, Aubert, & Herman, 1998). In 2004, about 3.4 million people died from consequences of abnormal fasting blood glucose (Yen-Chang, Trần Dương, Shu-Yin, & Pung-Ling, 2014). More than 80% of deaths caused by diabetes occur in low and mid income countries (Mathers & Loncar, 2006). WHO estimated that diabetes would be the 7th major cause of death in 2030 (Lopez & Mathers, 2006). Diabetes mellitus is a metabolic disorder of multiple etiologies characterized by chronic hyperglycaemia with disturbances of carbohydrate, fat and protein metabolism resulted from defects in insulin secretion, insulin action, or both (Geert et al., 2006). Diabetes mellitus may present with characteristic symptoms such as thirst, polyuria, blurring of vision, and weight loss (Intekhab & Barry, 2006). The long term effects of diabetes mellitus include progressive development of the specific complications of retinopathy with potential blindness, nephropathy that may lead to renal failure, and/or neuropathy with risk of foot ulcers, amputation, Charcot joints, and features of autonomic dysfunction, including sexual dysfunction (Intekhab & Barry, 2006).

Diabetes mellituswas divided into two major types, Type 1 (Insulin Dependent Diabetes Mellitus) and Type 2 (Non-Insulin Dependent Diabetes Mellitus), and about 90% of diabetes patients in the world are Type 2 diabetes (Lilliooja et al., 1993). Daily injection of insulin, the main treatment method employed to management of Type 1 diabetes, is administered and also brings great pain to patients. Alpha-glucosidase inhibitors focusing on reducing the digestion of carbohydrate are the most common and efficaciousagents utilized for the treatment of Type 2 diabetes. Since alpha-glucosidase inhibitors prevent the hydrolyzation of carbohydrates into glucose, a large amount of carbohydrates remain in the intestine. Therefore the bacteria will digest the carbohydrates, which may cause gastrointestinal side effects such as flatulence and diarrhea.

In recent years, more and more researches are focused on anti-diabetic effect brought by natural plants. For example, American Ginseng (Xie et al., 2004), green tea (Chen, Zhang, & Xie, 2005), astragalus (Zhou, Wu, & Ouyang, 2005) are reportedly used as anti-diabetic agent. *Actinidia kolomikta* (*A. kolomikta*), belong to genus *Actinidia* (over fifty-eight species) which is a locally renowned traditional medicine for diabetes (Guan et al., 2011). Anti-tumor activity, immunomodulatory activity (Guan et al., 2011), anti-proliferative activity (Liu et al., 2010), anti-oxidant activity (Sun et al., 2013) have been also evaluated gradually. Moreover, anti-diabetic activity of aqueous extracts from *A. kolomikta* roots was investigated. However, in this study, ethanol extracts and aqueous extracts from roots and leaves were further executed *in vitro* and *in vivo* through alpha-glucosidase inhibitory activity, oral glucose tolerance test and long term hypoglycemic activity.

2. Materials and Methods

2.1 Materials and Reagents

Glucose, sucrose, maltose, acarbose, glucose assay kit, dimethyl sulfoxide and ethanol (99.5%) were purchased from Wako Pure Chemical Industries. (Osaka, Japan). α-glucosidase from intestine of rat was provided from Oriental yeast Co., LTD. (Tokyo, Japan). Streptozotocin and glibenclamide were purchased from Sigma-Aldrich Co., LTD. (St. Louis, MO, USA). All other chemicals and solvents were analytical grade and used without further purification.

2.2 Preparation of Different Extracts

Roots (200 g) and leaves (200 g) of *A. kolomikta* were used in this experiment. Raw materials were washed with distilled water and dried in a conventional oven (50 °C, 72 h). Dried roots and leaves were well powdered by a dry pulverizer (National MX-152S, National, Co., Ltd., Japan).

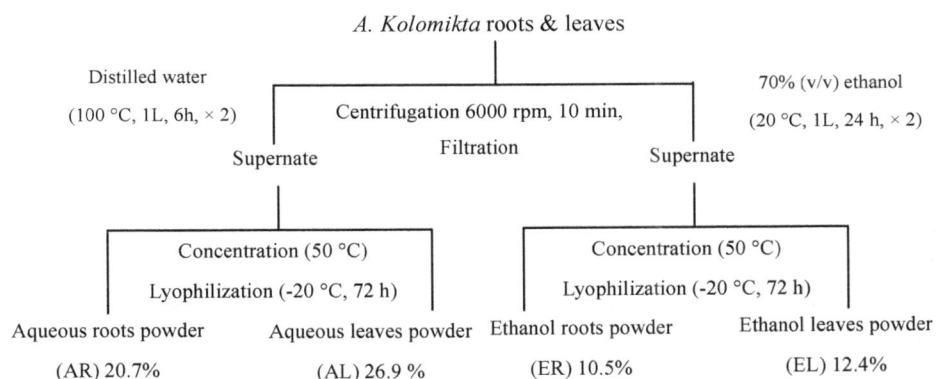

2.3 Animals and Experimental Design

10 weeks old of Sprague Dawley rats weighting about 250 g were obtained from Laboratory Animal Resource Center, University of Tsukuba (Japan). All mice were randomly separated into 5 groups consisting of six per cage and fed standard laboratory chow with 12-h dark/light cycle conditions for 1 week before the start of the experiments with a constant temperature of 20 ± 2 °C and humidity, 60 ± 5%. All laboratory feed pellets and bedding was autoclaved and supported by Laboratory Animal Resource Center, University of Tsukuba (Japan).

Type 2 diabetes was induced (Masiello et al., 1998) with slight modifications, by using standardized dose of STZ. Intraperitoneal injection of freshly prepared STZ (55 mg/kg) in 0.1 M citrate buffer (pH 4.5) in a volume of 1 mL/kg was injected to overnight-fasted normal rats, 15 min after intraperitoneally injected administration of nicotinamide (210 mg/kg). Hyperglycemia was confirmed by elevation in blood glucose levels, determined at 96 h after the STZ-nicotinamide administration. Rats with a fasting blood glucose range of 11-14 mmol/L were considered DM2 and subsequently used for the study.

2.4 Assay for Alpha-Glucosidase Inhibitory Activity of Extracts From A. kolomikta

Alpha-glucosidase inhibitory activity was determined using the method introduced a literature (Brain et al., 1997). Alpha-glucosidase solution (1.52 UI/ml) obtained by mixing 1 mg powder (76 UI) with 50 ml phosphate buffer (pH 6.9) was stored at -20 °C. 0.1 mL of gradient concentrations (0.094, 0.187, 0.375, 0.75, 1.5 and 3 mg/ml) of extracts then was mixed with 0.35 ml of sucrose (65 mM) and maltose solution (65 mM), respectively. After preheated (37 °C, 5 min), 0.2 mL of alpha-glucosidase solution was added into the preheated system and then reacted at 37 °C for 15 min. The reaction was arrested by heating the system in 100 °C water bath for 2 min. Acarbose was used in this experiment as the positive control and a group without an addition of extracts was served as negative control. The treatment of control was same as the treatment of extracts.

The activity of alpha-glucosidase was expressed as the glucose production level in the experiment. 0.2 ml of testing solution was combined with the solution got from alpha-glucosidase inhibitory test then 3 ml color reagent was added into the reactive system. Later the system was heated at 37 °C for 5 min and the absorption of the solution was checked at 505 nm. One test without the addition of the sample was served as blank control in this determination. The inhibitory activity of each sample was calculated as follows:

$$I = [1 - \frac{(As - Ab)}{(An - Ab)}] \times 100\% \tag{1}$$

Where I is the inhibitory activity, As is the absorbance of samples, An is the absorbance of negative control and Ab is the absorbance of blank control. All the data were expressed as mean ± S.D.

2.5 Oral Glucose Tolerance Test in Non-Diabetic Rats

The oral glucose tolerance test was performed in non-diabetic rats by using the similar method in literature (Barik et al., 2008). The rats were fasted overnight (16 h) before the test. Fasting blood glucose level in each rat was tested before the test. Rats were divided into five groups, and each group contained 6 animals. Control group was provided with an equal volume of distilled water. Group 2, 3 and 4 rats were administered aqueous extracts from the leaves of *A. kolomikta* orally at doses of 0.2, 0.4 and 0.8 mg/g body weight. Group 5 were fed with acarbose at a dose of 5 mg/kg body weight. Glucose (2 g/kg body weight) was fed 30 min after the administration of extracts. Blood was withdrawn from the retro barbital plexus at 30, 60, 90 and 120 min of extracts administration and plasma glucose level was determined by blood glucose meter. All the data were expressed asthe average level in 6 experimental animals in one group.

2.6 Long Term Anti-Diabetic Activities of Extracts From A. kolomikta

Group I was diabetic control. Group II and III were the extracts treated group with ethanol extracts from roots of *A. kolomikta* at the concentration of 0.2 and 0.4 mg/g body weight, respectively, and Group IV was treated with glibenclamide. The final group (Group V) was the normal control. Each group included 6 animals.

Extracts and glibenclamide were intragastric administered continuously daily for 28 days. The groups of diabetic and normal controls were administered with the same amount of distilled water. All the animals were free for feed and tap water during the experimental period. After the last dose of the extracts of the drugs, rats were fasted overnight for 16 h and the blood samples were obtained from the heart of animals into plain centrifuge tubes.

Blood samples were permitted to stand for 1 h and centrifuged at 3000 rpm for 15 min to obtain serum. The clear serum was used to biochemical assays.

Blood glucose level, plasma uric acid level, blood urea nitrogen, high-density lipoprotein, total cholesterol, triglyceride and the activities of alanine and aspartate aminotransferases in serum were analyzed by suing the Fuji Dri-Chem 7000 (Fujifilm Corporation, Tokyo, Japan) blood automatic analyzing system.

2.7 Statistical Analysis

All values are means of at least three replicates ± S.D. Differences in mean values between groups were analyzed by a one-way analysis of variance (ANOVA) and Duncan's multiplerange test using SPSS statistical software (version 16.0 for Windows, SPSS Inc., Chicago, IL, USA) to identify significant differences among means (P < 0.05).

3. Results and Discussion

3.1 Alpha-Glucosidase Inhibitory Activities of Ethanol and Aqueousextracts From A. kolomikta

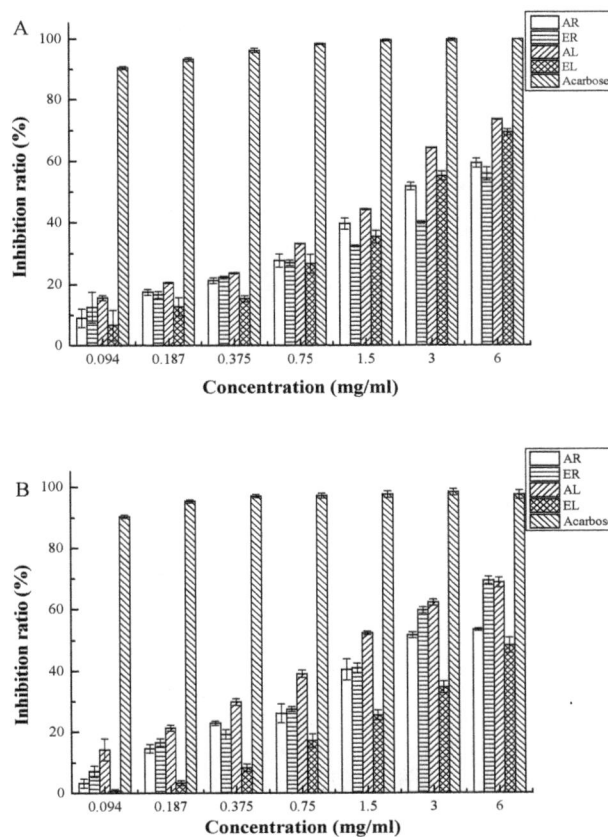

Figure 1. Alpha-glucosidase inhibitory activity of extracts from *A. kolomikta*. (A): Sucrose was the substrate. (B) : Maltose was as the substrate. AR: Aqueous extracts of roots; ER: Ethanol extracts of roots; AL: Aqueous extracts of leaves; EL: Ethanol extracts of leaves. Acarbose was invoked as negative control. Data are expressed as means ± S.D. of three independent experiments. (p < 0.05 in comparison with control)

Alpha-glucosidase inhibitory activities of aqueous and ethanol extracts with geometric gradient concentrations from both roots and leaves of *A. kolomikta* were illustrated in Figure 1. Besides, sucrose as the substrate was presented in Figure 1A and maltose as the substrate was displayed in Figure 1B. Acarbose,the alpha-glucosidase inhibitor can reduce the digestion of carbohydrate in the upper part of the small intestine, thus ameliorating postprandial hyperglycemia and enabling the beta-cells to make up the first phase insulin secretory defect in Type 2 diabetes mellitus (Furnary, Wu, & Bookin, 2004).

According to Figure 1A, it was demonstrated that each extract showed alpha-glucosidase inhibitory activity, which was in a dose-dependent manner. Moreover when treated with the concentration of 6 mg/ml, AL had the highest inhibitory activity (73.5%) among four extracts, which was in accordance with the previous results (Shai et al., 2010). Referred to Figure 1B, results including alpha-glucosidase inhibitory activity of each extract and AL possessing the higher inhibitory activity at each concentration were almost same with Figure 1B, however a slight difference of that was the ER manifested a similar inhibitory activity with aqueous extracts from leaves in higher concentrations, whose best inhibitory activity was 74.2% at the concentration of 6 mg/ml. Nevertheless, compared with the commercial drug acarbose, no precedence of extracts possessed higher inhibitory activities than acarbose in this test. Although the consequence that lower activities of both aqueous and ethanol extracts from *A. kolomikta* were obtained. It could be implicated that a high potential of medicine utilization and a novel alpha-glucosidase indicator instead of acarbose extracted from *A. kolomikta* after purification was completed feasibility.

3.2 Oral Glucose Tolerance Test in Non-Diabetic Rats

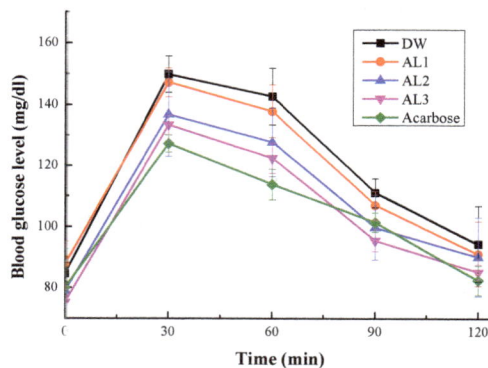

Figure 2. Effect of aqueous extracts from leaves of *A. kolomikta* on oral glucose tolerance test in normal rats. DW: distilled water. AL1: Aqueous extracts, 0.2 mg/g; AL2: Aqueous extracts, 0.4 mg/g; AL3: Aqueous extracts, 0.8 mg/g; Acarbose: 5 mg/kg. n = 6 for all groups. Data are expressed as means ± S.D. of three independent experiments. ($p < 0.05$ in comparison with control)

A long term hyperglycemiais was a major factor in the development of the complications of diabetes mellitus (Luzi et al., 1998). The oral glucose tolerance test was to investigate the effect of ALon glucose metabolism (Ariful et al., 2009). Distilled water and acarbose were invoked as the negative and positive control, respectively. The effect of aqueous extracts from the leaves of *A. kolomikta* on oral glucose tolerance test was administrated (Figure 2). Animals treated with extracts (0.8 mg/g, 133.4 mg/dl) and acarbose (127.3 mg/dl) showed a decrease in blood glucose level when compared with the control group (149.8 mg/dl) in 30 min after the administration of glucose, which was to simulate huge amounts of carbohydrates, proteins and lipid were transferred into glucose through gastrointestinal digestion and entered into the blood circulation. At the time of 60 minutes, a perceptible decrease was appeared in five groups which meant insulin and drugs had played a role in converting glucose into glycogen stored in the liver and muscles. During 60 min to 90 min, it is noteworthy that more effective hypoglycemic capability was performed in aqueous extracts which certificated itself efficacy possessed persistence. After 120 min, blood glucose level was back to the starting point (84.6 mg/dl) with treatment of either acarbose (82.5 mg/dl) or AL (AK1, 91.2 mg/dl; AK2, 90.2 mg/dl; AK3, 85.2 mg/dl) without causing a hypoglycemic state, which further proved that *A. kolomikta* had an ability of lowering blood glucosefor the treatment of Type 2 diabetes mellitus, especially in normoglycemic rats.

3.3 Effects of Ethanol Extracts From Roots of A. kolomikta on Blood Glucose Level

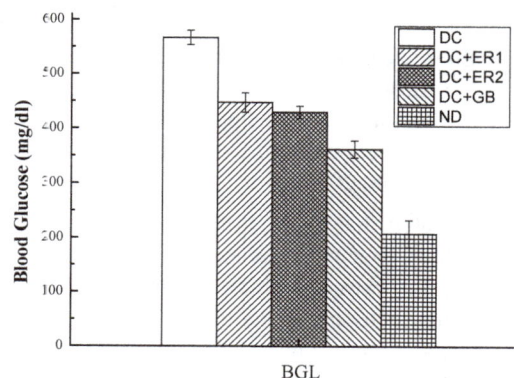

Figure 3. Effects of ethanol extracts from roots of *A. kolomikta* on blood glucose level in diabetic rats after 28 days of administrations. DC: Diabetic control; ER1: 0.2 mg/g; ER2: 0.4 mg/g; GB: Glibenclamide; ND: Non-diabetic control; BGL: blood glucose level. n = 6 for all groups. Data are expressed as means ± S.D. of three independent experiments. ($p < 0.05$ in comparison with control)

For the sake of evaluation of the long term anti-hyperglycemic activity of the extracts from *A. kolomikta* in streptozotocin-induced diabetic rats, a repeated oral glucose tolerance test for extracts from roots of *A. kolomikta* in diabetic rats was carried out. Sreptozotocin, which is a nitrosourea compound produced by Streptomyces achromogenes that induces DNA strand breakage in beta-cells causing diabetes mellitus (Kumar & Prakash, 2010; Arunachalam & Parimelazhagan, 2012; Kumar et al., 2011) by high doses of intraperitoneal injection was to produce Type 1 diabetes and a low level of streptozotocin injection can be used for the induction of Type 2 diabetes (Palsamya, Subramanian, & Veratrol, 2008). As described in the Figure 3, there was no doubt that the highest BGL appeared on the diabetic rats and the lowest BGL emerged from normal rats. After 28 days of administration, a decrease of blood glucose level was discovered in the group administrated with extracts (ER1, 447.0 mg/dl; ER2, 429.5 mg/dl) and glibenclamide (361.8 mg/dl). The group with a higher concentration (0.4mg/g) of extracts showed more effective on decreasing of the blood glucose level after 28 days of administration notwithstanding not as good as the reference drug, which may be ascribed to the crude nature of the plant extracts.

3.4 Effects of Ethanol Extracts From Roots of A. kolomikta on Protein Metabolism

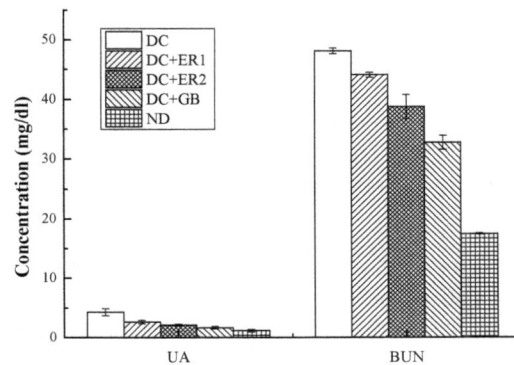

Figure 4. Effects ofethanol extracts from roots of *A. kolomikta* on protein metabolism in diabetic rats after 28 days of administrations. DC: Diabetic control; ER1: 0.2 mg/g; ER2: 0.4 mg/g; GB: Glibenclamide; ND: Non-diabetic control; UA: uric acid; BUN: blood urea nitrogen. n = 6 for all groups. Data are expressed as means ± S.D. of three independent experiments. (p < 0.05 in comparison with control)

It was manifest from Figure 4 that ethanol extracts from roots of *A. kolomikta* impacted protein metabolism by means of reduction of serum uric acid and blood urea nitrogen which was provided a theoretical basis for the possible protective effect of *A. kolomikta* against diabetes-induced renal dysfunction in the animals. More precisely, after 28 days administrations, uric acid (4.27 mg/dl) and blood urea nitrogen (48.1 mg/dl) was raised in the diabetic control group when compared with normal control (UA, 1.15 mg/dl; BUN, 17.47 mg/dl). The extracts reduced the level of uric acid (ER1, 2.62 mg/dl; ER2, 2.10 mg/dl) and blood urea nitrogen (ER1, 44.10 mg/dl; ER2, 38.75 mg/dl) in a dose-dependent manner after 28 days of administration and the treatment with glibenclamide reduced the uric acid (1.65 mg/dl) and blood urea nitrogen (32.75 mg/dl) level as well.

3.5 Effects of Ethanol Extracts From Roots of A. kolomikta on Lipid Profile

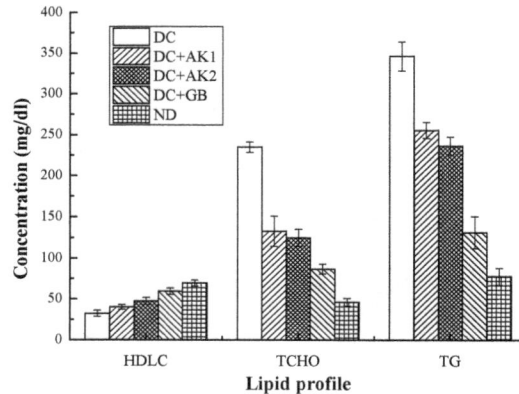

Figure 5. Effects of ethanol extracts from roots of *A. kolomikta* on the lipidprofilein diabetic rats after 28 days of administrations. DC: Diabetic control; AK1: 0.2 mg/g; AK2: 0.4 mg/g; GB: Glibenclamide; ND: Non-diabetic control; HDLC: High-density lipoprotein; TCHO: total cholesterol; TG: triglycerides. n = 6 for all groups. Data are expressed as means ± S.D. of three independent experiments. (p < 0.05 in comparison with control)

Diabetes poses a risk factor for coronary heart disease and the chance of coronary heart disease in diabetic patients is as four times as non-diabetic patients which may be one of the elements of coronary heart disease diabetes heart disease. In other words, patients who suffered diabetes can further complicated by coronary heart disease. Abnormal levels of total cholesterol and triglyceride are major important coronary risk factors (Temme et al., 1990), whereas previous studies showed that an increase in high-density lipoprotein is associated with a decrease in coronary risk. Most of the drugs that reduce total cholesterol also reduce high-density lipoprotein (Wilson, 1990). As can be observed in Figure 5, plasma lipid profile including total cholesterol, triglycerides and high-density lipoproteins were influenced entirely that the administration of extracts not only lowered the total cholesterol and triglyceride but also enhanced the level of high-density lipoprotein which indicated that the application of extracts from *A. kolomikta* may definitely reduce the incidence of coronary events occurred in the patient with diabetes.

3.6 Effects of Ethanol Extracts From Roots of A. kolomikta on Liver Indicators

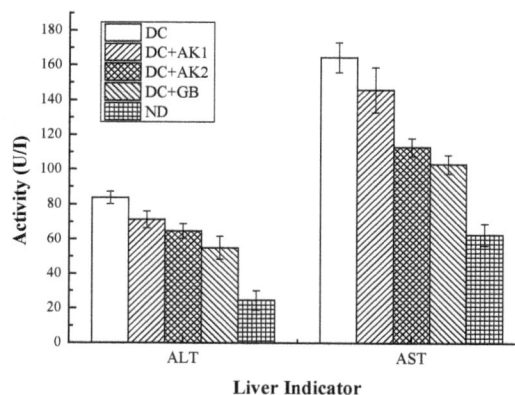

Figure 6. Effects of ethanol extracts from roots of *A. kolomikta* on liver indicators in diabetic rats after 28 days of administration. DC: Diabetic control; AK1: 0.2 mg/g; AK2: 0.4 mg/g; GB: Glibenclamide; ND: Non-diabetic control; ALT: alanine aminotransferase; AST: aspartate aminotransferase. n = 6 for all groups. Data are expressed as means ± S.D. of three independent experiments. (p < 0.05 in comparison with control)

Aspartate aminotransferase and alanine aminotransferase are reliable marker enzymes for liver function in animals. These two enzymes can be detected in the liver, cardiac muscle, skeletal muscle, kidney, brain, pancreas , lungs leukocytes and erythrocytes (Breitling, Arndt, & Drath, 2011). The increased activities of these two enzymes in blood indicate an increased permeability and damage or necrosis of hepatocytes (Adjroud, 2011). The effects of ethanol extracts from roots of *A. kolomikta* on liver indicators are shown in Figure 4. Diabetic rats showed increases in the liver indicators such as aspartate aminotransferase and alanine aminotransferase. On the contrary, a dose-dependent decrease of ethanol extracts after the treatment was detected in the rats and when treated with 0.4 mg/g per day, ethanol extract 2 had a similar regulatory effect (112.83 U/I) on aspartate aminotransferase as glibenclamide (103 U/I) after 28 days.This observation indicates the extracts from roots of *A. kolomikta* can protect the liver from oxidative pressure due to the high level of blood glucose in diabetic rats.

4. Conclusions

Anti-Diabetic activity of extracts from *A. kolomikta* was explored *in vitro* and *in vivo*. High inhibitory activity of aqueous extracts was exhibited in the alpha-glucosidase inhibitory activity test. In addition, blood glucose level was also reduced without causing a hypoglycemic state in the blood tolerance test *in vivo*. Moreover, some blood indicators such as protein metabolism, lipid profile and liver indicators caused by complications of diabetes were controlled effectively. These consequences indicated that *A. kolomikta* could be a potential and sustainable bio-resource to utilization for therapy of diabetic. However, further studies are in progress on isolation, purification, characterization and toxicity detection of the extracts.

References

Adjroud, O. (2011). The toxic effects of nickel chloride on liver, erythropoiesis, and development in Wistar albino preimplanted rats can be reversed with selenium pretreatment. *Environmental Toxicology, 28*(5), 290-298. http://dx.doi.org/10.1002/tox.20719

Amos, A., McCarty, D., & Zimmet, P. (1997). The rising global burden of diabetes and its complications: estimates and projections to the year 2010. *Diabetic Medicine, 14*(5), S1-S85. http://dx.doi.org/10.1002/(SICI)1096-9136(199712)14:5+<S7::AID-DIA522>3.0.CO;2-R

Ariful, M., Afia, A., Sarow, H., & Khurshid, A. (2009). Oral glucose tolerance test in normal control and glucose induced hyperglycemic rats with *Coccina cordifolla* L. and *Catharanthus roseus* L. *Pakistan Journal of Pharmaceutical Sciences, 22*(4), 402-404.

Arunachalam, K., & Parimelazhagan, T. (2012). Antidiabetic activity of aqueous root extract of Merremia tridentate Hall. F. instreptozotocininduced diabetic rats. *Asian Pacific Journal of Tropical Medicine, 5*(3), 175-179. http://dx.doi.org/10.1016/S1995-7645(12)60020-0

Barik, R., Jain, S., Qwatra, D., Joshi, A., Tripathi, G. S., & Goyal, R. (2008). Antidiabetic activity of aqueous root extract of Ichnocarpus frutescens in streptozootocin-nicotinamide induced type 2 diabetes in rats. *Indian Journal of Pharmacology, 40*(1), 19-22. http://dx.doi.org/10.4103/0253-7613.40484

Brain, R., Margaret, C., Jian, K., Ramesh, K. G., & John, H. M. (1997). Strain differences in susceptibility of streptozotocin-induced diabetes: effect on hypertriglyceridemia and cardiomyopathy. *Cardiovascular Research, 34*(1), 199-205. http://dx.doi.org/10.1016/S0008-6363(97)00045-X

Breitling, P., Arndt, V., & Drath, C. (2011). Liver enzymes: interaction analysis of smoking with alcohol consumption or BMI, comparing AST and ALT to γ-GT. *Public Library of Science, 6*(11), 27-51. http://dx.doi.org/10.1371/journal.pone.0027951

Chen, H. X., Zhang, M., & Xie, B. J. (2005). Components and antioxidant activity of polysaccharide conjugate from green tea. *Food Chemistry, 90*(1), 17-21. http://dx.doi.org/10.1016/j.foodchem.2004.03.001

Defronzo, R. A., Bonadonna, R. C., & Ferrannini, E. (1992). Pathogensis of NIDDM. A balanced overview. *International Textbook of Diabetes Mellitus, 15*(3), 318-368. http://dx.doi.org/10.2337/diacare.15.3.318

Furnary, A. P., Wu, Y., & Bookin, S. O. (2004). Effect of hyperglycemia and continuous intravenous insulin infusions on outcomes of cardiac surgical procedures: the Portland diabetic project. *Endocrine Practice, 10*(2), 21-33. http://dx.doi.org/10.4158/EP.10.S2.21

Geert, J. B., Salka, S., Eric, B., Carol, B., & Philip, S. (2006). Risk of dementia in diabetes mellitus: a systematic review. *The Lancet Neurology, 5*(1), 64-74. http://dx.doi.org/10.1016/S1474-4422(05)70284-2

Guan, D., Zhang, Z. Y., Yang, Y. N., Norio, S., Hu, H. H., Xing, G. Q., & Liu, J. Q. (2011). Antioxidant and antitumor activities of water extracts from the root of *A. kolomikta. Experimental and Therapeutic Medicine,*

2(1), 33-39. http://dx.doi.org/10.3892/etm.2010.163

Guan, D., Zhang, Z. Y., Yang, Y. N., Xing, G. Q., & Liu, J. Q. (2011). Immunomodulatory activity of polysaccharide from the roots of *Actinidia kolomikta* on macrophages. *International Journal of Biology, 3*(2). http://dx.doi.org/10.5539/ijb.v3n2p3

Intekhab A., & Barry G. (2006). Diabetes mellitus. *Clinics in Dermatology, 24*(4), 237-246. http://dx.doi.org/10.1016/j.clindermatol.2006.04.009

King, H., Aubert, R., & Herman, W. (1998). Global burden of diabetes, 1995-2025. Prevalence, numerical estimates and projections. *Diabetes Care, 21*(9), 1414-1431. http://dx.doi.org/10.2337/diacare.21.9.1414

Kumar, D., Kumar, S., Kohli, S., Arya, R., & Gupta, J. (2011). Antidiabetic activity of methanolicbark extract of Albizia odoratissima Benth. In alloxan induced diabetic albino mice. *Asian Pacific Journal of Tropical Medicine, 4*(11), 900-903. http://dx.doi.org/10.1016/S1995-7645(11)60215-0

Kumar, S., Kumar, V., & Prakash, O. (2010). Antidiabetic and anti-lipemic effects of Cassia siamealeaves extract in streptozotocin induced diabetic rats. *Asian Pacific Journal of Tropical Medicine, 3*(11), 871-873. http://dx.doi.org/10.1016/S1995-7645(10)60209-X

Lilliooja, S., Mott, D. M., Spraul, M., Ferraro, R., Foley, J. E., & Ravussin, E. (1993). Insulin resistance and insulin secretory dysfuction as precursors of non-insulin-dependent diabetes. Prospective Study of Pima Indians. *The New England Journal of Medicine, 329*(27), 1988-1992. http://dx.doi.org/10.1056/NEJM199312303292703

Liu, J. Q., Zhang, Z. Y., Xing, G. Q. Hu, H. H., Sugiura, N., & Intabon, K. (2010). Potential antioxidant and antiproliferative activities of a hot water extract from the root of Tonh khidum. *Oncology Letters, 1*(2), 383-387. http://dx.doi.org/10.3892/ol_00000068

Lopez, A. D., & Mathers, C. D. (2006). Measuring the global burden of disease andepidemiological transitions: 2002-2030. *Annals of Tropical Medicine & Parasitology, 100*(5), 481-499. http://dx.doi.org/10.1179/136485906X97417.

Luzi, L. (1998). Pancreas transplantation and diabetic complications. *The New England Journal of Medicine, 339*, 115-117. http://dx.doi.org/10.1056/NEJM199807093390210

Masiello, P., Broca, C., Gross, R., Roye, M., Manteghetti, M., Hillaire-Buys, D., … Ribes, G. (1998). Development of a new model of type 2 diabetes in adult rats administered with streptozotocin and nicotinamide. *Diabetes, 47*(2), 224-229. http://dx.doi.org/10.2337/diab.47.2.224

Mathers, C. D., & Loncar, D. (2006). Projections of global mortality and burden of disease from 2002 to 2030. *Plos Medicine, 3*(11), 442-444. http://dx.doi.org/10.1371/journal.pmed.0030442

Palsamya, P., Subramanian, & Veratrol, S. (2008). Resveratrol, a natural phytoalexin, normalizes hyperglycemia in streptozotocin-nicotinamide induced experimental diabetic rats. *Biomed Pharmacother, 62*(9), 598-605. http://dx.doi.org/10.1016/j.biopha.2008.06.037

Shai, L. J., Masoko, P., Mokgotho, M. P., & Magano, S. R. (2010). Yeast alpha glucosidase inhibitory and antioxidant activities of six medicinal plants collected in Phalaborwa, South Africa. *South African Journal of Botany, 76*(3), 465-470. http://dx.doi.org/10.1016/j.sajb.2010.03.002

Sun, H. Y., Wang, S. F., Li, S. H., Yuan, X., Ma, J., & Zhang, Z. Y. (2013). Antioxidant activity and immunomodulatory of extracts from roots of *Actinidia kolomikta*. *International Journal of Biology, 5*(3). http://dx.doi.org/10.5539/ijb.v5n3p1

Temme, E., Van, H., Schouten, E., & Kesteloot, H. (1990). Effect of a plant sterol-enriched spread on serum lipids and lipoprotein in mildly hypercholesterolaemic subjects. *Acta Cardiologica, 57*(2), 111-115.

Willis, J. A., Scott, R. S., Brown, L. J., Forbes, L. V., Schmidli, R. S., & Zimmet, P. Z. (1996). Islet cell antibodies and antibodies against glutamic acid decarboxylase in newly diagnosed adult-onset diabetes mellitus. *Diabetes Research and Clinical Practice, 33*(2), 89-97. http://dx.doi.org/10.1016/0168-8227(96)01281-8

Wilson, F. (1990). High density lipoprotein, low density lipoprotein and coronary heart disease. *American Journal of Cardiology, 66*(6), A7-A10. http://dx.doi.org/10.1016/0002-9149(90)90562-F

Xie, J. T., Wu, J. A., Mehendale, S., Aung, H. H., & Yuan, C. S. (2004). Anti-hyperglycemiceffect of the polysaccharides fraction from American ginseng berry extract in ob/ob mice. *Phytomedicine, 11*(2-3),

182-187. http://dx.doi.org/10.1078/0944-7113-00325

Yen-Chang, L., Trần Dương, T., Shu-Yin, W., & Pung-Ling, H. (2014). Type 1 diabetes, cardiovascular complications and sesame. *Journal of Traditional and Complementary Medicine, 4*(1), 36-41. http://dx.doi.org/10.4103/2225-4110.124817

Zhou, Y. F., Wu, Y., & Ouyang, J. P. (2005). Effects of Astragalus polysaccharide on insulin signal transduction in renal tissue of Type 2 diabetic rats. *Medical Journal of Wuhan University, 26*(2), 139-142.

Zimmet, P. (2000). Globalization, coca-colonization and the chronic disease epidemic: can the doomsday scenario be averted? *Journal of Internal Medicine, 247*(3), 301-310. http://dx.doi.org/10.1046/j.1365-2796.2000.00625.x

Permissions

All chapters in this book were first published in IJB, by Canadian Center of Science and Education; hereby published with permission under the Creative Commons Attribution License or equivalent. Every chapter published in this book has been scrutinized by our experts. Their significance has been extensively debated. The topics covered herein carry significant findings which will fuel the growth of the discipline. They may even be implemented as practical applications or may be referred to as a beginning point for another development.

The contributors of this book come from diverse backgrounds, making this book a truly international effort. This book will bring forth new frontiers with its revolutionizing research information and detailed analysis of the nascent developments around the world.

We would like to thank all the contributing authors for lending their expertise to make the book truly unique. They have played a crucial role in the development of this book. Without their invaluable contributions this book wouldn't have been possible. They have made vital efforts to compile up to date information on the varied aspects of this subject to make this book a valuable addition to the collection of many professionals and students.

This book was conceptualized with the vision of imparting up-to-date information and advanced data in this field. To ensure the same, a matchless editorial board was set up. Every individual on the board went through rigorous rounds of assessment to prove their worth. After which they invested a large part of their time researching and compiling the most relevant data for our readers.

The editorial board has been involved in producing this book since its inception. They have spent rigorous hours researching and exploring the diverse topics which have resulted in the successful publishing of this book. They have passed on their knowledge of decades through this book. To expedite this challenging task, the publisher supported the team at every step. A small team of assistant editors was also appointed to further simplify the editing procedure and attain best results for the readers.

Apart from the editorial board, the designing team has also invested a significant amount of their time in understanding the subject and creating the most relevant covers. They scrutinized every image to scout for the most suitable representation of the subject and create an appropriate cover for the book.

The publishing team has been an ardent support to the editorial, designing and production team. Their endless efforts to recruit the best for this project, has resulted in the accomplishment of this book. They are a veteran in the field of academics and their pool of knowledge is as vast as their experience in printing. Their expertise and guidance has proved useful at every step. Their uncompromising quality standards have made this book an exceptional effort. Their encouragement from time to time has been an inspiration for everyone.

The publisher and the editorial board hope that this book will prove to be a valuable piece of knowledge for researchers, students, practitioners and scholars across the globe.

List of Contributors

Ievgeniia Zhylkova
Centre of Human Reproduction "Clinic of Professor Feskov A. M.", 61098 Kharkov, Yelizarova Str. 15, Ukraine

Olexandr Feskov
Centre of Human Reproduction "Clinic of Professor Feskov A. M.", Kharkov, Ukraine

Iryna Feskova
Centre of Human Reproduction "Clinic of Professor Feskov A. M.", Kharkov, Ukraine

Olena Fedota
Kharkov National University named after V. N. Karazin, Kharkov, Ukraine

Vladyslav Feskov
Centre of Human Reproduction "Clinic of Professor Feskov A. M.", Kharkov, Ukraine

Marie-Claire Cammaerts
Faculté des Sciences, Université Libre de Bruxelles, Bruxelles, Belgium
DBO, CP 160/12, Université Libre de Bruxelles, 50, Av. F.D. Roosevelt, 1050 Bruxelles, Belgium

Zoheir Rachidi
Faculté des Sciences, Université Libre de Bruxelles, Bruxelles, Belgium

Geoffrey Gosset
Faculté des Sciences, Université Libre de Bruxelles, Bruxelles, Belgium

Alaa Alhazmi
Department of Biology, Lakehead University, 955 Oliver Road, Thunder Bay, ON, P7B 5E1, Canada

Johnathan Warren Stevenson
Department of Biology, Lakehead University, 955 Oliver Road, Thunder Bay, ON, P7B 5E1, Canada

Samuel Amartey
Division of Biology, Imperial College of Science, Technology and Medicine, London, E8 1PQ, UK

Wensheng Qin
Department of Biology, Lakehead University, 955 Oliver Road, Thunder Bay, ON, P7B 5E1, Canada

Ifat Ara Begum
Department of Biochemistry, Dhaka Medical College, Dhaka, Bangladesh
Asset Avalon, Flat-B/2, House-23, Road-1, Sector-6, Uttara, Dhaka, Bangladesh

Melva Silitonga
Biology Education Department, Faculty of Mathematic and Science, State University of Medan, Medan, Indonesia

Syafruddin Ilyas
Biology Department, Faculty of Mathematic and Science, University of North Sumatra, Medan, Indonesia

Salomo Hutahaean
Biology Department, Faculty of Mathematic and Science, University of North Sumatra, Medan, Indonesia

Herbert Sipahutar
Biology Education Department, Faculty of Mathematic and Science, State University of Medan, Medan, Indonesia

Nuzhat A. Akram
Department of Genetics, University of Karachi, University Road, Karachi 75270, Pakistan

Shakeel R. Farooqi
Department of Genetics, University of Karachi, University Road, Karachi 75270, Pakistan

Efrem-Fred A. Njau
School of Life Science and Bioengineering, The Nelson Mandela African Institution of Science and Technology, P.O.Box 447, Arusha, Tanzania

Jane Alcorn
College of Pharmacy and Nutrition, University of Saskatchewan, 105 Clinic Place, Saskatoon, SK, Canada

Patrick Ndakidemi
School of Life Science and Bioengineering, The Nelson Mandela African Institution of Science and Technology, P.O.Box 447, Arusha, Tanzania

Manuel Chirino-Trejo
Department of Microbiology, Western College of Veterinary Medicine, University of Saskatchewan, SK, Canada

Joram Buza
School of Life Science and Bioengineering, The Nelson Mandela African Institution of Science and Technology, P.O.Box 447, Arusha, Tanzania

Aregbesola Oladipupo Abiodun
Department of Microbiology, Faculty of Science, Obafemi Awolowo University, Ile-Ife 220005, Nigeria

Oluduro Anthonia Olufunke
Department of Microbiology, Faculty of Science, Obafemi Awolowo University, Ile-Ife 220005, Nigeria

Fashina Christina Dunah
Department of Microbiology, Faculty of Science, Obafemi Awolowo University, Ile-Ife 220005, Nigeria

Famurewa Oladiran
Department of Microbiology, Ekiti State University, P.M.B. 5363, Ado-Ekiti 36001, Nigeria

Dan Zhu
Graduate School of Life and Environmental Sciences, University of Tsukuba, Ibaraki, Japan

Hongyi Sun
Graduate School of Life and Environmental Sciences, University of Tsukuba, Ibaraki, Japan

Shuhong Li
Graduate School of Life and Environmental Sciences, University of Tsukuba, Ibaraki, Japan

Xuansheng Hu
Graduate School of Life and Environmental Sciences, University of Tsukuba, Ibaraki, Japan

Xi Yuan
Graduate School of Life and Environmental Sciences, University of Tsukuba, Ibaraki, Japan

Chao Han
Graduate School of Life and Environmental Sciences, University of Tsukuba, Ibaraki, Japan

Zhenya Zhang
Graduate School of Life and Environmental Sciences, University of Tsukuba, Ibaraki, Japan

K. P. Ivanov
Institute of Physiology, Russian Acad. Sci., Sankt-Petersburg, Russia

Kwang Hyun Ko
Hanyang University, Korea

Doaa Kirat
Department of Physiology, Faculty of Veterinary Medicine, Zagazig University, Zagazig, Egypt

Nora E. Abdel Hamid
Department of Physiology, Faculty of Veterinary Medicine, Zagazig University, Zagazig, Egypt

Wafaa E. Mohamed
Department of Physiology, Faculty of Veterinary Medicine, Zagazig University, Zagazig, Egypt

Mohamed Hamada
Department of Physiology, Faculty of Veterinary Medicine, Zagazig University, Zagazig, Egypt

Shimaa I. Shalaby
Department of Physiology, Faculty of Veterinary Medicine, Zagazig University, Zagazig, Egypt

David Cammaerts
Faculté des Sciences, Université Libre de Bruxelles, Av. F. Roosevelt, Bruxelles, Belgium

Sunday O. Okoh
Department of Chemical Fibre and Environmental Technology, FIIRO, Lagos, Nigeria
Department of Chemistry, University of Fort hare, Eastern Cape, South Africa

Ade O. Oyewole
Department of Chemical Fibre and Environmental Technology, FIIRO, Lagos, Nigeria

Ruth O. Ishola
Department of Chemical Fibre and Environmental Technology, FIIRO, Lagos, Nigeria

Adenike D. Odusote
Department of Chemical Fibre and Environmental Technology, FIIRO, Lagos, Nigeria

Omobola O. Okoh
Department of Chemistry, University of Fort hare, Eastern Cape, South Africa

Chima C. Igwe
Department of Chemical Fibre and Environmental Technology, FIIRO, Lagos, Nigeria

Gloria N. Elemo
Department of Food Technology, FIIRO, Lagos, Nigeria

Yu Liu
Graduate School of Life and Environmental Sciences, University of Tsukuba, Ibaraki, Japan

Delin Cheng
Taisei Kogyo Co. Ltd., Tokyo, Japan